Biomedical Engineering

Biomedical Engineering

J. H. U. BROWN, PH.D.

*Associate Director, National Institute of
General Medical Sciences, National Institutes
of Health, Bethesda, Maryland*

JOHN E. JACOBS, PH.D.

*Executive Director, Biomedical Engineering
Center, Northwestern University, Evanston,
Illinois*

LAWRENCE STARK, M.D.

*Professor of Physiological Optics and of
Engineering Science, University of California,
Berkeley, California*

F. A. DAVIS COMPANY · PHILADELPHIA

Preface

It is admittedly difficult to write a textbook in a field as broad and diversified as biomedical engineering. We excuse our temerity on the grounds of necessity.

There is a wealth of material available on each of the topics included in this book and some in fact are taught as separate courses in many institutions. In this light we would like to emphasize at the outset that we are not trying to produce a treatise on biomedical engineering, a textbook for the advanced student, or even a manual for the guidance of the well initiated.

Rather, this is a book designed to acquaint the novice and beginning student both in engineering and in biology with some of the basic principles and approaches of this rapidly developing field. While we have attempted to prepare material which requires a minimal mathematical and life science background, we would not advise a student with a preparation in neither or in only one area to attempt the course outlined.

We believe that before a student plunges into a problem area he should be taught the basic principles of the science underlying it through exposure in breadth rather than in depth. With this in mind, the book has been arranged so as to provide an introduction to basic principles of modeling, system theory, control, biological systems analysis, instrumentation, the development of instruments, devices, and systems, and engineering in the delivery of health services.

Since the book is general in nature and most of the present programs in the field have been oriented towards particular faculty interests, we hope it will serve as a general expository reading guide to be supplemented by the teacher. Most institutions are not equipped to conduct laboratories in all areas of biomedical engineering, but we cannot stress too strongly our conviction of the importance of an accompanying laboratory, even though it may deal with only a few aspects of the many problems. References have been provided which will allow the student to go further in depth at any point he wishes. In many cases, references which have not been keyed to the text have been included for general information. It is certain that anyone reading the book will find sins of both omission and commission and for these we apologize. We assume responsibility for the balance and direction of the contents based largely upon our personal experiences and prejudices regarding important topics in biomedical engineering.

J. H. U. B.

J. E. J.

L. S.

Introduction

A gulf has existed between the medical and the physical and engineering sciences. The mathematical treatment of problems by the engineer has been traditionally considered by the physician as far removed from the pragmatic treatment of disease, while the engineer has regarded the patient-oriented approach of the physician as empirical and nonscientific. A new multidisciplinary science is now arising which bridges this gulf and draws upon both the life sciences and the physical sciences for help and support. This area is biomedical engineering.

The problems of combining two widely divergent sciences into a new discipline are not easy to solve. They do not lie in the establishment of a *think tank,* but must provide complete integration of the biological sciences from cell biology to medicine with the engineering sciences, including all forms of engineering, into a new way of dealing with mutual problems.

This does not mean that we can neglect the theoretical underpinnings of biomedical engineering while we develop a highly practical science. Rather, since biomedical engineering is a developing science, it is even more critical that it have a strong theoretical base upon which to erect a structure of problem-oriented research and development. This is particularly true since we know only a very little about the systems of information, communication, and control in the human body, particularly outside the central nervous system and still less about the application of engineering skills to the measurement and control of bodily processes.

Too often the biological scientist has designed an experiment or piece of apparatus to measure a particular biological parameter, only to find the results totally unsatisfactory because he did not understand the engineering principles involved. The engineer, on the other hand, may have been capable of designing a perfectly adequate piece of apparatus or of modeling a physiological system, only to fail miserably because he did not understand the complexities and interrelations within the body.

Biomedical engineering can be approached from the standpoint of its degree of theoretical development as outlined above or from the standpoint of function. From the latter viewpoint, three distinct phases are clearly visible. One is the theoretical treatment which deals with the applications of mathematics and physics to information, communication, and control in the human body. The engineering here is largely in the area of research model making and mathematical treatment of general physiological principles. By necessity, this phase occurs largely within the academic environment where freedom of research endeavor permits wide-ranging investigations to take place.

The second phase of biomedical engineering occurs in the overlap between government, the university, and industry. This is the phase of development. The university investigator is usually interested in the development of instrumentation to the prototype stage, and the industrial complex is interested in production—somewhere in between lies the stage of development at which the device must be engineered for maximum performance and ease of production. The university scientist sees this phase as one of rather mundane testing and does not care to undertake it. Industry sees it as a period of expensive cut-and-try and does not care to expend the large funds which might be required to get a device on the market. It is here that the government can provide a major input. The use of the National Laboratories such as Oak Ridge to develop and test prototypes, the provision of funds to underwrite development costs, and the use of government contracts for underwriting the costs of the first models may be necessary.

Finally, when the theoretical development has been completed and a prototype instrument has been introduced and tested, the final stage of biomedical engineering must be entered—the stage of delivery of health services. With a combination of funds from hospitals and private foundations, the government, and industry, systems must be created to deliver health services at maximum efficiency and minimum cost.

Regardless of whether we approach biomedical engineering from the standpoint of development of a science or from the standpoint of a complex of academia, government, and industry for the delivery of health services, we must first understand the problems. The human body is a complex system of various closely interrelated subsystems, each of which affects the action of the other. We cannot adequately discuss the circulation without a thorough understanding of the nervous control of blood vessels, the hormonal control of the force of the heartbeat, the consideration of the heart as a pump, the relation of respiration to circulation and of the structure of the body to the flow of blood to various organs. The same interrelations exist with any of the many subsystems of the body. Each must be thoroughly understood from both the physiological and the engineering viewpoint.

From the above very brief discussion, it is apparent that no overview of the subject of biomedical engineering exists at present. There are several excellent monographs on control systems of the body. Papers exist which discuss in experimental detail the individual elements of the body, and occasional symposia have pointed up problems in biomedical engineering as a whole.

The authors believe that, in order for a student to understand biomedical engineering, he should be presented at some point in his training with material which gives a general overview of the subject and which points out where further information can be obtained as needed. This book attempts to serve that end. The book has been divided into several major areas which comprise the field of biomedical engineering.

In line with these objectives, this book presents a general theoretical overview of the subject as an introduction. The chapter by Yates is concerned with the use of mathematics to solve biological problems by systems analysis. The work by Bekey on control theory discusses control of various systems of the body in relatively general terms. Bugliarello discusses problems of fluid flow, while Stark points out the adaptation of systems, using the eye as an example. Agarwal's section deals with a detailed analysis of a specific fluid flow system as an example of the general problems discussed by Bugliarello.

Related Appendices contain a mathematical treatment of control theory by Bekey and a section on analog computation by Mitchell for those individuals who wish to pursue these subjects in depth, as well as a section on pattern recognition and a few problems for those interested in practical applications.

Following the theoretical introduction are chapters concerned with the application of theory to specific organ systems. The nervous system, the fluid balance, the kidney system, and the endocrine system are used as examples. We realize that many other systems such as the cardiovascular system, the respiratory system, and the nervous system (from the standpoint of single fibers) could have been included here, but these were deliberately omitted for the sake of brevity and because sufficient information exists in the literature and is well documented in the references.

Following the section on theoretical implications and the section on illustrations of body systems in operation, we decided to include a section on instrumentation. It could be logically argued that this should follow a section on theoretical treatment, but here again we have elected to go directly from theory to example and to delay methods of measurement to a later portion of the book. In this section we have made no real attempt to go into design of specific amplifiers or transducers. We discussed the general theory of design with application to illustrate the principles. A chapter on miniaturization has been included because it affords an opportunity to discuss a rapidly developing field and to point out the problems of telemetry in living systems.

Since most problems connected with measurement of parameters in the living organism must of necessity involve the introduction of foreign materials, a section has been included on the design of materials for implantation and on tissue reactions to implanted materials.

The book ends with a section which discusses in part the problems of development of instrumentation and the delivery of health services, since these after all are the important results of good biomedical engineering. Theory is of no value to mankind unless it can be applied to the solution of practical problems in the real world.

This, then, is the format of BIOMEDICAL ENGINEERING. We have attempted to explore the gamut from theory to development to practical application and to point out the role of the biomedical engineer in these areas. We have deliberately avoided complex mathematics, experimental results, and detailed explanations. The book, we hope, is readable and provides a broad survey of a rapidly developing discipline which may well be the science of tomorrow.

THE AUTHORS

Contributors

PETER H. ABBRECHT, Ph.D., M.D.
Department of Physiology, University of Michigan Medical School, Ann Arbor, Michigan

GYAN C. AGARWAL, Ph.D.
Department of Systems Engineering, University of Illinois at Chicago Circle, Chicago, Illinois

GEORGE A. BEKEY, Ph.D.
Departments of Electrical Engineering and Computer Science, University of Southern California, Los Angeles, California

DANIEL BERTAUX
Ingenieur Principal de l'Air, Centre d'Etudes et de Recherches en Automatisme, Cite Fournier 78, Velizy-Villacoublay, Paris, France

J. H. U. BROWN, Ph.D.
Associate Director, National Institute of General Medical Sciences, National Institutes of Health, Bethesda, Maryland

GEORGE BUGLIARELLO, Sc.D.
Dean of Engineering, University of Illinois at Chicago Circle, Chicago, Illinois

PETER J. DALLOS, Ph.D.
Professor of Audiology and Electrical Engineering, Northwestern University, Evanston, Illinois

JAMES F. DICKSON, III, M.D.
Program Director, Biomedical Engineering, National Institute of General Medical Sciences, National Institutes of Health, Bethesda, Maryland

DONALD S. GANN, M.D.
Professor and Director, Department of Biomedical Engineering, Case-Western University, Cleveland, Ohio

EVAN GREENER, Ph.D.
Professor and Chairman, Department of Biological Materials, Northwestern University Medical and Dental Schools, Chicago, Illinois

TIN-KAN HUNG, Ph.D.
Biotechnology Committee, Department of Engineering, Carnegie-Mellon University, Pittsburgh, Pennsylvania

JOHN E. JACOBS, Ph.D.
Director, Biomedical Engineering Center, Northwestern University, Evanston, Illinois

WEN H. KO, Ph.D.
Engineering Design Center, Case-Western Reserve University, Cleveland, Ohio

EUGENE LAUTENSCHLAGER, Ph.D.
Associate Professor, Department of Biological Materials, Northwestern University Medical and Dental Schools, Chicago, Illinois

JAMES MITCHELL, M.D.
M.D. Anderson Hospital, Houston, Texas

LYLE F. MOCKROS, Ph.D.
Technological Institute, Northwestern University, Evanston, Illinois

GEORGE MOORE, Ph.D.
Department of Electrical Engineering, University of Southern California, Los Angeles, California

CARLOS E. QUEVEDO, Ph.D.
Carnegie-Mellon University, Pittsburgh, Pennsylvania

LAWRENCE STARK, M.D.
> *Professor of Physiological Optics and of Engineering Science,* University of California, Berkeley, California

M. H. WEISSMAN, Ph.D.
> *Biotechnology Committee,* Carnegie-Mellon University, Pittsburgh, Pennsylvania

MATTHEW B. WOLF, Ph.D.
> *Departments of Electrical Engineering and Physiology,* University of Southern California, Los Angeles, California

F. EUGENE YATES, M.D.
> *Professor of Biomedical Engineering,* University of Southern California, Los Angeles, California

Contents

PART FOUR. *Biomaterials*

PART FIVE. *Engineering and Health Care*

APPENDICES

PART ONE
Engineering Concepts in Biomedicine

1
Systems Analysis in Biology

Any serious attempt to treat the biological systems from the physical point of view forces examination of uncertainties concerning physics, mathematics, logic, and human knowledge itself. An easy and casual imposition by a "bioengineer" of convenient mathematical techniques or physical models upon biological systems will do no more than lead to a pseudoquantitative description of biological systems that may be less useful than the purely qualitative and highly verbal descriptions now used by biologists. To escape from such a casual approach requires an acknowledgement of the present state of logical systems, and of physics.

The Defects of Mathematics

Mathematics has been called the language of science largely because of its great usefulness in physics; an almost mystical isomorphism between certain observable features of the physical universe and the properties of certain differential equations has cast a spell over those who become familiar with them. For example, that great and profound quantity of classical thermodynamics, entropy, appears not as a physical concept implying the ultimate demise of order in the universe, nor as giving unidirectional sequencing to macrophenomena, but instead

3

it slips in without fanfare as an integrating factor for an inexact differential equation.

Biologists do not believe that differential equations specifically, or mathematics generally, are the language of their science, because they have not seen that science advanced through techniques that are primarily mathematical, except in the philosophically limited (though practically powerful) sense that biologists benefit as do other scientists from advances in computational methods for treatment of data. In spite of past skepticism regarding the usefulness of mathematics as the language of biological science, the biologist now must consider "applied" physics and its associated "applied" mathematics (the vacuousness of the distinction between "pure" and "applied" has been colorfully noted by Hammer[1]) in the advancement of biology.

If he begins by assuming that science is elegant in proportion to its mathematical content, he will not be encouraged by the foundations of mathematics itself. Jacob Bronowski has summarized with admirable clarity the severe limitations of logical systems, as we know them, that have been revealed by the logical theorems of Godel, Turing, Church, and Tarski. He has written[2]:

> It follows in my view that the unwritten aim that the physical sciences have set themselves since Isaac Newton's time cannot be attained. The laws of nature cannot be formulated as an axiomatic, deductive, formal and unambiguous system which is also complete. And if at any state in scientific discovery the laws of nature did seem to make a complete system, then we should have to conclude that we had not gotten them right. Nature cannot be represented in the form of what logicians now call a Turing machine—that is, a logical machine operating on a basic set of axioms by making formal deductions from them in an exact language. There is no perfect description conceivable, even in the abstract, in the form of an axiomatic and deductive system.

> Of course, we suppose nevertheless that nature does obey a set of laws of her own which are precise, complete and consistent. But if this is so, then their inner formulation must be of some kind quite different from any that we know; and at present, we have no idea how to conceive it. Any description in our present formalisms must be incomplete, not because of the obduracy of nature, but because of the limitation of language as we use it. And this limitation lies not in the human fallibility of language, but on the contrary in its logical insufficiency.

> This is a cardinal point: it is the language that we use in describing nature that imposes (by its arrangement of definitions and axioms) both the form and the limitations of the laws that we find.

The appeal to mathematics rests, in the light of the above, upon its demonstrated usefulness in expressing and defining certain functional relationships and in providing the only means we now have for extracting, from sets of such relationships selected, more compact and more useful relations than are provided by the initial set. Another mathematician, Richard Bellman, has also commented very explicitly upon the limitations of mathematics[3]:

> Mathematics is an excellent tool in certain idealized situations. Fortunately or unfortunately, depending on one's point of view, the real world escapes the simple axiomatic structure needed for the application of mathematics. Essentially, the principal difficulty of mathematics is that it always insists upon supplying more information than is needed, answers to questions that are never asked.

> We are generally not interested in all continuous functions, nor all differential equations, nor all situations in which the statement "A implies B, B implies C, therefore A implies C" is valid. What is desired are sets of rules for working with various subclasses of functions, differential equations, and logical statements. In the parlance of mathematical physics, we need closure. We require ways of modifying general methods to handle specific questions.

> Another way of putting this is that we require methods for extracting the essential information from the sea of data in which we float. This is an extraordinarily difficult problem, or class of problems. We can assert that the ability of the human being that is remarkable is not so much the ability to handle large amounts of information as the ability to discard large amounts and to concentrate upon the essentials.

> It is quite probable that the methods used by the brain-mind are entirely different from those of classical mathematics. One way of seeing this is to consider the fact that humans make mistakes in recognition and are susceptible to illusions of all types. Analysis of the kinds of mistakes and the possible causes of these errors would be extremely valuable in connection with the understanding of human processes. In any case, it seems safe to say that at present the mathematical methods we now possess are not adapted to study significant decision-making processes.

> Frequently, mathematical physics is stymied by the task of describing common phenomena, such as turbulence, stress and strain, elastic creep, and so on. In these cases we use either extensive experimentation or an analogue computer of some type to carry out the process of obtaining numerical answers. We have emphasized in the foregoing that mathematics and digital computers are presently incapable of handling the problem of carrying out decision-making in situations where there is insufficient information along classical lines. This means that if we wish to obtain numerical results in processes of this nature we must employ an analogue computer of some type. The human being is this analogue computer, and it is the only one we possess. Consequently, the role of the human as principal decision-maker in economic, political, and sociological processes is secure for as far as we can see into the future.

It seems that the biologist must abandon any hope he may have had for a comprehensive *mathematical* representation of biological systems, which are the most complex we know. We will be constrained, as Bellman notes that the mathematicians are constrained, "by the futility of attempting to consider all aspects of a class of problems simultaneously," and by the demonstrated "utility of constructing theories based upon levels of problems" (hierarchies).

The vigor of physics as a theoretical as well as an experimental science has been manifested in the persistence of metaphysical arguments in philosophy that have interacted with work in physics itself to give it new directions. Perhaps biology will be mature when a "metabiology" develops and similarly interacts with *biology* which so far is rich in content but poor in concept. Preliminary explorations in metabiology have been undertaken by Iberall and his colleagues[7-9].

Apart from the defects at the foundations of physics, the biologist quickly discovers other limitations of physics—even in its sturdiest form, classical mechanics. The idealizations of mechanical systems that are treated effectively by the mathematical physicist simply do not encompass much of what is relevant in biology. As Iberall has remarked, the very details of noise, uncertainty, error, and other limiting influences in real systems that are ignored in classical mechanics and linear network theory become the primary concern of the physical biologist, who is interested in the design of biological systems as given, and not as they might have been designed by him.

The physicist who looks into biology has to undergo a deep transformation of his habits: he must confess the limited ability of idealizations in physics to describe the performance of observables in the range of the physical dimensions occupied by biological systems such as man. Thus, the biologist cannot look either to mathematicians or to physicists for his philosophy, although they in contrast can do much to perfect their philosophies by looking at biological science. These remarks acknowledge that self-constructing historical systems such as evolved living systems could never have persisted without coming to terms at every step with the nonideal aspects of physical reality.

The Defects of Biology

A system is considered "animate" when it reveals the capacity to sustain itself in a state in which observable processes occur without causing the system itself to follow a path toward the most general mechanical, chemical, thermal, and electrical equilibrium for all the processes observed. According to this view, some animate systems of biological interest are composed of connected inanimate systems, in which instance the animate state, or "life" itself, is revealed as a systems property. In other instances the system of interest is composed of connected animal subsystems, in which case the overall system is a "higher form" of life. "Life" in either case is ultimately a system attribute, and not an attribute of molecular components.

Biological science has from its very beginning been largely descriptive. Its sweeping concepts, even today, are very few; the list below is essentially exhaustive:

1. The single cell is the smallest system that unambiguously has all the attributes that define a system as "living" (even though we cannot precisely define the term "living").
2. Living systems are self-constructing, seemingly "goal-directed," self-reproducing hereditary systems whose characteristics may change in time through adaptation of the phenotype to the environment, and through the play of chance, the environment, and mating habits upon the genotype and the gene pool.
3. *Omne cellula e cellula* (even though a precellular stage of "life" is acknowledged to have existed initially. Whether life is being continuously created on earth now from abiological precursors is not known, but it is usually assumed not to be occurring, either because the conditions are not right or because the widespread existing living forms would consume and use the primordial precursors of life before they could be organized into new life forms).
4. The information transferred during cell replication and used during subsequent self-construction is stored in its entirety in nucleic acid polymers, and is realized by catalyzed transcription of the code and translation into structural and functional proteins. The resultant system has functional potentials, many of which will not be specifically realized without influence from certain physical and chemical features of the environment.
5. The storage forms of the Gibbs free energy used in the support of synthesis, secretion, repair, maintenance, and motion in biological systems are few in number, and adenosine triphosphate (ATP) is the archetype of these compounds.
6. Biological systems obey the known laws of thermodynamics, or of physics and chemistry in general.

Each of the six assertions above could be made the subject of animated debate. Biology is handicapped by the logical difficulties and paradoxes inherent in self-reference, and some of its major explanations are incomplete, curiously anthropomorphic, or even tautological. The phrases "survival of the fittest" or "the fitness of the environment," which in past years had a certain vogue in biology, smell very strongly of their human origin. The peculiar aroma of biology, as distinct from that of physics, has been well described by Pattee[10]:

> What is it then about the fundamental facts of living matter which are still not easily understood in terms of physical theory? To answer this question we must have some idea of what it means to a physicist to understand his observations in terms of the theory. Biology and physics have long had different traditions with respect to the relation between observations and theories, so that it is not unexpected that this difference in style often causes misunderstanding. Biologists have always had to struggle with an enormous variety of facts and with great complexity and lability in the simplest living units. Consequently molecular biology appears from this perspective as a set of relatively simple unifying observations about all life, and the ability to reproduce in the complexity of even the simplest living cell.

> But the physicist sees a different problem. He has already learned a set of unifying foundations or theories which quite accurately describe an enormous range of experience. The physicist also collects facts, but it is the overwhelming concern of the physicist to find out whether or not these facts can be predicted by or "reduced" to theory. Moreover his rules for "reduction" require relatively high experimental and formal precision compared to that expected in biology. Historical experience has also taught physicists that elementary concepts are clarified not by applying them to more and more complicated systems, but rather by asking deeper questions about the elementary concepts themselves. For example, the laws of classical dynamics would not have evolved into the more profound relativistic or quantum theories simply by applying these classical laws to systems with more and more degrees of freedom. For this reason it does not seem reasonable to expect the current interpretation of heredity in terms of the Watson-Crick template model and the formal description of the coding and synthesis process to develop into a more profound theory of life simply by applying it to more and more complex organisms or by the exhaustive description of, say, one strain of bacteria.

> If living matter differs from non-living matter because of its "hereditary" property, which allows for evolution by natural selection, then the central question for the physicist is to explain how hereditary structures would arise and persistently propagate and evolve in a relatively disorderly environment. In other words, if the nature of heredity is "the secret of life" then surely we must know something of the physical significance of this concept if we are to claim we understand this secret.

The defects in biological science can be seen in the paucity of theory, and in the nonquantitative character of such theories as do exist. Even the great triumphs of molecular biology and biochemistry are merely demonstrations that certain physicochemical transfer processes are present; these demonstrations usually do not answer any quantitative questions about these processes, e.g.: How fast do they go? How much material or energy flows? What partitioning of substrates will occur at branch points in biochemical chains? How strong or how weak are the coupling relations among the chains? In short, we have a chemical anatomy of biological systems, a road map for the transport routes, but little ability to predict actual traffic patterns. The descriptive stage of biological science must of course precede the quantitative state, and the description is a noteworthy achievement; it is a necessary but insufficient foundation for a theoretical biology.

In summary, biology suffers from an overcommitment to analytical and reductionistic approaches that lead to descriptions of subcellular processes, and an undercommitment to synthesis. This present situation has been described by Gray as follows[11]:

> Analysis strives to define the behavior of components isolated from their systems; synthesis strives to deduce the behavior of systems from a knowledge of their components and organization. Thus, analysis moves downward through the hierarchy while synthesis moves upward. But in order to understand a system, at whatever level of organization, one must synthesize as well as analyze. Both are necessary, yet neither is sufficient. Analysis properly comes first, but synthesis must follow.

> Yet there is a field, traditionally unrelated to physiology, which has long concerned itself with systems—with their analysis and synthesis as well as with theory and principles. It had the good fortune to be able to evolve naturally from the simple to the complex as a growing body of theory and methodology provided the catalyst. That field is engineering.

Systems Science

Systems science is the body of knowledge and techniques that permit quantitative description, prediction, and mathematical representation of the performance characteristics of a complex system of components with their functional relationships that comprise the signal pathways, system topology, and connectivity. In the simplest cases, the components are understood to be noiseless, and to accept inputs and produce outputs with an input/output relationship that can be adequately described for arbitrarily chosen purposes by a linear set of ordinary differential equations with constant coefficients. In such a restricted case, the solution of the system problem (e.g.: What will be the trajectories of the stated variables in time following any perturbation or "forcing" of the system at some input or at some internal point?) is straightforward and can be carried out by a variety of standard analytical techniques. The global system then has no emergent properties not fully predictable from knowledge of the components and their connections.

Incompletely known systems pose difficulties. If the transfer characteristics or information-processing characteristics of the components and connections (signal flow channels) are not completely known, then the whole system may have "emergent" properties not anticipated from the partial knowledge gained by analysis of components and connections.[12] The larger a system, and the greater the number of hierarchies of components and signal flow loops within it, the more likely are emergent properties to appear. If, in addition to incompletely specified complexity, the system also follows nonlinear mechanics, then the existing body of knowledge concerning systems in the concrete or in the abstract fails to serve for predictions. Thus, it is the restricted science of linear systems that flowered first in engineering practice, and that is now being explained to biologists.[13-15]

The justification for urging application of the principles of analysis of linear systems to biological problems is that the resulting models then become analytically tractable. Furthermore, for some purposes linear approximations may suffice. However, if the purpose of an investigation is the understanding of a given biological design, then nonlinearities must be accepted and no general method of analysis exists to deal with them. In order to proceed, an investigator must turn to simulation.

Model Building and Simulation

The nature of the contribution of model building and simulation to science has been well described by Grodins,[16] and some of his views are paraphrased in this section. Model building is intrinsic to all science, and it becomes extremely explicit in systems science. Facts (observations) and the concepts (laws, models) used to relate them must be taken together to define a science; neither facts nor concepts are sufficient alone.

A collection of observations, without concepts to relate them, may serve as raw data in the development of science. Organization of such raw data into nosological or taxonomic systems represents a primitive form of science, because classification requires some guidelines, and these guidelines are skeletal concepts. A branch of science has reached a mature stage of development when a set of observations is at hand and is being compared to testable implications extracted by deduction from a conceptual scheme (model). The conceptual scheme may be embodied in physical devices such as homologs, which emphasize structure, or in analogs, which emphasize function. It may also be expressed symbolically—by words, pictures, mathematics, or special computer programs. Simulation is the act of deducing observable implications from a conceptual scheme cast in any of the above forms.

If the process of deduction does not involve an assumption concerning the real structure relating the observations, then the conceptual scheme is "empirical" or "formal." If, however, a structure is assumed, then the model is explanatory, and it attempts to approach the isomorphic ideal in which there is a one-to-one quantitative correspondence between functional relations in the real world and those in the abstracted model. The aim of advanced simulation in theoretical biology is the codification of understanding in explanatory models that approach the isomorphic ideal. In pursuing this aim, the biologist uses as functional relations those unit processes made available to him from the sciences of physics and chemistry. At the present time, it is almost universally assumed that biology has no unit processes or elemental, basic functional relations of its own, distinct from those of physics and chemistry. Vitalism is thoroughly dead. It is the emergent properties of biological systems, those properties that stem from incompleteness of knowledge of the elemental components and causal relations, that give biology its peculiar aspect of residual vitalism. What remains to be seen is whether study of biological systems will force exposure of an incompleteness of physics and chemistry, and thus lead to recognition of new unit processes in these sciences. However, even if this happens, biology will concede these processes to those other sciences to which it is cognate, and no vitalistic principle outside of physics and chemistry is ever again likely to be invoked as an explanation in biology.

The successful study of physical reality by means of models and simulation requires intuition as well as knowledge and luck, and the whole activity is intensely personal. Therefore, it is to be expected that no general strategy can be offered as a guideline for all purposes. In fact, what steps are taken in simulation studies depends very much on the purpose of the investigator conducting the experiments. His aim may be to predict performance of a system under very limited conditions only, in which case a highly empirical, formal model may suffice for the study. If, in contrast, he wishes to explore an hypothesis that attempts to specify the design of a biological system, he will accept the isomorphic ideal, and seek the unit processes out of which the model may be constructed. Examples of both kinds of modeling in physiology and biochemistry may be found in several recent collections.[17,18]

In what follows we shall suppose that the scientist studying a biological system is interested in an abstract representation of the given design of that system, i.e., that he wants to achieve a simulation based upon a dynamic model which attempts to approach the isomorphic ideal. Some of the views to be presented have appeared elsewhere in slightly different form.[19,20] The major objectives of the systems approach in biology are (a) analysis: the determination of those properties and arrangements of components and of unidirectional signal pathways that give rise to the conspicuous functional attributes of a higher level system; and (b) synthesis: the rationalization and, ultimately, the mathematical abstraction of the dynamic system attributes, using component properties and pathways of connection and coupling.

The modern concern about system dynamic properties goes beyond the historical interest in function, though the two are of course related. They differ in that the historical concern was mainly for the observation of function and its qualitative description. Quantitative description sometimes followed, but, if so, the emphasis was strongest upon steady-state quantification. In modern biology the emphasis includes the dynamic properties of system arrangements as revealed by their transient responses, and with this changed emphasis comes a more explicit strategy of research and modeling. It is the purpose of the analytical stages of biological work to identify and characterize the mechanism—the relevant unit processes—of a system of interest. It is the purpose of the synthetic stage to demonstrate how the dynamic system properties at the highest level of interest arise from the lower level unit process characteristics and couplings. Such a demonstration requires the development of a dynamic model of the system.

Models have served well in physics and chemistry and are now becoming important in biology. This approach deserves emphasis, for those who have never engaged in computer simulation of biological system performance may not have discovered for themselves the powerful demands for intellectual precision and clarity that simulation imposes upon a scientist. Computer models provide a physical embodiment of hypothesis; and shortcomings, contradictions, or failures, as well as the need for new data of previously unsuspected importance, are made startlingly evident during simulation.

Successful simulation codifies facts efficiently, predicts some new aspects of system performance accurately, and represents what is meant by "understanding" of complex systems. Successful simulation leads to the assignment of weighting factors to unit processes, and suggests points of attack on systems that may be exploited by experiment, by disease, or by rational therapy. Successful simulation also defines the stability regimes for biological systems, relates these to the average operating point of the system, and in so doing reveals modes of oscillation and instability that may appear when systems are overdriven outside their "normal" ranges, as may occur in disease.

Forms of Models

Techniques of quantitative modeling currently available for biological work are summarized in Table 1-1. The heavy demand placed on simulation has already provoked some development of computation equipment and in the future may be expected to have a more profound effect. Furthermore, the distributed character of biological systems, their marked and essential nonlinearities, and their time-varying parameters and memory for their recent past history may invigorate the development of general systems theory. At the present time, a deterministic systems approach is mathematically limited by the intractable properties of nonlinear differential equations. For that reason, the use of computers in biological modeling is absolutely essential. In fact,

Table 1-1. Forms of Quantitative (Metric) Models

A. Sets of equations
    1. Continuous deterministic
    2. Piecewise continuous deterministic
    3. Stochastic deterministic (e.g., statistical mechanics)
    4. Finite-state discontinuous (e.g., Boolean)
    5. Stochastic nondeterministic
B. Physical homolog
C. Physical analog
    1. Direct (special purpose analog computer)
    2. Indirect (general purpose analog computer)
D. Digital computer programs
    1. Direct simulation by programmed equation sets
    2. Digital-analog simulation programs
    3. Continuous-system modeling programs
    4. Finite-state programs
E. Hybrids
    1. Hardware hybrids
    2. Equation set and program hybrids

computer use is superior to the use of classical mathematics for modeling highly nonlinear functions such as those of neural elements and brains.

The use of the machines shifts the attention of the investigator away from an overenthusiastic development of mathematical expressions of transfer functions with limited applicability toward mechanisms as manifested in unit processes. The use of linear electrical models for the simulation of biological processes has such severe limitations as to constitute a real danger to the novice. The nonlinearities inherent in real chemical, thermal, fluid mechanical, mechanical, and biological systems are not so conspicuous in electrical systems, and the seductive formalisms of linear control theory and feedback theory that have formerly prospered so well in electrical engineering usually will not serve for more than crude approximations in biological work. Thus, the demands of biology may be expected to elevate the art and science of modeling itself, for, after all, biological systems are the most complex systems yet known to man.

Hierarchies and Time Domains

Whatever form is used to model a biological system, the multilevel character of that system eventually taxes the ingenuity and intuition of the modeler. Almost every biological system has a lower-level unit process structure, each unit process of which is itself part of a subsystem, while at the same time the system of interest is a subsystem of a still higher level of organization. Inevitably, theoretical system analysis has to emphasize a particular level of organization, but in so doing it must acknowledge the analytical results available at the next lower level of organization, and the coupling relationships that tie into the next higher level.

The hierarchies of structure are associated with hierarchies of functional time domains. How modeling proceeds depends as much upon the choice of time domain as upon the structural level of interest. To illustrate the multilevel approach, we shall consider briefly some of the attributes of the adrenocortical system that have been discussed elsewhere in more detail,[21] as well as in Chapter 7 of this book.

The adrenal cortex may be considered as two glands, weakly coupled. One gland produces a class of hormones called glucocorticoids, and the other produces a class called mineralocorticoids. The glucocorticoid gland operates a synthetic chain of enzymatically catalyzed steps that convert cholesterol to a final product such as cortisol.

The conversion progresses at a rate determined by an input signal provided by corticotropin. At each subcellular enzymatic step, in the time domain of milliseconds, Michaelis-Menten kinetics give the detailed relation between the reaction rate and substrate concentrations. Such kinetics yield an equation for the steady state that is a rectangular hyperbola in form.

These kinetic considerations provide the basis for a saturation effect in the chain of enzymatic steps for, as soon as any one step enters the state described by the asymptotic region of the hyperbolic curve, it becomes a zero-order reaction and is rate limiting in the chain. Zero-order reactions in biological systems are more commonly rate limiting than are irreversible reactions (the other possible type of rate determinant), and if an irreversible reaction precedes a zero-order reaction in a chain, no steady state is possible. Since the adrenal glucocorticoid chain apparently has a steady state open to it, the rate-determining element is probably that of a reaction at zero-order kinetics provided by Michaelis-Menten theory.

When the adrenal glucocorticoid gland is studied in the time domain of minutes to hours, then the detailed kinetics at the millisecond level are not prominent. Nevertheless, the saturation characteristic that they provide still emerges as a prominent feature of the whole chain of steps. At the level of the whole gland, the Michaelis-Menten feature can be modeled empirically with a rectangular hyperbola generated without dynamics, because the empirical function has already been rationalized at a lower level of structure and time domain by the use of reaction kinetics as unit processes. Thus, this empiricism is an informed post hoc empiricism, rather than a blind ad hoc formalism.

In a comprehensive model of the adrenal glucocorticoid gland, the Michaelis-Menten element appears as a low-level component of a more complex structure.[22] However, in a still more comprehensive model of the whole adrenal glucocorticoid control system, the adrenal glucocorticoid gland itself appears as only one component. Furthermore, the global adrenal glucocorticoid control system is only one of many endocrine control systems that regulate chemical processing in animals such as humans, and in the time domain of several hours to days the contribution of the adrenocortical glucocorticoid control system to the regulation of blood sugar in man can appear as an arbitrary function with zero-order dynamics, or can be ignominiously lumped into a single parameter of a higher-level model.[23] The overwhelming importance of consideration of hierarchies in the understanding of biological systems has already been used as the basis of a philosophical treatise by the novelist Arthur Koestler.[24]

Dynamic Testing of Models with Multiple Forcing Functions

Proper application of systems analysis in biology requires that both the steady-state (static) and transient (dynamic) behavior of biological systems and their analogs be explored. An excellent example of the value of dynamic testing can be found in the work of Clynes,[25] who examined the response of the pupillary area to flashes of light, and then inverted the stimulus and presented negative flashes (pulse and step decreases in light intensity) to the eye. The unexpected finding was that the pupil constricted to both positive and negative flashes. This result indicated that the receptors may have a unidirectional rate-sensitive property previously unemphasized because the appropriate dynamic tests had not been performed.

Since almost all biological systems probably involve dynamic asymmetries and nonlinearities, it is essential that numerous different forcing functions be applied and that several different initial operating conditions be established in experiments in the laboratory and with the model. If a model successfully predicts the response of a biological system to step forcing from zero initial conditions, it may nevertheless fail to predict the response to any other forcing function, or even to the same one applied with nonzero initial conditions.

The standards of model testing have so far been rather low in biology, and the scientific community is often asked to consider models that have been documented by only one forcing function, under one set of initial conditions. Unless the model is known to be fully linear (in which case it may be trivial), further testing is necessary. In any case, whether or not the model is linear, the real biological system must be subjected to extensive dynamic testing before its characteristics can be sufficiently known to justify or validate a model. Appropriate forcing functions for dynamic testing include step, pulse, ramp, exponential, and sinusoidal functions. All but the last can be achieved, even in chemical systems, by programmed pumps. Sinusoidal forcing of chemical systems is also possible under special conditions.[21]

Signals and Noise: Continuous and Discrete Systems and Models

A signal is a time record of a measured variable. Both the measurement method and the variable itself will be associated with some noise. Noise may be regarded as changes detected in the measurement record of a variable of interest which are not associated with causal changes in the variable itself in response to particular causal events of interest to the observer. Thus, noise is relative, and what is noise to one observer may be the signal of primary interest to another observer. For example, variations of arterial blood pressure with respiration may be of immediate interest to the investigator concerned with circulatory-respiratory coupling, whereas to the person studying the response of the baroreceptors to pulse pressure cycles, the superimposed respiratory cycle may be a noisy nuisance.

Whatever the point of view of the observer, he will be confronted with some uncertainty in his methods, as well as with the possibility that the biological receptors of interest to him may be unable to detect every change in a continuous input variable. They may quantize continuous variables. The question then arises whether or not a continuous-system model which is highly deterministic should be used to simulate a biological system which may have intrinsic statistical properties different from those of the deterministic model.

Occasionally, separate finite-state models and continuous-system models of the same biological system have been developed,[19],[26] and both have been helpful. Of course, in engineering, hybrid approaches are possible. For example, in the case of a prosthetic device, an activator was designed to give a continuous controlled motion from an input which could assume only four distinct states.[27] However, the possible usefulness of combined discrete- and continuous-system modeling programs has not yet been demonstrated, though the technique has been considered.[28] At present the relative merits of continuous-system models, finite-state discrete-system models, or hybrid models remain unsettled, and the application of the systems approach in biology through simulation will benefit from further work on the comparisons of these different types of modeling programs. Meanwhile, the biologist who has believed that the system of interest to him is a continuous system might profit from considering in more detail the statistical properties of his measurement techniques, and of the system itself. His confidence in the appropriateness of continuous-system models may be shaken by the experience.

Stability

The traditional respect paid by physiologists to Claude Bernard's brilliant recognition that highly organized, multicellular animal systems maintain internal environments that resist the vicissitudes of their external environments has led to an overemphasis on the static stability of many of the state variables in biological systems. Iberall has suggested that the revered concept of homeostasis be supplanted by that of homeokinesis, in which regard is given to the probability that the stability achieved by biological systems is a marginal stability of ensembles of coupled, nonlinear, limit-cycle oscillators.[29] This view of biological systems leads logically to an emphasis upon spectral analysis of functional systems, an emphasis that has not heretofore been sufficiently prominent in biological research.

Stability is achieved in some cases by a simple regulation based upon the establishment of two independent functional relations, of opposite slope when graphically displayed, between a pair of variables. For example, when blood sugar increases independently, insulin secretion rate is dependently increased by one mechanism, whereas when insulin injection rate is independently increased, blood sugar is dependently decreased by a different mechanism. With respect to these two processes, blood sugar is stabilized at the point represented graphically by the intersection of the two functional relations. This point is determined by the parameters of the two relations, and not by set points, reference inputs, or servoaction within a control system. Riggs has introduced an analysis of such simple regulation in biological systems,[30] and characterized the effectiveness of the regulation by a term he calls the "homeostatic index," which is given by the negative product of the absolute values of the first derivatives of the two independent functions.

At a more elaborate level, stability or accurate following can be achieved by the servo mode of action of closed-loop, negative-feedback systems. Just how much use is made by biological systems of such techniques is still in doubt. Comparator action to generate error signals may be uncommon, except in organ systems such as brains, where an algebraic subtraction of variables expressed in the same measurements units is at least conceptually possible.

In engineering systems, stability is sometimes achieved by use of the Invariance Principle, which provides that a disturbance can itself directly initiate a control action to cancel out its effects, without recourse to comparison of the controlled variable with a reference value, as is necessary in negative-feedback control.[31] The Invariance Principle can be combined with negative-feedback control for better performance, but it is not yet known whether biological systems achieve stability in this manner.

The question of how biological systems achieve stability is far from trivial, because in the answer lies the basis for the philosophical foundations of theoretical biology itself, and systems analysis in biology cannot escape concern with the answer.

References

1. Hammer, P. C.: *The role and nature of mathematics.* The Math. Teach. 57:521, 1964.

2. Bronowski, J.: *The logic of the mind.* Amer. Sci. 54:1, 1966.

3. Bellman, R.: *Mathematical models of the mind.* Math. Biosci. 1:287, 1967.

4. Bohm, D., and Bub, J.: *A proposed solution of the measurement problem in quantum mechanics by a hidden variable theory.* Rev. Mod. Phys. 38:453, 1966.

5. Lindsay, R. B.: *Physics—to what extent is it deterministic?* Amer. Sci. 56:93, 1968.

6. Eyring, H.: *This changing world.* AAAS Bulletin, September, 1965.

7. Iberall, A. S., and McCulloch, W. S.: *1967 behavioral model of man—his chains revealed.* NASA Contractor Report CR-858, July, 1967.

8. Iberall, A. S., and McCulloch, W. S.: *The organizing principle of complex living systems.* J. Basic Engrg. In press.

9. Iberall, A. S.: *Introduction to the content of a general systems science.* US Army Research Office Contract Report DAHC 1967 C0027.

10. Pattee, H. H.: *The physical basis of coding and reliability in biological evolution.* Biophys. Lab. Rep. #193, W. W. Hansen Laboratories of Physics, Stanford University, May, 1967.

11. Gray, J. S.: *A physiologist looks at engineering.* Science 140:463, 1963.

12. Ashby, W. R.: *An Introduction to Cybernetics.* Science Editions, John Wiley, New York, 1963.

13. Milsum, J.: *Biological Control Systems Analysis.* McGraw-Hill, New York, 1966.

14. Grodins, F. S.: *Control Theory and Biological Systems.* Columbia University Press, New York, 1963.

15. Milhorn, H. T.: *The Application of Control Theory to Physiological Systems.* W. B. Saunders, Philadelphia, 1966.

16. Grodins, F. S.: *Theories and Models in Regulatory Biology.* In press, 1969.

17. Mesarovic, M. E. (ed.): *Systems Theory and Biology.* Springer-Verlag, New York, 1968.

18. Stacy, R. W., and Waxman, B. D.: *Computers in Biomedical Research.* Vol. II, Academic Press, New York, 1965.

19. Yates, F. E., Brennan, R. D., Urquhart, J., Dallman, M. F., Li, C. C., and Halpern, W.: *A continuous system model of adrenocortical function,* in Mesarovic, M. D. (ed.): *Systems Theory and Biology.* Springer-Verlag, New York, 1968, p. 141.

20. *A view of systems physiology.* Physiologist 11:115, 1968.

21. Yates, F. E., Brennan, R. D., and Urquhart, J.: *Adrenal glucocorticoid control system.* Fed. Proc. 28:71, 1969.

22. Urquhart, J., Krall, R. L., and Li, C. C.: *Analysis of the Koritz-Hall hypothesis for the regulation of steroidogenesis by ACTH.* Endocrinology 83:390, 1968.

23. Gatewood, L. C., Ackerman, E., Rosevear, J. W., and Molnar, G. D.: *Tests of a mathematical model of the blood-glucose regulatory system.* Comput. Biomed. Res. 2:1, 1968.

24. Koestler, A.: *The Ghost in the Machine.* Macmillan, New York, 1967.

25. Clynes, M.: *Unidirectional rate sensitivity: a biocybernetic law of reflex and humoral systems as physiologic channels of control and communication.* Ann. N.Y. Acad. Sci. 92:946, 1961.

26. Gann, D., Schoeffler, J. D., and Ostrander, L.: A finite state model for the control of adrenal corticosteroid secretion, in Mesarovic, M. D. (ed.): *Systems Theory and Biology.* Springer-Verlag, New York, 1968, p. 185.

27. Tomovic, R., and McGhee, R. B.: *A finite state approach to the synthesis of bioengineering control systems.* IEEE Trans. on Human Factors in Electronics, Vol. HFE-7, 1966, p. 65.

28. Brennan, R. D.: *Simulation Programming Languages.* Proc. of the IFIP Working Conference on Simulation Programming Languages. North-Holland, Amsterdam 1968, p. 371.

29. Iberall, A. S., and Cardon, S. Z.: *Regulation and control in biological systems.* Proc. of IFAC Tokyo Symposium on Systems Engineering for Control System Design, 1965, p. 463.

30. Riggs, D. S.: *The Mathematical Approach to Physiological Problems.* Williams and Wilkins, Baltimore, 1963.

31. Preminger, J., and Rootenberg, J.: *Some considerations relating to control systems employing the Invariance Principle.* IEEE Trans. on Automatic Control, July, 1964, p. 209.

2
Control Theory
in Biological Systems

Regulation and control in biological systems were of interest to biologists long before Cannon coined the word "homeostasis" to describe the tendency of living organisms to regulate certain variables in their internal environment.[1] It is common knowledge that variables such as body temperature, blood pressure, composition of body fluids, blood pH, the partial pressures of certain gases carried by the bloodstream, and the concentrations of a large number of chemical species in body tissues are closely controlled. Therefore, it is not surprising that we should investigate the applicability of the tools developed for the study of automatic control systems in engineering to the improvement of our understanding of biological control systems. During the past twenty-five years, a vast body of knowledge has accumulated concerning the analysis and synthesis of control systems. In fact, our everyday life surrounds us with examples of controllers and regulators constructed by man, such as devices which maintain the temperature of our hot water, adjust the gain of our radios, operate the power-assist devices in our automobiles, and regulate a myriad of processes in modern industry. The purpose of this chapter is to provide a brief introduction to the tools and methodology of control systems theory, and to indicate the applicability of this theory to a range of biological problems.

The chapter will be concerned with such topics as the importance of feedback in control systems, the effect of feedback on system sensi-

21

tivity to disturbances, the investigation of system stability, and transient response. The viewpoint of control theory will be applied to a series of example systems, including those concerned with the regulation of blood pressure, muscular force, respiration, and body water content.

The mathematical conceptual descriptions of the biological systems discussed are in fact simplified representations of the actual living organism. The mathematical tools introduced will be only those appropriate to the study of uncoupled linear dynamic systems. Hence, the biological problems that we examine will be simplified until they can be treated using these tools. And yet, as any student of biology knows, biological systems are characterized by their nonlinearity; by the variability which one observes, both in the repetition of the same experiment with a given organism and with different members of the same species; and by the fact that specific cause-and-effect relationships in living organisms are extremely difficult to isolate. It may then be asked whether the material in this chapter is simply an exercise in simplification, rather than a useful introduction to biological control systems. A dual answer to this criticism may be offered. First of all, even the basic elementary notions on linear control theory provide a point of view into the behavior of biological systems which leads to a level of insight quite different from that obtained from anatomical or verbal description. Secondly, there are tools and techniques in the field of control theory which are applicable to the study of time-varying, nonlinear, cross-coupled, and interacting systems, even in the presence of random disturbances. However, as may be expected, a considerably higher level of mathematical sophistication is required for the utilization of these tools. Hence, this chapter is an introduction, both in the techniques of control theory and in the description of biological control systems. As the student grows in knowledge and sophistication, he should be able to describe the biological systems with which he is concerned with greater and greater accuracy, and utilize more advanced techniques for their description and interpretation.

Open-Loop Versus Closed-Loop Control:
The Tradeoffs

In order to illustrate the advantages and disadvantages of the notion of feedback in control systems, let us consider two alternative designs for a temperature control system. The function of the system is to raise the temperature of a typical home in the winter time from an initial temperature of 50°F to a desired temperature of 70°F in as short a time as possible. Two alternative designs of such a system are shown in Figure 2-1.

Consider first Figure 2-1a. The controller indicated in this figure has the capability of turning the fuel supply on or off, based on a previously computed control strategy. Assume that the heating plant is capable of producing a 1° Fahrenheit temperature rise for each minute of operation. Thus, we can describe the plant by its "transfer function"

$$\frac{T_c}{u} = 1°F/min \tag{1}$$

Given the initial temperature $T_i = 50°F$, and the desired or reference temperature $T_{REF} = 70°F$, it is evident that a possible controller function is described by

$$u = 1 \text{ for } 0 \leq t \leq 20 \text{ min}$$
$$u = 0 \text{ for } t \geq 20 \text{ min} \tag{2}$$

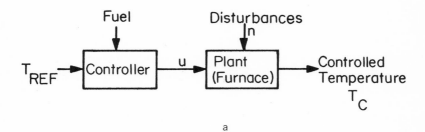

a

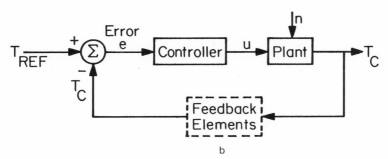

b

Figure 2-1. Alternative mechanizations of temperature control systems: a. open-loop, b. closed-loop.

It is evident that this system will be capable of producing the desired final temperature only under certain conditions, namely, if the initial temperature is indeed 50°F, if the plant is indeed described by Eq. (1), and if no external heat inputs or outputs (such as open windows, for example) are allowed. Extraneous inputs to the system are indicated in the figure as disturbances. This type of control system is sometimes referred to as "open-loop control." Let us summarize its problems: (a) the plant and controller characteristics must be known accurately, (b) variations in the environment and system parameters may produce erroneous results, and (c) disturbance inputs may produce erroneous results.

Figure 2-1b shows the same system with the addition of negative feedback, so that at each instant of time the room temperature is compared to the desired or reference temperature, thus producing an error which in turn becomes the input to the controller. In this case the controller input e is described by

$$e(t) = T_{REF} - T_c(t) \tag{3}$$

A typical controller function for this type of situation might be chosen by deciding to turn the fuel supply on when the error is greater than some temperature error threshold, and leaving the fuel off otherwise, that is, let

$$\begin{aligned} u &= 1 \text{ for } e(t) > 0.1°F \\ u &= 0 \text{ for } e(t) \leq 0.1°F \end{aligned} \tag{4}$$

Several features of this design become immediately apparent. First of all, it is evident that this system must incorporate some way of measuring the actual room temperature, thus increasing the complexity of the system over that shown in Figure 2-1a. On the other hand, if the initial temperature is not 50°, the furnace will simply stay on for more or less time, until the temperature error drops below 0.1°F. If the furnace characteristics change with time, e.g., from an accumulation of deposits, so that it eventually produces a temperature rise of only 0.8°F per minute, this too will be corrected by simply leaving the furnace on for a longer period of time. Let us summarize some of the differences between this "closed-loop" control and the open-loop control above:

1. Closed-loop control can compensate for plant and controller variations.
2. Closed-loop control can compensate for environmental disturbances.

3. Closed-loop control requires sensors (to measure the values of control variables) and some form of comparison process to obtain error quantities.

It will also be shown in Appendix 1 that the advantages obtained from closed-loop control exact an additional price from the designer, namely, that as a result of feedback, such systems may under certain conditions enter into oscillation.

Biological control systems invariably involve one or more feedback loops. The resulting closed-loop control enables the organism to function almost normally over a wide range of environmental conditions (disturbances) and in the face of variations in the biological parameters responsible for its adaptation to a variety of bodily states (e.g., exercise, postural changes, emotional factors, illness, etc.).

We can divide these control systems into two classes, voluntary and involuntary. Such tasks as steering a vehicle, following a moving object with the eyes, or reaching for a glass of beer are examples of voluntary control in which visual and other sensory inputs provide the necessary feedback. On the other hand, blood pressure, chemistry of the blood, and water content of the body are all involuntarily controlled, with the feedback coming from sensors in the body which are sensitive to these variables. These specific examples will be discussed in more detail.

In order to study linear control systems, we need to review the mathematical description of system components, both in the time domain and in the frequency domain. This will enable us to introduce the notion of open-loop and closed-loop transfer functions. With these basic tools, it will be possible to compare open-loop and closed-loop systems on the basis of sensitivity and stability. These mathematical concepts are discussed at length in the Appendix. The student should completely understand these concepts before proceeding further in this chapter.

Examples of Biological Control Systems

This chapter presents brief discussions of four biological systems, with emphasis on the analytical tools and techniques described in the Appendix. Other general treatments of biological control and regulation can be found in such books as those by Grodins,[6] Milhorn,[7] Yamamoto and Brobeck,[8] and Milsum.[9] Specific references for further study of particular systems are given in each of the following sections.

Respiratory Control of Blood Chemistry

It is appropriate to begin with a discussion of the manner in which the respiratory system operates to regulate blood chemistry, since this biological control system was one of the first to which control techniques were applied.[6,10] This controller, sometimes referred to as the "respiratory chemostat," functions to maintain blood concentrations of the gases CO_2 and O_2 and of the hydrogen ion within limits compatible with life. Variations from normal in the concentrations of these chemical species in the blood caused by metabolic changes, as in exercise or by breathing abnormal gas mixtures, are detected by specialized sensors (chemoceptors) in the body (Fig. 2-2). Chemoceptors respond to changes in blood chemistry by modifying the rate at which they transmit nerve impulses to the central nervous system (CNS). The CNS processes this information and in turn controls the muscles involved in breathing to change the ventilation rate ($\dot{V}_A$), which is the product of the rate and depth of breathing. $\dot{V}_A$ is the final control signal, since it affects the concentrations of the chemical species in the blood through altering the rates of removal of gases from the body and the resultant changes in blood acidity.

The elements of the respiratory chemostat for CO_2 control may be viewed in standard control system form in Figure 2-3. Here we assume that the chemoceptor simply senses the concentration difference $[CO_2]_e$ between the actual blood concentration $[CO_2]_B$ and the normal blood

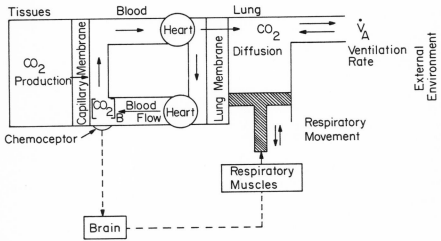

Figure 2-2. Schematic diagram of the respiratory chemostat.

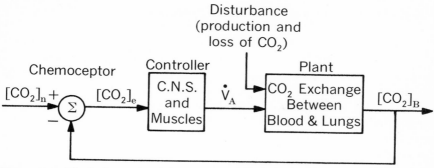

Figure 2-3. Block diagram of the respiratory chemostat.

concentration $[CO_2]_n$ and relays this information to the CNS. For the controller we may write that

$$\dot{V}_A = \dot{V}_{A_{ss}} + K[CO_2]_e \tag{5}$$

where $\dot{V}_{A_{ss}}$ = the value of $\dot{V}_A$ at $[CO_2]_n$

$$[CO_2]_e = [CO_2]_n - [CO_2]_B$$

and $\qquad\qquad\qquad K$ = a constant

Eq. (5) assumes that the dynamic response of the controller is much faster than that of the plant. The equation describing the dynamics of the plant can be derived from a CO_2 mass balance. Assuming $[CO_2]$ is the same everywhere in the bloodstream and is in equilibrium with the lung gas, then from Figure 2-2 we have

$$V_B \frac{d}{dt}[CO_2]_B = \dot{i}_{CO_2} - \dot{V}_A[CO_2]_B \tag{6}$$

where V_B = the blood volume (liters)

$\dot{i}_{CO_2}$ = metabolic production rate of CO_2 (liters/hr)

Eq. (6) states that the time rate of change of the quantity of CO_2 in the bloodstream is equal to the metabolic production rate of CO_2 in the body minus the rate at which CO_2 is eliminated from the body. Since $\dot{V}_A$ is a variable, Eq. (6) is nonlinear, and Laplace transforms do not enable us to obtain a transfer function for the plant. However, by assuming for the present that $\dot{V}_A$ is constant, we may take a heuristic look at the stability of this system. With this assumption, from Eq. (6) the transfer function of the plant is

$$\frac{[CO_2]_B(s)}{\dot{i}_{CO_2}(s)} = \frac{1}{V_B(s + \dot{V}_A)} \tag{7}$$

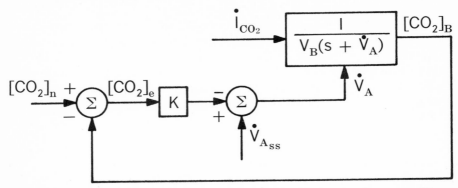

Figure 2-4. Simplified linear representation of the respiratory chemostat (illustrating parametric feedback).

Figure 2-4 shows the final form of the control system. We see that the final control signal $\dot{V}_A$ appears as a parameter in the plant transfer function (parametric feedback). Due to this nonlinearity, stability analysis is possible only by using computer simulation. However, there are some interesting conclusions which can be reached from the present approach.

For a step increase in $\dot{I}_{CO_2}$ (heavy exercise), we see from Figure 2-4 that $[CO_2]_B$ responds with an exponential rise to a new value. The speed of response and new steady state value are both determined by the magnitude of $\dot{V}_A$.

As $[CO_2]_B$ increases, the error is sensed by the chemoceptor and relayed to the CNS, which increases $\dot{V}_A$, thus tending to further increase the speed of response, but actually decreasing the new steady state magnitude of $[CO_2]_B$. Thus, this control system actually adapts to a disturbance by altering its dynamic response to accommodate changing physiological requirements.

Although the simplified representation of the respiratory chemostat is always stable in response to changes in CO_2 production rate since $\dot{V}_A$ is always positive, this is not true for the actual system in the organism under all conditions. It has been observed that there is an abnormal oscillatory breathing pattern (Cheyne-Stokes breathing) in some individuals. Milhorn and Guyton[11] postulate that this type of breathing is produced by transport delays which originate from the finite transfer rates of CO_2 between the tissues and the lung by way of the bloodstream. In our simplified analysis, the blood flow rate was implicitly assumed to be very fast in relation to the CO_2 diffusion rates into and out of the bloodstream.

Grodins and associates[12] have studied the respiratory control system in great detail using computer simulation. One of the aims of the simulation was to postulate a form for the controller variable ($\dot{V}_A$) to yield transient behavior of the system which matches experimental data. Computer simulation allows the inclusion of nonlinearities and complexities that cannot be handled by conventional linear control theory.

Control of Arterial Blood Pressure

The regulation of arterial blood pressure in the body is accomplished by means of the so-called "baroreceptor reflex," which is shown schematically in Figure 2-5, and in block diagram form in Figure 2-6. This control system functions in such a way that blood pressures in the large arteries supplying the head and brain are regulated in the presence of disturbances, for example those arising from postural changes, which can cause the blood mass to move due to the force of gravity. The actual blood pressure is detected by means of specialized sensors known as "baroreceptors" (or simply baroceptors) which are located primarily in the arterial wall of the aortic arch, near the bifurcation of the carotid arteries (in the carotid sinus) and in the branching of other arteries. These sensors, which are stretch or deformation sensitive, emit a train of nerve impulses along their output fibers at a rate related to the arterial pressure and its variations. The central nervous system processes this information with regard to the "reference" or "normal" pressures of the particular organism. Deviations from this reference level result in changes in the output or control signals in both the sympathetic and parasympathetic nervous systems that affect the resistance of the peripheral vascular bed, the heart rate, and the strength of contraction of the ventricles. Specifically, sympathetic activity acts upon the smooth muscles of the peripheral vascular system to cause vasoconstriction or vasodilation, thus raising or lowering the resistance to blood flow. Signals along sympathetic fibers also influence the contractility of the heart and the rate of heartbeat. Finally, signals originating in the vagus (parasympathetic) nerve also affect the heart rate.[13]

The sequence of events which follows a blood pressure disturbance may be sketched as follows: Assume that pressure at the baroceptors falls. The reflex then acts to constrict the peripheral vasculature, increase the heart rate, and increase cardiac output (by strengthing the contractile force of the ventricles). All three of these effects tend to

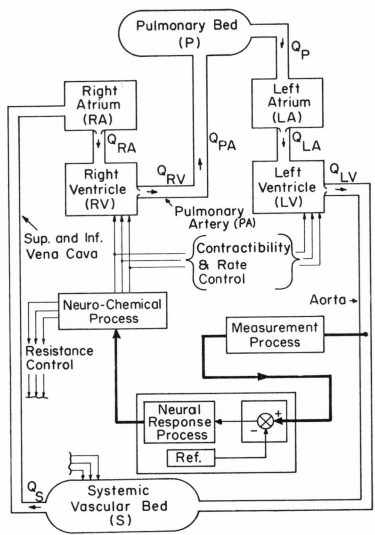

Figure 2-5. Schematic diagram of blood pressure control system.

raise the central arterial pressure, thus returning it to normal. It is evident that this is a closed-loop system with negative feedback.

The existence of this control system and the physiological role of the baroceptors have been known for over forty years, and portions of the system have been subjected to intensive investigation. In order

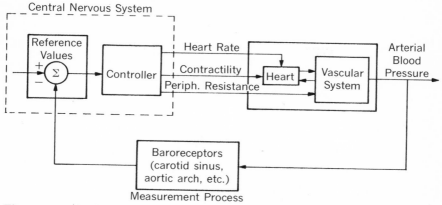

Figure 2-6. Block diagram of blood pressure control system.

to obtain mathematical expressions which characterize various blocks in the system, the loop is opened, as indicated in the diagram in Figure 2-7. With this configuration, it is possible to obtain experimentally a transfer function which approximately represents the baroreceptors, e.g., by applying a sinusoidal pressure waveform to the carotid sinus. Under these conditions, the carotid sinus baroreceptors in dogs have been characterized by[14]

$$G_{br}(s) = \frac{N(s)}{P_{cs}(s)} = \frac{K(1 + .036s)}{(1 + .0018s)(1 + as)} \tag{8}$$

where $P_{cs}(s)$ is the Laplace transform of the applied sinusoidal pressure wave, $N(s)$ is the Laplace transform of the impulse rate $n(t)$ along the carotid sinus nerve, and the constant K is used to account for the

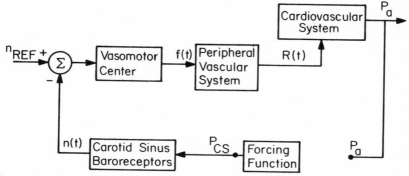

Figure 2-7. Open-loop measurements on blood pressure control system.

difference in units between pressure (in mm of Hg) and nerve impulse rate (in impulses per sec). The unspecified pole, located at $s = -1/a$, is needed when pressure input frequencies exceed 15 Hz, since experiments have shown that the gain of the receptors drops rapidly at high frequencies. For input frequencies below 15 Hz, $a = 0$.

Unfortunately, the transfer function of (8) only approximates the results of experimental observations. Even with a sinusoidal pressure wave, the nerve impulse rate is not sinusoidal and the transfer function represents a fit only to the fundamental component of receptor response. The harmonic distortion (when inputs in the frequency range of 0.5 to 15 Hz were used) was about 10 percent, indicating the presence of some nonlinearity in the receptor characteristic.

It is also interesting to examine the transfer function for low frequencies, i.e., when $a = 0$. In this case the zero is located at $s = -1/.036$, while the pole is located at $s = -1/.0018$. A sketch of the amplitude portion of the Bode diagram (Fig. 2-8a) illustrates the fact that the baroreceptor behaves as a high-pass filter in the vicinity of 10 Hz.

The overall open-loop frequency response of the pressure control system has been studied by Grodins,[6] Levison and his co-workers,[15] Scher and Young,[16] Sagawa,[17] Warner,[18] and others. In this case, the carotid sinus is forced with a sine wave of pressure superimposed on a mean or steady-state pressure $\overline{P}_{cs}$. The output (systemic arterial pressure) then contains a mean as well as a pulsating component. The fundamental component of the output oscillation can be related to the input oscillation by the linear differential equation

$$K_1 p_a(t) + K_2 \dot{p}_a(t) = K_3 p_{cs}(t - \tau) \tag{9}$$

where $p_a(t)$ and $p_{cs}(t)$ refer to the sinusoidal components of the arterial and carotid sinus pressures, respectively, τ is a time delay, and the K's are constants. However, the mean output pressure, $\overline{P}_a$, depends in a nonlinear way on the input frequency. Increasing the frequency of carotid sinus pressure wave form leads to a decrease in $\overline{P}_a$. This phenomenon, known as "frequency-dependent rectification"[16] makes linear analysis of the overall response impossible. Nevertheless, the oscillatory portions of the response alone, as indicated in eq. (9), lead to the Bode diagram (Fig. 2-8b) from which it can be seen that the overall system functions at frequencies much lower than those at which the receptor gain is highest.

To investigate the source of the disparity in frequency response characteristics shown in Figure 2-8, additional experiments have been used. Scher and Young[16] have studied the relation between sympathetic

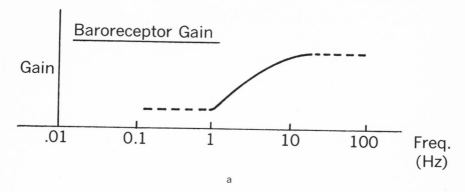

a

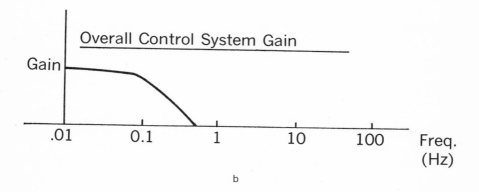

b

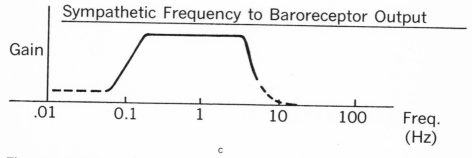

c

Figure 2-8. Bode diagrams for portions of blood pressure control system.

nerve impulse rate f(t) and the baroceptor firing rate n(t). This relation can be approximated by

$$\Delta f(t) = K_4 n(t - \tau_1) \tag{10}$$

where $\Delta f(t)$ is the change in sympathetic impulse rate and τ_1 represents a delay due to CNS processing of the information. Relation (10) is valid for frequencies in the vicinity of 1 Hz. (Units must be considered carefully. Thus, the slope of a pressure sine wave is measured in mm of Hg per sec, while the slope of a nerve impulse rate sine wave is measured in impulses per sec per sec.) For the frequency range in which Eq. (10) is valid, its corresponding transfer function is

$$\frac{\Delta F(s)}{N(s)} = K_4 e^{-\tau_1 s} \tag{11}$$

Since the complex exponential $e^{-\tau_1 s}$ (the Laplace transform of the time delay operator) has a magnitude characteristic of unity, it contributes only to the phase portion of the Bode diagram. The amplitude portion is shown in Figure 2-8c. The experimentally known attenuation of the CNS processing portion of the system near 0.1 and 10 Hz is also shown. Eqs. (10) and (11) represent only the straight-line portion of the diagram.

   If one examines the three Bode diagrams of Figure 2-8, it is evident that the generally low-pass nature of the overall system characteristics is due neither to the baroreceptors nor to CNS processing. Hence, we can hypothesize that the frequency response of the total loop is limited by the dynamic response of the smooth muscle of the resistance vessels in the cardiovascular system.[14] It is interesting to note that this hypothesis follows directly from a control systems approach to the study of blood pressure regulation.

Neuromuscular Control System

   The neuromuscular system of the body consists[19,20] of the skeletal muscles, the prime movers of limbs; the nerves, which lead to and stimulate the muscles to contract and move a limb; sensors, located within the muscle fibers, which are sensitive to changes in length or tension; and the central nervous system (CNS), which integrates all the input information from sensors and higher centers in the brain. Output or control signals, in the form of nerve impulses, are transmitted along nerve axons eventually to stimulate movement in order to accomplish a desired muscular task. To understand the dynamic response of an individual in a given control task, such as steering a car or docking a space vehicle, we must investigate and interrelate each of the elements of this system.

The desire to move a limb initiates a series of nerve impulses generated from higher centers in the brain. These travel down nerve trunks in the spinal cord to ultimately connect (synapse) with large nerve cells (α motoneurons) at various levels of the cord. The α motoneuron is stimulated to generate nerve impulses along its axon at an increased rate, which in turn causes the muscle fibers it innervates to contract or shorten.

The movement of a limb involves the coordinated action of at least two individual muscles (Fig. 2-9). One of these (the agonist) will be stimulated by its α motoneuron to shorten, whereas the other(s) will be inhibited from shortening. This inhibition results from a decrease in the rate of firing of the α motoneurons, due to the inhibitory influence of an intermediate nerve cell (interneuron) in the spinal cord which has, itself, been stimulated by feedback signals from sensors in the muscle.

As the agonist muscle changes length, sensors (muscle spindles) in the muscle modify the rate of discharge of nerve impulses along their

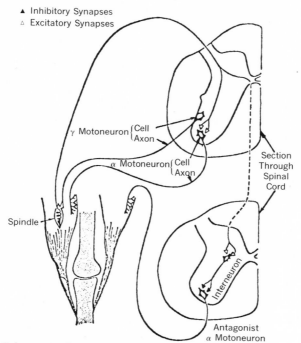

Figure 2-9. Schematic diagram of neuromuscular system.

axons which lead back to the spinal cord, and synapse with both the α motoneuron and the nerve cell bodies of the interneurons, thus completing the feedback pathway. Figure 2-10 shows these interactions in block diagram form. There is an additional loop shown in the diagram involving the small or γ motoneurons. These may also be stimulated from higher centers and in turn cause positional changes of the spindles, thus providing a bias adjustment to the sensor.

This system may be simplified and diagrammed in more conventional control system form as shown in Figure 2-11. Here we look at the net or average result of the two or more muscles in moving the limb. The muscle spindle responds to an error in muscle position $(\gamma_c - x)$ by changing the frequency (Δf_{sp}) at which it discharges nerve impulses along its axon leading back to the spinal cord. Stimulation from higher centers sums with the spindle output to give a net change in the rate of α motoneuron firing (Δf_α). The muscle then responds by a change in length according to the magnitude of the load and the degree to which it is stimulated.

A transfer function for the sensor (muscle spindle) has been derived[21] from a hypothetical mechanical model, in the form

$$\frac{\Delta f_{sp}(s)}{(\gamma_c - x)(s)} = \frac{K_{sp}(T_{sp}s + 1)}{(\beta T_{sp}s + 1)} \tag{12}$$

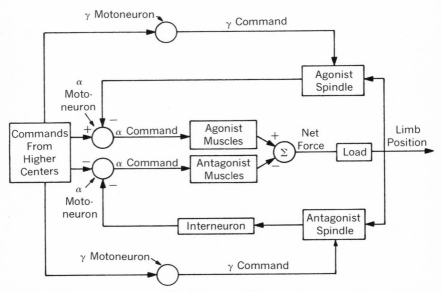

Figure 2-10. Block diagram of neuromuscular system.

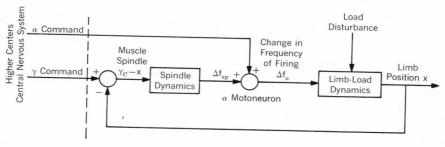

Figure 2-11. Limb position control system.

where K_{sp} is a constant and T_{sp} is the time constant of the spindle. The transfer function (12) is of the lead-lag form, which gives added stability to the neuromuscular system. The limb-load dynamics (the plant) have been represented by a number of mathematical models.[21,22] Figure 2-11a shows one of these.[21] A transfer function may be derived from this model for limb position (x) as a function of differential firing rate (Δf_α)

$$\frac{X(s)}{\Delta F_\alpha(s)} = \frac{C \dfrac{(K_M + K_L)}{M}}{s^2 + \dfrac{B_M + B_L}{M} s + \dfrac{K_M + K_L}{M}} \tag{13}$$

where C = a constant
M = the mass of the muscle, limb, and load
B_M = the viscous damping of the muscle
B_L = the viscous damping of the load
K_M = elastic spring constant of muscle
K_L = elastic spring constant of load

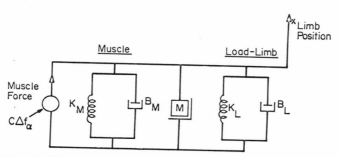

Figure 2-11a. Model of muscle-load dynamics.

The muscle parameters B_M and K_M are in reality functions of the tension developed in the muscle and the degree of stimulation. Muscles change their physical state when they are stimulated in such a way that both K_M and B_M increase. Thus, parametric feedback is present here as in the respiratory chemostat system, although these parameters do not act as the final control signal.

Load disturbances such as variations in K_L cause changes in limb position which must be compensated by changes in muscle tension by way of the feedback loop through the muscle spindle. This involuntary reflex mechanism (stretch reflex) is the main control means by which precise movements are obtained. Thus, the question of stability of this reflex is of prime importance in all coordinated muscular movements.

Regulation of Body Water Content

The total volume of water in the body (TBW) is regulated around a normal value for each individual by a combination of nervous and hormonal influences acting upon intake and excretion of fluids. The

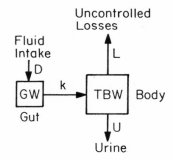

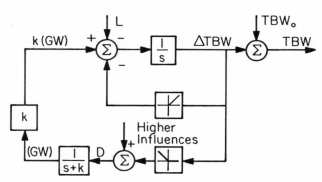

Figure 2-12. Two-compartment model of total body water control system.

TBW is maintained between 58 and 62 percent of an individual's body weight, in spite of an uncontrolled daily water loss (in feces, perspiration, etc.) of about 2 percent of the body weight. Excesses in fluid intake are compensated by increased quantities of fluid excretion by the kidneys (urine), while fluid deficits are made up from increased fluid intake.

Equations can be derived for the input and output of fluids from the body from the simplified model shown in Figure 2-12:

$$\frac{d}{dt}(GW) = D - k(GW) \tag{13}$$

$$\frac{d}{dt}(TBW) = k(GW) - U - L \tag{14}$$

where (GW) = volume of water in the gut
    D = rate of water intake into the gut
    k = water absorption constant
    U = rate of urine output
    L = rate of uncontrolled water loss

The naive assumptions that drinking occurs when TBW is less than some normal value TBW_0, and that urine excretion takes place when $TBW > TBW_0$, yield the control system of Figure 2-12, which even at this simplified state of analysis involves nonlinearities, making conventional control analysis difficult. By inspection of the control system, we see that, at steady state, the uncontrolled rate of water loss (L) would be compensated by a steady rate of drinking (D_{ss}), with ΔTBW being

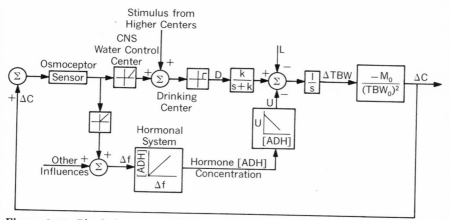

Figure 2-13. Block diagram of TBW control system illustrating osmoceptor activity and hormonal influences.

sufficiently negative to achieve the value of D_{ss}. Thus, a conclusion has been reached of a steady state drinking rate which is contrary to our experience and physiological knowledge. Obviously, refinements must be made in the model to match the known physiology.

It is known that there are specialized receptors (osmoreceptors) in the body which are sensitive to changes in effective solute concentration (osmotic pressure) of the body fluids from the normal. It can be shown that changes in the volume of the body fluids (ΔTBW) due to water addition or loss are equivalent to changes in the solute concentration. The solute concentration may be expressed as

$$C = \frac{M}{TBW} \tag{15}$$

where C = the solute concentration in milliosmoles/liter of fluid
       M = the quantity of solute in milliosmoles

Small changes in the concentration (ΔC) may be expressed by a Taylor series expansion of Eq. (15):

$$\Delta C \approx - \frac{M_0}{(TBW_0)^2} (\Delta TBW)$$

where M_0 = the normal quantity of solute in the system
      TBW_0 = the normal quantity of water in the system

The osmoceptors relay the information concerning ΔC to a higher center in the brain responsible for water control in the body. The center must coordinate the intake of water through drinking and the output through urine production. To a reasonable approximation, drinking occurs as an on-off phenomenon at a constant rate. This is equivalent to the heater in the thermostat system discussed in the first part of the chapter. Urine excretion, on the other hand, is under direct hormonal influence. The hormone (ADH or antidiuretic hormone) is produced by the pituitary gland upon stimulation from the water control center when solute concentration increases above normal. In its absence, urine excretion rate (U) is maximum, and when ADH is at its highest concentration. U is minimum. The overall system is shown in Figure 2-13.

In summary, we may review qualitatively the action of this control system. If L takes a step increase (e.g., due to heavy exercise), then TBW decreases and ΔC increases. ADH production is increased, which increases [ADH], resulting in a decrease in U (a water-saving mechanism). After ΔC exceeds a threshold (3 to 4 percent increase over normal), drinking begins at a constant rate D and continues until TBW returns to the normal range. U also returns to normal. If, on the other hand, drinking is stimulated from higher centers due to emotional or other factors, then the opposite set of events occurs with an increased urine production and a turning off by the drinking center of any stimulation to drink. Reeve and Kulhanek[23] have simulated a control system for body water regulation similar to this one, and have found that a limit-cycle oscillation occurs in TBW with a period and amplitude critically dependent upon the drinking threshold.

Concluding Remarks

Hopefully, by now, the reader will have noted the degree of usefulness of control theory to the study of biological systems. There is no question that the normal functioning of the body is dependent upon controlling processes, from the cellular level up to and including the organ system level. We have found that viewing these systems in conventional block diagram form has provided a great deal of qualitative insight into their function, as well as illustrating the areas in which quantitative data are necessary to fill in our gaps of knowledge. However, conventional control theory based upon the assumption of linearity is not in itself adequate to explain the workings of the inherently nonlinear biological control system. Computer simulation must then be used to finish what the control system approach has started. Some may go directly to simulation as their tool, but many others will go through the processes described here to study these and other biological systems, and will benefit from the experience.

References

1. Cannon, W. B.: *The Wisdom of the Body.* Norton, New York, 1936.

2. Aseltine, J. A.: *Transform Method in Linear System Analysis.* McGraw-Hill, New York, 1958.

3. Kuo, B. C.: *Automatic Control Systems.* Prentice-Hall, Englewood Cliffs, N.J., 1962.

4. Dorf, R. C.: *Modern Control Systems.* Addison-Wesley, Reading, Mass., 1967.

5. Pipes, L. A.: *Applied Mathematics for Engineers and Physicists.* ed. 2. McGraw-Hill, New York, 1958.

6. Grodins, F. S.: *Control Theory and Biological Systems.* Columbia University Press, New York, 1963.

7. Milhorn, H. T.: *The Application of Control Theory to Physiological Systems.* W. B. Saunders, Philadelphia, 1966.

8. Yamamoto, W. S., and Brobeck, J. R. (eds.): *Physiological Controls and Regulations.* W. B. Saunders, Philadelphia, 1965.

9. Milsum, J. H.: *Biological Control Systems Analysis.* McGraw-Hill, New York, 1966.

10. Grodins, F. S., Gray, J. S., Schroeder, K. R., Norins, A. L., and Jones, R. W.: *Respiratory responses to CO_2 inhalation. A theoretical study of a nonlinear biological regulator.* J. Appl. Physiol. 7:283, 1954.

11. Milhorn, H. T., Jr., and Guyton, A. C.: *An analog computer analysis of Cheyne-Stokes breathing.* J. Appl. Physiol. 20:328, 1965.

12. Grodins, F. S., Buell, J., and Bart, A. J.: *Mathematical analysis and digital simulation of the respiratory control system.* J. Appl. Physiol. 22:260, 1967.

13. Burton, A. C.: *Physiology and Biophysics of the Circulation.* Yearbook Medical Publishers, New York, 1965.

14. Spickler, J. W., Kezdi, P., and Geller, E.: Transfer characteristics of the carotid sinus pressure control system, in Kezdi, P. (ed.): *Baroreceptors and Hypertension.* Pergamon Press, New York, 1967, pp. 31–40.

15. Levison, W. H., Barnett, O., and Jackson, W. D.: *Nonlinear analysis of the baroreceptor reflex system.* Circ. Res. 18:673, 1966.

16. Scher, A. M., and Young, A. C.: *Servoanalysis of carotid sinus reflex effects on peripheral resistance.* Circ. Res. 12:152, 1963.

17. Sagawa, K.: Relative rate sensitive and proportional elements of the carotid sinus during mild hemorrhage, in Kezdi, P. (ed.): *Baroreceptors and Hypertension.* Pergamon Press, New York, 1967, pp. 97–105.

18. Warner, H. R.: *Use of analog computer in the study of control mechanisms in the circulation.* Fed. Proc. 21:87, 1962.

19. Katz, B.: *Nerve, Muscle and Synapse.* McGraw-Hill, New York, 1966.

20. Stark, L.: Neurological feedback control systems, in Alt, F. (ed.): *Advances in Bioengineering and Instrumentation.* Plenum, New York, 1966, pp. 289–385.

21. McRuer, D. T., Magdaleno, R. E., and Moore, G. P.: *A neuromuscular actuation system model.* IEEE Trans. on Man Machine Systems, September, 1968, p. 61.

22. Houk, J. C.: *A mathematical model of the stretch reflex in human muscle systems.* M.S. Thesis, M.I.T., Cambridge, 1963.

23. Reeve, E. B., and Kulhanek, L.: Regulation of body water content: a preliminary analysis, in Reeve, E. B., and Guyton, A. C. (eds.): *Physical Bases of Circulatory Transport.* W. B. Saunders, Philadelphia, 1967.

24. Koshikawa, S., and Suzuki, K.: *Study of osmo-regulation as a feedback system.* Med. Biol. Engrg. 6:149, 1968.

3
Fluid Flow

Flows play an essential role in life processes. The physiological conditions in the basic biological organism, the cell, are characterized by the fluidity of the cellular material. Stimulation, anesthesia, and death are associated with changes in fluidity. Increasingly sophisticated flow processes have made possible the evolution of multicellular organisms, up to the complexity of the air-breathing homothermal mammal.

This chapter presents a brief overview of some of the principal methods available today for achieving a quantitative mechanical description of biological flow processes.

A fluid is a substance whose deformation under the action of a force is fully or partly irreversible. It is indeed the irreversible part that constitutes the flow. The force causing the flow may be mechanical or the result of more generalized agencies involving chemical and electrical potentials, temperature gradients, etc. Figure 3-1 exemplifies the complexity of flow processes in a biological system, caused by factors such as multiphasic nature of the fluids, flexible boundaries through which mass, heat, and momentum exchanges may take place, complex geometry with branches and irregularities, and flow conditions varying with time.

In the analysis of a biological flow system such as that in Figure 3-1, two aspects are of interest: (a) the bulk relationships between inputs and outputs and (b) the details of the flow. To determine the former,

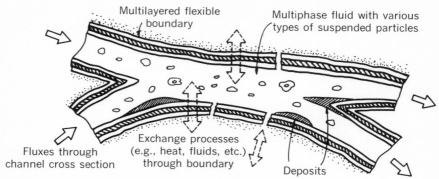

Figure 3-1. A schematic of a biological flow system indicating some of the parameters.

a "black-box" approach (one-dimensional analysis) is used, while for the latter the differential equations are solved which describe the flow characteristics of an infinitesimal fluid element. These two approaches are related, since the bulk characteristics result from integration of the detailed ones.

Mechanical analysis of a flow problem requires choice of a convenient coordinate system, and establishment and solution of equations describing the balance of forces in the flow and the characteristics of the fluid. The solution can be reached through several avenues. Analytic solutions are the most desirable, because of their parametric nature with a number of parameters describing a family of flows. Complete analytical solutions for three-dimensional flows are seldom feasible at present. Solutions are available only for certain simple cases in which the complexity of a problem is reduced by simplifying the original geometry, or by analyzing the corresponding two- or one-dimensional case. Numerical analysis provides an alternate approach to the solution of complex flow problems. Solutions can also be obtained by resorting to analogs, i.e., to other more easily studied phenomena (mechanical, electrical, etc.) governed by the same equations.

Experimental measurements are needed to provide the data necessary to describe the phenomena, to specify coefficients in the equations, and to check the correctness of various types of solutions.

Complex biological flow phenomena can be analyzed through model studies designed to maintain similarity of corresponding entities (geometrical configurations, velocities, and forces) in the model and the living organism (the prototype). Thus, if L_M, V_M, and F_M are respectively a geometrical dimension, a velocity, and a force in the model, and if L_P, V_P, and F_P are the corresponding quantities in the organism, each of the ratios L_M/L_P, V_M/V_P, and F_M/F_P must be constant for homologous lengths, velocities, and forces (geometric, kinematic, and dynamic similarity, respectively).[1]

Conservation of Mass

Mathematical analysis of biological flow systems is based on the differential equations stating the general principles of conservation of mass, momentum, and energy. In a continuous flow field, the rate of exchange of mass within any closed surface must be equal to the net rate of mass flux (mass flux being defined as mass times velocity) through the surfaces. Fluids can be considered incompressible whenever the ratio of the velocity V of the fluid to the sound velocity c in the fluid (a ratio known as the Mach number $M = V/c$) is small (<0.4), as is the case in most biological situations. In the Cartesian coordinates (x_1, x_2, and x_3), the continuity equation for an incompressible fluid is $\nabla \cdot \vec{V}$, that is

$$\frac{\partial u_1}{\partial x_1} + \frac{\partial u_2}{\partial x_2} + \frac{\partial u_3}{\partial x_3} = 0 \tag{1}$$

where u_i is the velocity component of the fluid in the x_i direction.

Constitutive Equations

In general terms, the constitutive equation of a substance is a relation between a forcing function and a response function. A broad spectrum of material behavior can be described by different functional relationships. At one end of the spectrum are solid elastic materials whose chief characteristic is reversible deformation under external forces; at the other end are materials traditionally defined as fluids and characterized by continuous irreversible deformation when sheared. Between the two extremes stand substances partaking in varying degree of the characteristics of both solids and fluids (viscoelastic substances).

In a one-dimensional flow of a linear (i.e., Newtonian) incompressible fluid characterized by a velocity distribution u(y), the shear stress τ between two fluid layers is linearly proportional to the velocity gradient du/dy (Fig. 3-2a):

$$\tau = \mu \frac{\partial u}{\partial y} \tag{2a}$$

where μ is called the dynamic viscosity of the fluid. More generally, in the three-dimensional flow of such a fluid, the viscous stress component τ_{ij} is a linear function of the rate of angular deformation (or shear rate)

$$\tau_{ij} = \mu \left(\frac{\partial u_i}{\partial x_j} + \frac{\partial u_j}{\partial x_i} \right) \tag{2b}$$

where τ_{ij}, oriented in the j-direction, is on a plane normal to the i-direction (Fig. 3-2b). When $i = j$, the viscous stress component

$$\tau_{ii} = 2\mu \frac{\partial u_i}{\partial x_i} \tag{2c}$$

is normal to such a plane (Fig. 3-2c) and is superimposed on the pressure p.

However, biological fluids are usually nonlinear and time-dependent. Experimental information leading to a rigorous determination of their constitutive characteristics is at present very scarce. It is thus necessary to resort to empirical or semiempirical constitutive equations, which are of limited validity outside of the range of conditions under which they were obtained. For biological fluids, among the simplest and more widely used descriptions of this kind, derived for the case of a fluid sheared between two parallel plates with shear rate du/dy (Fig. 3-2d), are the following models.[1]

The Bingham Plastic Model (Fig. 3-2e).

$$\tau = \tau_y + \mu \frac{du}{dy} \tag{3}$$

where τ_y is the stress that must be exceeded before flow can occur (yield stress), and the slope of the curve is the dynamic viscosity μ.

The Power-Law Model (Fig. 3-2f).

$$\tau = C_1 \left(\frac{du}{dy} \right)^n \qquad n > 0 \tag{4}$$

where C_1 is a constant. For $n = 1$, the relation between τ and rate of deformation du/dy is linear, i.e., the fluid is Newtonian, and C_1 becomes the dynamic viscosity μ. For $n < 1$, the apparent viscosity, i.e., the ratio of τ to du/dy ($= C_1 (du/dy)^{n-1}$), decreases with higher shear rates du/dy (pseudoplastic, or shear-thinning behavior), whereas for $n > 1$, it increases (dilatant or shear-thickening behavior).

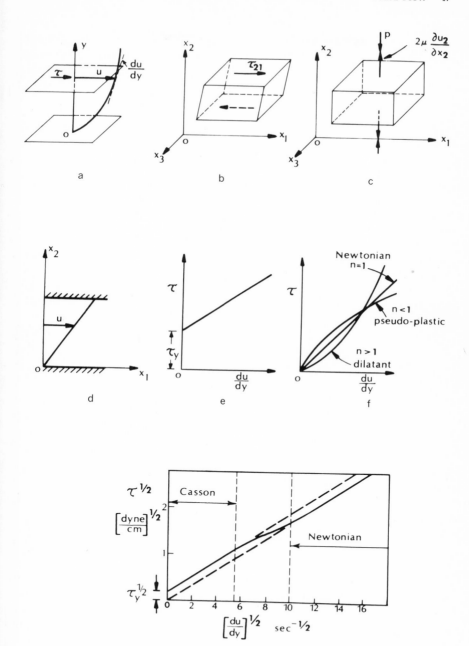

Figure 3-2. Flow in biological system. Explanations in text.

The Casson Model.

$$\tau^{1/2} = \tau_y^{1/2} + C_2\left(\frac{du}{dy}\right)^{1/2} \tag{5}$$

where C_2 is a constant and, as for the Bingham Plastic Model, τ_y is a yield stress.

At times the behavior of a biological fluid is better described by a combination of different models. Thus, for normal human blood, a model has been suggested consisting of three regions[2]: a Casson region for low shear rates (<31.5 sec^{-1}), a Newtonian region for high shear rates (>105 sec^{-1}), and an intermediate transition region (Fig. 3-2g).

Equations of Motion for Linear Fluids
(The Navier-Stokes Equations)

From Newton's Second Law, the product of mass and acceleration (the time rate of change of linear momentum, or the so-called inertial force) of any arbitrary volume in space is equal to the sum of the surface and body forces acting on the element. Surface forces are due to the stress vector, while body forces are field forces acting throughout the element. Examples of body forces are gravity (which is significant, e.g., when the density varies with position) and electromagnetic forces (important, e.g., in membrane transport and in biomedical instruments such as the electromagnetic flow meter).

For linear and incompressible fluids, the equations of motion are

$$\rho\left(\frac{\partial u_i}{\partial t} + \sum_{j=1}^{3} u_j \frac{\partial u_i}{\partial x_j}\right) = \rho X_i - \frac{\partial p}{\partial x_i} + \mu \sum_{j=1}^{3} \frac{\partial^2 u_i}{\partial x_j^2} \quad \text{for i = 1, 2, 3} \tag{6}$$

(DENSITY × ACCELERATION)   BODY   (PRESSURE + VISCOUS)
     INERTIAL FORCE   FORCE   (FORCE   FORCE)
                                   SURFACE FORCE

Eq. (6) shows that the flow acceleration (or deceleration) in the i-direction is due to the pressure gradients $\partial p/\partial x_i$, the body force ρX_i, and the gradients $\mu \sum_{j=1}^{3} \partial^2 u_i/\partial x_j^2$ of the viscous stress components. If the body force is due to a gravitational potential Ω (such that $X_i = -\partial\Omega/\partial x_i$), and if the x_2-axis is oriented in the vertical direction, then $X_1 = X_3 = 0$, $X_2 = g$, the gravitational acceleration, and $\Omega = -gx_2$. In the confined flow of a homogeneous fluid, the gravitational effects can be included in p.

Solution (integration) of the equations of motion leads to a description of the flow field. The solution is determined by the boundary and initial conditions which characterize a given flow phenomenon. For example, a boundary condition for most fluids is that the velocity of the fluid at the boundary (in both the tangential and the normal components) must equal the velocity of the boundary:

$$(u_i)_{\text{wall}} = \left(\frac{\partial \zeta_i}{\partial t}\right)_{\text{vessel}} \qquad \text{for } i = 1, 2, \text{ and } 3 \tag{7}$$

where ζ_i is the i-component of the displacement of the vessel wall. When the vessel wall is fixed, $(u_i)_{\text{wall}} = 0$. If the flow is confined by elastic walls, the equations of motion of the fluid and of the elastic wall must be solved simultaneously. Other examples of boundary conditions are the pressures at the entrance and outlet of a blood vessel. When the flow is unsteady, in addition to the boundary conditions, it is necessary to give a complete description of the flow at some prescribed instant (initial conditions).

Peristaltic Flows

Pumping by periodic motion of a flexible boundary is a characteristic biological mechanism for imparting energy to fluids (as in the heart, veins, intestine, and ureters†). The motion of a homogeneous Newtonian fluid, moving in a vessel under the action of waves propagating longitudinally along the vessel wall (peristaltic flow), has been analyzed under the simplifying assumptions of two-dimensional or axisymmetric flow and constant-wall wave celerity c.

Small Amplitude-to-Wavelength Ratio: The Linear Case. For a two-dimensional case, a coordinate system (x^*, y^*) set to travel with the wave makes the wave pattern stationary, and the flow velocity (u^*, v^*) independent of time (since the local acceleration vanishes). For a wavelength $\lambda \gg A$, the convective inertia effects are negligible and so is the pressure gradient normal to the longitudinal (x) axis. Thus, Eq. (6) reduces to the "quasi-one-dimensional" relationship

$$\frac{dp}{dx^*} = \mu \frac{\partial^2 u^*}{\partial y^{*2}} \tag{8}$$

where p includes the gravity effects. Integration with the boundary conditions $\partial u^*/\partial y^* = 0$ at $y^* = 0$, and $u^* = -c$ at $y^* = h$ (where h,

---

†In the case of the ureter, the wave celerity c may be typically of the order of 3 cm/sec, and the kinematic viscosity ν of the urine 0.007 cm²/sec. The cross section of the normal ureter varies from nearly round, when fully distended, to roughly star-shaped when contracted. In the latter configuration, each arm of the star may be treated approximately as a two-dimensional channel. In the normal ureter, the wave amplitude is generally not too small compared to the lumen, whereas under pathological conditions it may be small.

the lumen, is a function of x only) gives the longitudinal velocity component u* and the discharge q* (per unit width) with respect to the moving coordinate system, while integration of the continuity equation gives the radial velocity component v* $(= -\int_0^{y*}$ $(\partial u*/\partial x*)\, dy*$, with v* = 0 at y* = 0). Figure 3-3 provides an example of longitudinal velocity profiles u(y) (with respect to the fixed coordinates) due to a sinusoidal wall-wave:[3]

$$h = R + A\sin\frac{2\pi}{\lambda}(x - ct) \tag{9}$$

where A is the wave amplitude. The transport ratio θ $(=\bar{q}/Ac$, ratio of the actual time-mean flow $\bar{q}$ $(=q* + Rc)$ to the product Ac of wave amplitude and celerity) is related to the pressure difference Δp_λ per wavelength λ, to the wave-amplitude ratio ϕ $(=A/R)$, celerity c, wavelength λ, and fluid viscosity μ by

$$\theta = \frac{3\phi}{2 + \phi^2} = 1 - \Delta p_\lambda \frac{2R^2(1 - \phi^2)^{5/2}}{9\mu c\lambda\phi^2} \tag{10}$$

The linear relationship shows that, when Δp_λ is positive, the amount of fluid transported by peristaltic action increases with viscosity, wavelength, and celerity, but decreases as Δp_λ increases.

In the ureter, work done by the peristaltic wall-wave to move the urine from the kidney to the bladder increases with bladder pressure. The required pumping power could be generated in the ureter by increasing the number of pumping steps between kidney and bladder, i.e., by a larger number of waves of shorter length. At present, however, physiological information to substantiate this hypothesis is insufficient.

Spacial integration of the velocity profiles (such as those in Fig. 3-3a) yields the stream function $\psi*$ $(= \int_0^y *u*\, dy*).\ddagger$ The configuration of the streamlines (Fig. 3-3b) changes with the transport ratio θ and the amplitude ratio ϕ. For complete occlusion ($\phi = 1$), all the fluid is transported with the wave celerity c (so that $\theta = 1$) and the $\psi*$ streamlines form two vortices. The vortices persist over a portion of the cross section as ϕ and θ decrease to values not much smaller than unity.

The path of fluid particles in the channel (path lines) can be obtained by integrating the particle velocities over time. A sample path line, when the net flow is small (Fig. 3-3c), indicates that under these conditions, though the bulk of the flow moves downstream, some fluid near the wall can move upstream. This has been suggested as a possible

---

$\ddagger$ $\psi*$ is defined by u* = $\partial\psi*/\partial y*$ and v* = $-\partial\psi*/\partial x*$. Thus, the difference $\Delta\psi*$ between two lines of constant $\psi*$ gives the discharge with respect to the moving reference frame (u* = u − c).

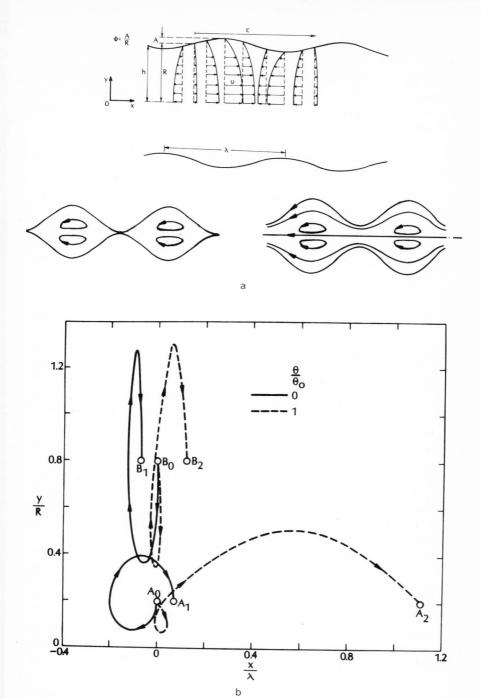

Figure 3-3. Characteristics of two-dimensional sinusoidal flow.[3]

mechanism for bacterial transport from the bladder toward the kidneys.[3]

The three-dimensional axisymmetric case has been analyzed with analogous procedure and results.[4]

Large Amplitude-to-Wavelength Ratio: The Nonlinear Case. When the ratio of the amplitude to the wave length A/λ is larger, as in the healthy ureter and in some heart-lung machines, the convective acceleration cannot be neglected. For the two-dimensional case, if in Eq. (6) the velocity components are expressed in terms of the stream function ψ, and if the pressure is eliminated by cross differentiation, the vorticity-transport equation is obtained:

$$\frac{\partial}{\partial t}\nabla^2\psi + \psi_y\nabla^2\psi_x - \psi_x\nabla^2\psi_y = \frac{1}{\mathbf{R}}\nabla^2\nabla^2\psi \tag{11}$$

where $\nabla^2\ (=\partial^2/\partial x^2 + \partial^2/\partial y^2)$ is the Laplacian operator, $\mathbf{R}(= c R/\nu)$ is a Reynolds's number, and $-\nabla^2\psi\ (= \partial v/\partial x - \partial u/\partial y)$ is the vorticity (the rotational velocity of a fluid particle in the x,y plane). A perturbation solution for a sinusoidal wall-wave in the absence of a hydrostatic pressure difference between inlet and outlet of the tube (so that there would be no flow if the walls were motionless) yields for the time-averaged longitudinal-velocity component (approximate to the second order of the amplitude ratio $\phi = A/R$)[4]

$$\frac{\overline{u(\eta)}}{c} = \frac{\phi^2}{2}\left[D - \mathbf{R}\frac{R}{\rho c^2}\left(\frac{\partial p}{\partial x}\right)(1 - \eta^2) - \frac{2\pi\mathbf{R}R^2}{200\lambda}\mathcal{F}(\eta)\right] \tag{12}$$

where $\eta = y/R$, and $\overline{(\partial p/\partial x)}$ is the time-mean pressure gradient generated by the peristaltic flow. When the time-mean pressure gradient exceeds a critical value $(\partial p/\partial x)_{cr}$, a reversal of the time-mean velocity $\overline{u}$ occurs in the neighborhood of the center line. The influence of the pressure gradient $\overline{(\partial p/\partial x)}$ on the shape of the velocity profile is shown qualitatively in Figure 3-4.

Flows in the Axial Plasmatic Gaps of Capillaries

If the flow of plasma in the axial gaps between the erythrocytes in a capillary is assumed to be axisymmetric and steady, and if the inertial terms are neglected (slow flow), the equations of motion can be solved numerically by a finite-difference method.[5] Sample results in Figure 3-5a show the motion of the plasma with respect to the erythrocytes (given by the streamlines ψ^*), and in Figure 3-5b the pressure

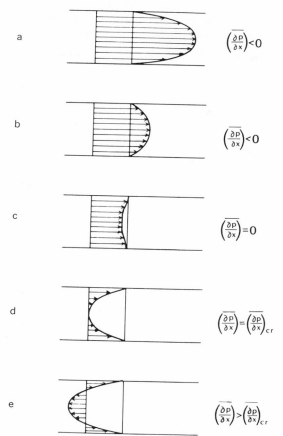

Figure 3-4. Two-dimensional peristaltic flow.

drop Δp in the axial direction (expressed as a ratio to the pressure drop Δp_0 for Poiseuille flow). Since $\Delta p/\Delta p_0$ decreases as the gap L/D increases, lower pressure gradients are required to convey the plasma if the erythrocytes travel in groups rather than individually.

Laminar Vortices

Vortices often occur when the boundary configuration changes abruptly. Figure 3-6 gives an experimental and numerical description of the steady flow in a circular steplike transition.[6] Experimental observations in vitro have shown a tendency to sedimentation in the

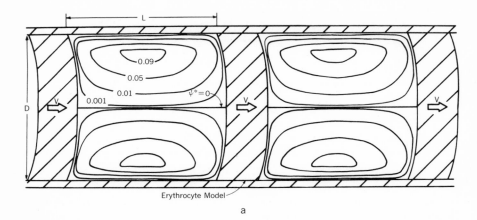

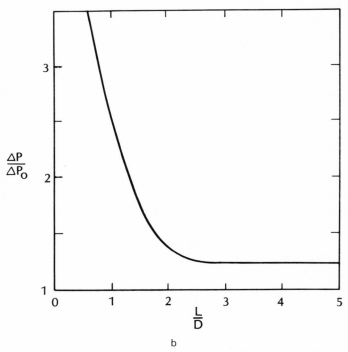

Figure 3-5. a. A model illustrating the relation of the flow of plasma to the
RBC. b. The relationship of pressure difference to the axial gaps
in the bloodstream.

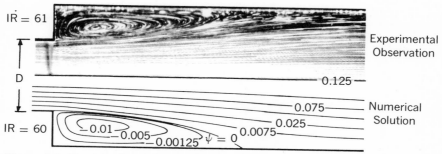

Figure 3-6. Vortices formed at a tube expansion.

vortex zone. The sediment can be driven away by making the flow highly turbulent. In vivo observations with blood have shown similar deposition of fibrin in the vortex zone of an abrupt expansion, leading over long time periods (of the order of one year) to a gradual streamlining of the expansion.[7]

The region of small velocities in the vortex behind an artificial aortic valve is believed to lead to the formation of thrombi. Similarly, weak eddies behind a stenotic aortic valve which cannot open fully may cause the deposition of fibrin or the formation of clots. Behind a normal aortic valve, on the other hand, the vortices are strong, preventing thrombi formation and attachment of the valve leaves to the aortic wall.

The Work-Energy Relationship and Bernoulli's Theorem

If the terms in the force equations are multiplied by the corresponding velocity components u_i, and if the resulting equations for each component are added, a scalar equation is obtained[9]

$$\frac{D}{Dt}\left(\rho \frac{V^2}{2}\right) = \sum_{i=1}^{3} \rho u_i X_i + \sum_{i=1}^{3} u_i \left[-\frac{\partial p}{\partial x_i} + \sum_{j=1}^{3} \frac{\partial \tau_{ji}}{\partial x_j}\right] \qquad (13)$$

CHANGE OF   WORK DONE   WORK DONE
KINETIC ENERGY   BY BODY   BY STRESSES
                FORCES

where D/Dt is the differential operator $(\partial/\partial t + \sum_{i=1}^{3} u_i \, \partial/\partial x_i)$. The work-energy relationship states that the rate of change of the kinetic energy $\rho V^2/2$, due to both time variation $(\partial/\partial t)$ and convection $(\sum_{i=1}^{3} u_i \, \partial/\partial x_i)$ of $\rho V^2/2$, is caused by the rate at which work on the unit element under consideration is done by the body forces, and by

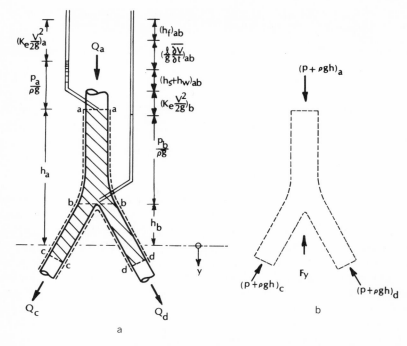

Figure 3-7. The energy/pressure relationships at the bifurcation of an artery.

the pressure and the viscous stresses acting on it. Part of the work done by the viscous stresses is transformed into heat and represents the rate of dissipation of energy.

If a vessel wall is flexible and moves (e.g., under pulsating flow), integration of Eq. (13) over the volume Ψ between two cross sections at a distance ℓ from each other (a and b in Fig. 3-8) yields the Bernoulli equation

$$\left(K_e\frac{V^2}{2g} + \frac{p}{\rho g} + h\right)_a$$
$$= \frac{Q_b}{Q_a}\left(K_e\frac{V^2}{2g} + \frac{p}{\rho g} + h\right)_b + \left(\frac{\ell}{g}\frac{\partial \overline{V}}{\partial t}\right)_{ab} + (h_f + h_s + h_w)_{ab} \quad (14)$$

where the index ab refers to the reach from a to b, and the expansion (or contraction) of the vessel between a and b makes $Q_b \neq Q_a$ (the discharges at sections b and a). All terms in Eq. (6) are energy fluxes divided by the mass flux ($\rho g Q_a$) at section a and constant g and are thus expressed in units of length (the height of a fluid column with respect to a reference level). The constant where $\overline{V}$ is the mean velocity

$$K_e = \frac{1}{A} \int \left(\frac{V}{\overline{V}}\right)^3 dA$$

is an energy coefficient resulting from an integration of the energy term,

$$\ell \frac{\overline{\partial V}}{\partial t} = \frac{1}{A} \int \frac{\partial V}{\partial t} d\forall$$

is the volume integral of the local inertia force divided by ρ and the mean cross-sectional area A, the Q's are discharges, the h's elevations of the centroid of the cross sections, h_f is the total energy dissipation, h_s the work done by viscous stresses at sections a and b and on the wall, h_w the energy transmitted to the wall by the fluid

$$h_w = \frac{1}{Q_a} \int_w B_w(V_w)_n \, dS$$

where B_w is the Bernoulli sum on the wall, $(V_w)_n$ is the normal velocity component on the vessel wall, and the integration is performed over the vessel wall w. For a perfect elastic vessel, h_w is conservative, being stored in the vessel wall when the wall expands (systole) and returned to the fluid when the wall relaxes or contracts (diastole). If the vessel is viscoelastic, only part of h_w is conservative and the rest will be dissipated. When energy is imparted to the fluid, e.g., by a contraction of the vessel wall caused by muscle action, h_w is negative.

If the vessel is rigid and stationary, h_w vanishes, the ratio Q_b/Q_a becomes unity, and h_s is due only to the work done by viscous stresses at the two end sections. When the flow is uniform at either end, $h_s = 0$.

One-Dimensional Analysis—Flow Through a Bifurcation

Information on the bulk characteristics of a flow can be obtained by a one-dimensional analysis (the "black-box" approach), employing equations derived from the integral forms of the equations of continuity, of motion, and of Bernoulli. For example, for the flexible and impermeable bifurcation in Figure 3-8, the equation of continuity becomes

$$Q_a = Q_c + Q_d + \frac{d\forall}{dt}$$

where Q_a is the inflow, Q_c and Q_d are the outflows, and $d\forall/dt$ is the rate of change of the vessel volume due to the deformability of the vessel wall. The equation of motion for the forces acting in the i-direction becomes

$$\sum_i F_i = \left[\rho V \frac{\overline{\delta u_i}}{\delta t}\right]_{ab} + \left[\rho V \frac{\overline{\delta u_i}}{\delta t}\right]_{bd+bc}$$

ALL SURFACE AND BODY FORCES	INERTIAL FORCE FOR REGION BETWEEN a AND b DUE TO LOCAL ACCELERATION	INERTIAL FORCE FOR BRANCHES bc AND bd DUE TO LOCAL ACCELERATION

$$+ [K_m \rho Q\bar{u}_i]_c + [K_m \rho Q\bar{u}_i]_d - [K_m \rho Q\bar{u}_i]_a$$

MOMENTUM FLUX AT c	MOMENTUM FLUX AT d	MOMENTUM FLUX AT a

$$+ \int \rho u_i \sum_{j=1}^{3} u_j \frac{\partial x_j}{\partial n} dS \quad (16)$$

MOMENTUM FLUX ON THE FLEXIBLE WALL

where $\bar{u}_i$ and $(\overline{du_i/dt})$ are, respectively, a mean velocity component and a mean component of the local acceleration, dS is a surface element, and

$$K_m = \frac{1}{A} \int_A \left(\frac{u_i}{\bar{u}_i}\right)^2 dS$$

is a coefficient resulting from the integration of the momentum term over the cross-sectional area A. The Bernoulli equation for the segment ab of the bifurcation is given by Eq. (14); a similar equation can be written for the branch bc:

$$\left[K_e \frac{\overline{V}^2}{2g} + \frac{p}{\rho g} + h\right]_b$$
$$= \frac{Q_c}{Q_b^*}\left[K_e \frac{\overline{V}^2}{2g} + \frac{p}{\rho g} + h\right]_c + \left(\frac{\ell}{g}\frac{\partial \overline{V}}{\partial t} + h_f + h_s + h_w\right)_{bc} \quad (17)$$

where Q_b^* is the fraction of Q_b flowing into the branch bc, and the ratio Q_c/Q_b^* is unity if the branch is rigid.

Pulsatile Flows[9–13]

Pulsatile flows are generated by a periodically time-varying pressure. A rigorous analysis of these flows is not feasible at present. However, approximate descriptions can be obtained under a variety of simplifying assumptions.

One-Dimensional Analysis. For a segment of artery of unit length, neglecting the nonuniform character of the flow, the one-dimensional equation of motion in the axial direction can be written

$$\left[\pi R^2\left(-\frac{\partial p}{\partial z}\right) - \pi R^2 R_f Q\right] = \rho\frac{\delta Q}{\delta t} \tag{18}$$

PRESSURE   BOUNDARY
FORCE   SHEAR FORCE

SURFACE FORCES   INERTIAL
FORCE

where R_f is a resistance parameter

$$R_f = \frac{2\mu}{RQ}\left(-\frac{dw}{dr}\right)_{r=R} \tag{19}$$

Eq. (19) shows that the pressure gradient $\partial p/\partial z$ required to move the fluid must overcome not only the shear resistance at the boundary $(\pi R^2 R_f Q)$, as in the case of steady flow, but also the inertial force $(\rho\delta Q/\delta t)$. The flow characteristics depend on the relative importance of these two terms, given by the frequency parameter α ($\alpha = R\sqrt{\eta\rho/\mu}$) which represents the ratio of an inertial force ρQn to a viscous one $\mu Q/R^2$.

The equation of continuity for an elastic and permeable vessel can be written in the form

$$-\frac{\partial Q}{\partial z} = \frac{\partial A}{\partial p}\frac{\partial p}{\partial t} + Q_{LP}p \tag{20}$$

where A is the cross-sectional area, $\partial A/\partial p$ characterizes the elasticity of the vessel wall, and Q_{LP} is the rate of leakage per unit length and per unit of transmural pressure p (the pressure difference across the wall thickness). Equations (18) and (19) can be solved by studying the flow in electric circuits whose elements are analogous to those of the flow problem (Table 3-1).

Axisymmetric Incompressible Flow in Rigid Tubes. Analysis of pulsatile flow in a long rigid tube reduces to the solution of the equation of the forces in the axial (z) direction

$$\rho\frac{\partial w}{\partial t} = -\frac{\partial p}{\partial z} + \mu\left(\frac{\partial^2 w}{\partial r^2} + \frac{1}{r}\frac{\partial w}{\partial r}\right) \tag{21}$$

with the boundary conditions. The system described by these three equations is linear, and can thus be represented by a Fourier series.

Table 3-1.   Fluid-electric Analogy

Tube Flow			Electric Cable		
Variable Name	Symbol	Dimensions	Variable Name	Symbol	Dimension
Pressure	P	dyne/cm^2	Voltage	V	volts
Discharge	Q	cm^3/sec	Current	I	amps
Density per unit area	ρ/A	gm/cm^5	Inductance per unit length	L	henries/cm
Resistance coefficient	R_f	$\dfrac{\text{dyne-sec}}{\text{cm}^6}$	Resistance per unit length	R	ohms/cm
Deformability	$\partial A/\partial p$	cm^4/dyne	Capacitance per unit length	C	farads/cm
Leakage	Q_{LP}	cm^4/dyne-sec	Conductance (leakage to ground) per unit length	G	amp/(volt-cm)
Impedance	Z	$\dfrac{\text{dyne-sec}}{\text{cm}^4}$	Impedance per unit length	Z	ohms/cm

Its input is the pressure gradient corresponding to the particular pulse considered and its output the velocity profiles. Given the linearity of the system, the N-harmonic of the input function contributes only to the N-harmonic of the output function, and the solution corresponding to any pulse is the summation of the solutions corresponding to each of the harmonics constituting the pulse.

The solution of Eq. (21) is a Bessel function of the first kind and of order zero.[8,9] A sequence of velocity profiles calculated from the measured pressure-gradient with a pulse-frequency n = 2.8 sec^{-1} in a femoral artery of the dog is shown in Figure 3-8. While the maximum forward velocity occurs on the axis, the maximum backward velocity occurs away from it (at r/D between 0.3 and 0.4).

Pressure Gradient, Phase, and Group Velocities. In the circulatory system, as well as in other biological flow systems, it is generally more difficult to determine the pressure gradient $\partial p/\partial z$ (by simultaneous measure of the pressure between two points) than the variation of pressure with time $\partial p/\partial t$ at a given point. Under the assumption that the pressure pulse does not distort as it travels, the former can be related to the latter by setting the total differential of p with respect to time equal to zero.

$$\frac{dp}{dt} = \frac{\partial p}{\partial t} + \frac{\partial p}{\partial z}\frac{dz}{dt} = \frac{\partial p}{\partial t} + c\frac{\partial p}{\partial z} = 0 \qquad (22a)$$

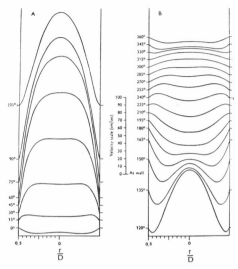

Figure 3-8. Pulsatile velocity profiles calculated from a measured $\partial p/\partial z$ in the femoral artery of a dog, Period $\cong 0.36$ sec. (1° corresponds to 1 msec) (From ref. [9])

whence

$$\frac{\partial p}{\partial z} = -\frac{1}{c}\frac{\partial p}{\partial t} \tag{22b}$$

where $c = (\partial z/\partial t)$ is the celerity of the pressure wave. For a sinusoidal wave traveling without reflection through an inviscid incompressible fluid inside a thin-walled elastic tube (of thickness $d \ll R$, and Young's modulus of elasticity E), the celerity is

$$c = \sqrt{\frac{Ed}{2R\rho_s}} \tag{23}$$

In the arterial tree, E increases from the root of the aorta toward the peripheral regions. The celerity must thus increase accordingly. For example, $c = 8.3$ m/sec for the thoracic aorta and 11 m/sec for the abdominal aorta of a dog (with $\rho_s = 1.08$ gm/cm^3 and $d/2R = 0.08$, and the E moduli for the two aortas equal respectively to 9.3×10^6 and 15.6×10^6 dynes/cm^2).

Equation (23) is based on highly idealized conditions. In reality, c is bound to be frequency-dependent if the viscosity of the fluid is not neglected.

In the case of the circulatory system, the frequency dependence of c makes each harmonic component of the pulse travel at a different celerity (*phase velocity*), and distorts the shape of the compound pulse as it propagates through the system. The apparent velocity of propagation of the compound pulse is called *group velocity,* to distinguish it from the phase velocities of its components.

Fluid viscosity dampens the amplitude of a wave and reduces its phase velocity. Phase velocities are smaller for low frequencies because of the predominance of the viscous forces.

The transmission of a pulse is further modified by reflections, which arise whenever the configuration of the boundaries along the path of the pulse is altered. For instance, in the circulatory system, reflections occur—in a magnitude dependent on the degree of alteration of the cross section—when a vessel is occluded (e.g., by an atherosclerotic plaque) or at bifurcation. In the latter case, since the cross-sectional area of the two branches is generally larger than that of the parent vessel, the reflection is stronger when the wave propagates from the branches toward the parent vessel than when it propagates in the opposite direction. Figure 3-9 shows measurements of pressure and velocity for different segments of the aorta of a dog.[9]

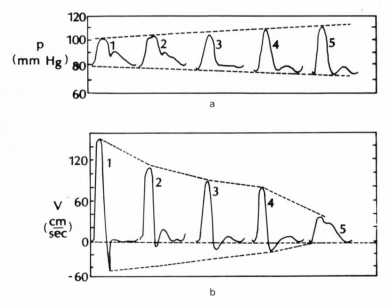

Figure 3-9. a. Pressure and b. velocity in the aorta of a dog at increasing distance from the heart.[9]

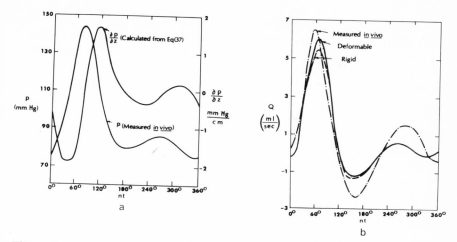

Figure 3-10. Deformable wall model for pulsating flow in femoral artery of dog. (From ref. [10])

Pulsatile Flow in Deformable Tube

Pulsatile flow in a deformable tube can be analyzed by coupling through the boundary conditions (Eq. (7)) at the inner surface of the tube wall, the equations of motion of the wall material and of the fluid in the tube. Results of an approximate analytical solution[10] are shown in Figure 3-10, where the pressure gradient $\partial p/\partial z$ (Figure 3-10a) has been calculated from *in vivo* measurements of $\partial p/\partial t$, using Eq. (22b) with a celerity c (-760 cm/sec) based on an expression that includes the elastic effect of the vessel wall. In Figure 3-10b the calculated flow rates for the deformable tube are compared with those for a rigid tube and with the discharge measured *in vivo*.

Boundary Layers, Entrance, and Curvature Effects

Under particular cirumstances (e.g., at high Reynolds numbers), the influence of a boundary on the velocity and other transport quantities is effectively confined to a fluid layer of limited dimension adjacent to the boundary—the boundary layer. The development of such a layer when flow enters a uniform rigid tube is similar to what occurs, e.g., in the circulatory system whenever there are sudden reductions in vessel diameter, as in the transition from the heart to the aorta or at the inlet of a bifurcation. At the entrance of the tube ($z = 0$), the velocity is uniformly distributed, except at the wall where the fluid has the same velocity as the boundary (Eq. 7). The difference in velocity between fluid and boundary generates vorticity (i.e., rotation of fluid particles) at the wall. Convection and viscous diffusion (analogous, e.g., to the transport of heat from a boundary) gradually transport the vorticity into the flow, altering the uniform ($dw/dr = 0$) velocity distribution. Fluid in a boundary layer of thickness δ is thus decelerated, acquiring velocity gradients $dw/dr \neq 0$. At the same time, it is obvious that, if the fluid is incompressible and the wall rigid, the fluid in the core must be accelerated to maintain a constant flow rate from section to section. The boundary layer thickness increases progressively downstream until it reaches the tube axis and becomes constant (fully developed flow, $\delta = R$). The distance required for full development is, in general, a function of the Reynolds number (and hence the flow regime) and of the nature of the boundary, being, e.g., equal to 0.25 **R** for laminar steady flow in a rigid tube. If the flow is unsteady, δ varies not only with position, but also with time. For a pulsatile flow due to an oscillatory uniform velocity at the inlet of a tube, the downstream velocity profiles and the longitudinal pressure gradients have been obtained from a solution of linearized Navier-Stokes equations.[11]

In biological situations, the distribution of the fluid entering the vessel is usually not uniform, and the oncoming flow may contain vortices which are carried into the vessel, affecting the boundary layer development.

Boundary layers develop in similar fashion around bodies immersed in a flow, such as a catheter inserted in a vessel or a spherical artificial heart valve. If the catheter is set along the axis of a straight tube, the boundary layer thickness increases continuously downstream of the leading edge of the catheter, until the velocity distribution in the annulus between the catheter and the vessel wall becomes fully developed (i.e., remains constant). In practice, however, the catheter and the vessel are seldom coaxial, eliminating the axial symmetry of the flow and imparting to it a spiral pattern. In the case of the spherical artificial heart valve (Fig. 3-11c), the assumption of axisymmetric flow can be better justified, but the divergence of the boundary usually causes the boundary streamline to separate from the wall in the rear of the sphere (boundary layer separation). Numerical simulations indicate that the vortex generated during a sudden acceleration (as during systole) may have a high rotating speed, large shearing stresses, and

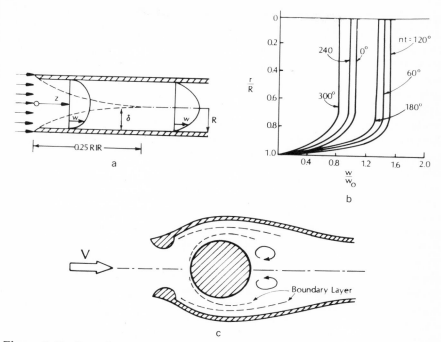

Figure 3-11. Boundary layer conditions during flow.

a large dissipation of energy, with a low pressure at the center.[10] The relation of these flow characteristics to hemolysis and the possible occurrence of gaseous embolism is still unexplored, but is likely to be significant.

Secondary flows characterized by circulatory motions in sections normal to the main (longitudinal) flow direction are often present in curved vessels (or around a curved catheter). The phenomenon is due to the lower velocities of the fluid layers near the boundary, which on a curve can be deflected more easily than the faster layers in the core of the flow. The combination of the main and the secondary flow results in a spiral type of motion.

Turbulent Flows

Local disturbances due to a variety of causes, such as boundary irregularities or suspended particles, are almost always present in a biological flow system. If the disturbances are damped by the additional viscous stresses resulting from the higher velocity gradients which they produce, the flow is stable; otherwise, it is unstable. Stability depends on the Reynolds number of the flow and on the characteristics of the disturbance. If the disturbances become unstable and are amplified, the fluid no longer flows in orderly layers with only molecular inter-changes of momentum, as in the laminar case, but exhibits the charac-teristics of a highly random mixing process (turbulent flow). Turbulence leads to much higher energy expenditures in the conveyance of the fluid, for the same flow rate, than if the flow were laminar. On the other hand, turbulence leads to higher mixing of mass (e.g., O_2, CO_2) and heat.

Other factors being equal, the transition from stability to instability, i.e., from laminar to turbulent flows, occurs when the Reynolds number exceeds a critical value $\mathbf{R}_{cr}$. For example, for steady flows in a straight circular rigid tube, $\mathbf{R}_{cr}$ ($= \rho DW/\mu$), where W is mean velocity, is ap-proximately 2000. In general, the value of $\mathbf{R}_{cr}$ for transition from laminar to turbulent conditions (as flow rates increase) is somewhat larger than that for transition from turbulence to laminar flow. For pulsatile flow in rigid tubes, $\mathbf{R}_{cr}$ decreases as the frequency parameter α increases; for a given α, the peak value of $\mathbf{R}_{cr}$ appears at a moderate value of the ratio of the amplitude of the periodic component of the discharge Q to the mean discharge $\bar{Q}$ (flow-amplitude ratio.)[12] The high mixing that charac-terizes turbulence tends to equalize the mean (time-average) distribu-tion of quantities transported by the flow (mass, heat, momentum, and hence velocity). Thus, in a tube the velocity profile for turbulent flow is considerably flatter than that for laminar flow.

High Reynolds numbers, and hence the possibility of turbulence, are present in several biological flow situations, e.g., when air is dis-charged from the nose into the atmosphere, in the chambers of the heart, or possibly during systolic peaks in the aorta. However, rapid mixing of fluid in a pulsatile flow does not necessarily indicate turbu-lence, since it may be caused by transient laminar vortices or secondary laminar flows. Furthermore, if the flow pulsates, there may not be sufficient time for the turbulence-inducing vorticity to be diffused and convected from the boundary into the core of the flow, even if the instantaneous Reynolds number is high (as at peak systole).

Diffusion and Heat Transport

An important function of many flows in biological organisms is conveyance and transfer of heat and chemical species. Thus, blood capillaries transfer O_2 to the surrounding tissue, in turn receiving CO_2

from them; hormones are conveyed by the bloodstream from one point to another of the arterial system; subcutaneous peripheral beds transmit heat from the body to the environment, etc. In each of these cases, heat or chemical species are conveyed in part by convection, i.e., by motion of the fluid containing the thermally excited or chemically different molecules, and in part by diffusion or thermal conductivity. A study of these processes is outside the scope of this chapter, but a brief mention of their characteristics is desirable.

Diffusion and thermal conductivity are, like viscosity, transport processes of physical quantities due to a gradient of corresponding field variables. The simplest description of these processes can be achieved by referring to the geometry of Fig. 3-12. For a unit area normal to the transfer direction, the relationship between the flux F(M) (the amount of property M being transported per unit time and unit area in the direction normal to the plates) and the gradient of M in the gap ds between the plates can be stated in the form of a constitutive equation:

$$F(M) = -\rho K \frac{dM}{ds} \qquad (24)$$

where the minus sign denotes a decrease in mass flux as the distance s increases, and constitutive coefficient K has the dimensions of $(\text{length})^2 (\text{time})^{-1}$ and, like the coefficient of viscosity, cannot generally be expected to be a constant.

If the quantity M is made to represent, respectively, momentum density (mass δm of a fluid particle times its velocity), heat density $(\delta m \times C_p \times T)$, where C_p is the specific heat at constant pressure), and density of molecular species present in concentration $C_A (\delta m \times C_A)$, Eq. (24) becomes:

1. For the transport of momentum:

$$\tau = \rho \nu \frac{du}{dy} \qquad \text{dynes per cm}^2 \qquad (2a)$$

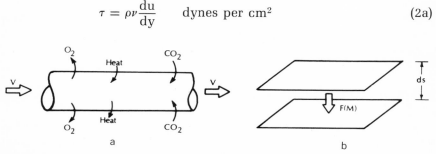

Figure 3-12. A model of diffusion.

2. For the transport of heat:

$$q = -\rho \alpha C_p \frac{dT}{dy} \qquad \text{calories per cm}^2 \qquad (25)$$

3. For the transport of mass of the chemical species A into chemical species B:

$$j = -\rho D_{AB} \frac{dC_A}{dy} \qquad \text{mass per sec-cm}^2 \qquad (26)$$

where $\nu(=\mu/\rho)$, α, and D_{AB} are, respectively, the coefficients of kinematic viscosity, thermal diffusivity, and molecular diffusivity.

If the diffusion processes are accompanied by mass flow, the total flux of the quantity transported, with respect to a fixed coordinate system, is the summation of the two processes. Thus, for the case of diffusion of chemical species A into chemical species B, the mass flux $\overline{F}_A$ (mass per second per unit area) is given by

$$\vec{F}_A = -\rho D_{AB} \vec{\nabla} C_A + \rho C_A \vec{V} \qquad (27)$$

where the first term on the right-hand side is the vectorial generalization of Eq. (24), and the second term represents the mass flow of species A, under the action of the velocity $\vec{V}$.

The equation of conservation of mass for the chemical species A is a generalization of the equation of continuity, stating that the next flux of mass A, through an infinitesimal reference volume, plus the rate of production P_A of A inside the volume (e.g., by biochemical reactions) equals the rate of mass accumulation of A inside the fluid element:

$$\frac{\partial \rho_A}{\partial t} + \nabla \cdot \vec{F}_A = P_A \qquad (28)$$

where P_A is in general a function of concentration, tissue, and location. From Eq. (27) and from $\rho_A = \rho C_A$, Eq. (28) can be written

$$\frac{\partial \rho C_A}{\partial t} + \rho C_A \nabla \cdot \vec{V} + \vec{V} \cdot \nabla \rho C_A = \rho D_{AB} \nabla^2 C_A + (\nabla C_A) \cdot (\nabla \rho D_{AB}) + P_A \qquad (29)$$

For an incompressible ($\nabla \cdot \vec{V} = 0$), homogeneous ($\rho = $ constant) fluid, and a dilute solution ($D_{AB} = $ const. $= \mathfrak{D}$), the equation reduces to

$$\frac{\partial C_A}{\partial t} + \vec{V} \cdot \nabla C_A = \mathfrak{D} \nabla^2 C_A + \frac{P_A}{\rho} \qquad (30)$$

where the left-hand term is the total derivative DC_A/Dt, composed of a local term and a convective term. For oxygen diffusing into blood, a typical value of $\mathfrak{D}$ is 9×10^{-6} cm² per sec. By introducing a dimensionless time $t^* = tV/D$ (where V and D are, respectively, a reference velocity and a reference length) and a dimensionless length $L^* = L/D$, Eq. (30) can be written

$$\frac{DC_A}{Dt} = \frac{1}{P}\nabla^{*2} C_A + P_A^*$$
(31)

where $\nabla^{*2}C_A$ and P_A^* are now dimensionless and $\mathbf{P}$ is the Peclet number (VD/D). Eq. (31) shows that the relative importance of the term encompassing the convective component (left-hand term) over the diffusive term (first right-hand term) in the transport of the chemical species A is determined by the magnitude of the Peclet number. If the field velocity V and the production rate P_A are both zero, Eq. (31) becomes the expression of Fick's Second Law:

$$\frac{\partial C_A}{\partial t} = \frac{1}{P}\nabla^{*2}C_A$$
(32)

Formally identical expressions are obtained for the heat transfer case by replacing C_A by T, $\mathfrak{D}$ by αC_P, and P_A by the rate of heat production.

As in the case of the momentum equation, the solution of Eq. (31) presents considerable difficulties, both analytically and numerically, and has been achieved only for simplified conditions. On the other hand, Eq. (32) is a parabolic partial differential equation which has been solved for large classes of initial and boundary conditions.

A Special Case

Unlike in most engineering fluid systems, mathematical modeling problems in biological fluid systems are extremely complex. The complexity in these systems arises from four factors: (a) distensibility of the fluid vessels, (b) non-Newtonian nature of biological fluids, the blood in particular, (c) numerous branchings of the vessels, and (d) pulsatile nature of pressures. Various neural controls which the central nervous system (CNS) exerts to regulate flow and distribution make the task of mathematical modeling even more difficult.

In this section we present a lumped parameter model of the cerebral fluid system using some of the concepts of systems analysis. Admittedly, this model is a grossly simplified portrait of the real biological system, but it does help to study the mechanisms through which physiological and pathological parameters of this system exert control on the cerebrospinal fluid (CSF) dynamics and cerebral blood flow rate (CBF).

Model of Cerebral Fluid System

A schematic representation of the cerebral fluid system is shown in Figure 3-13. The human CNS is well supplied with blood entirely by the internal carotid, spinal, and vertebral arteries. The normal blood flow averages about 750 ml per min for the whole brain.[1] The blood exits at the jugular bulb and spinal veins. A unique feature of the cerebral circulation is that it takes place within a relatively rigid container, the cranium. The dura mater, which is closely applied to the cranial vault, faithfully encloses the brain and is relatively inelastic. Since the brain tissue and the brain fluids are nearly incompressible, it follows that the combined volume of the brain tissue, cerebrospinal fluid, and intracranial blood must be nearly constant, and that the volume of any one of these components can be increased only at the expense of one or both of the others. This has long been accepted as the Monro-Kellie doctrine.

There is little question that the CSF is derived from blood which has entered the dural compartment, and that it rejoins the bloodstream to exit the compartment. Thus, the CSF flow forms a branch parallel to the blood flow along some segments of the vascular system. The normal CSF formation in humans averages about 0.3 ml per min.[2]

In a rigid capillary tube, during flow the pressure within its lumen drops from P_a at the input end to P_c at the outflow end. If an incompressible fluid and laminar flow are assumed, then the resistance to flow, which is distributed along the length of the tube, is derived from

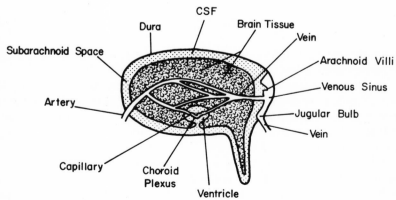

Figure 3-13. A schematic of the brain and its blood supply.

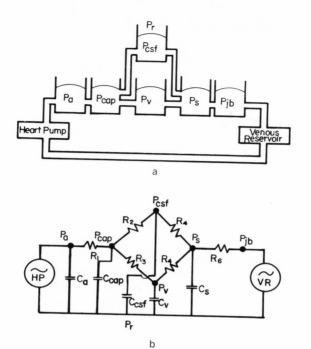

Figure 3-14. a. A lumped parameter model of the cerebrovascular system, and b. its electrical counterpart.

the Poiseuille-Hagen formula[3]

$$R_{ac} = \frac{8}{\pi} \cdot \eta \cdot \frac{L}{r^4} \tag{33}$$

where R_{ac} is the fluid resistance to flow from a to c, L the length, r the radius, and η the absolute viscosity of the fluid.

A lumped parameter model is assumed by defining a number of compartments for each part of the cerebral fluid system (Fig. 3-14). The resistance to flow due to a particular vessel type is lumped at the outflow of its compartment. Figure 3-14a is the electrical circuit representation of the fluid model.

The cerebral fluid system is represented by lumping the vascular tree into arterial, capillary, venous, venous sinus, jugular bulb, and cerebrospinal fluid compartments. All the vascular and cerebrospinal compartments are shown in Figure 3-15.

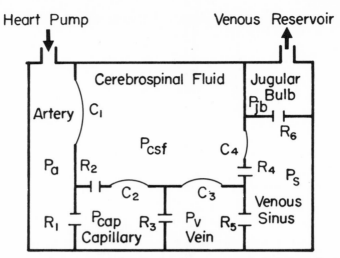

Figure 3-15. A schematic of the fluid compartments of the cerebrovascular
        system.

Volume Change and the Capacitative Analogy

Imagine a hydraulic system with an inflow at pressure P_a, outflow
pressure P_c, and a compartment with rigid walls and one or more
localized elastic membranes to permit volume change in the compartment; the pressure P_b throughout the compartment is assumed to be
uniform (Fig. 3-16). The volume of the compartment is a function of
the transmural pressure $(P_b - P_r)$ and the elastance of the membrane
and its geometry (P_r is the outside pressure). For the purpose of this
discussion, it is sufficient to represent this relationship as

$$V_b = V_{bo} = C_b(P_b - P_r)$$

where V_{bo} is the volume of the compartment when the transmural
pressure $(P_b - P_r)$ is zero, and C_b is the capacitance term composed
of the elastance and geometric factors of the membrane.

Any change in resistance values or pressures will cause changes
in flow rates and also changes in the volume of the compartment. The
electrical analogy is that the compliance of the membrane is represented as a capacitance, and thus change in the volume of the compartment is analogous to the charge stored on the capacitance.

If there are two compartments separated by a membrane enclosed
in a rigid container with a fixed volume (Fig. 3-17) under conditions

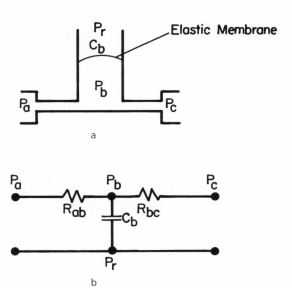

Figure 3-16. A simplified model of the pressure relationships in the brain.

of zero flow, the compartments b and c are at equal pressures and the transmural pressure across the membrane is zero. Thus, the membrane is not distorted. Under conditions of flow, the pressure at b exceeds the pressure at c by an amount equal to the product of the flow rate and the resistance. This pressure difference distorts the membrane. In this fixed-volume case, any change in the volume of compartment b must be equal to and opposite in sign to the change of the volume of compartment c.

Model of Cerebral Fluid System Revised

The vascular and CSF compartments within the dural sac are connected by resistance paths through which fluid flow occurs and capacitance paths in which membrane displacements simulate fluid flow. These paths intimately link the compartment volumes, resistances, and flow rates in a unique fashion. Figure 3-18 shows the overall model for all of the compartments in the intracranial fluid system considered here, with the compartments now enclosed in a fixed rigid container in the skull. An additional assumption made to simplify this model is that, since it is so easy to exchange volume between the CSF com-

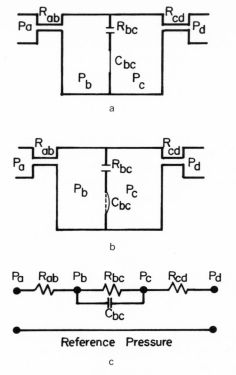

Figure 3-17. An expanded model of Figure 3-16 including further parameters.

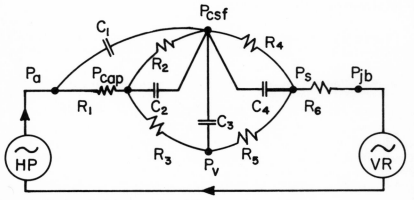

Figure 3-18. An electrical model of the complete system.

partment and the vascular compartments, the volume exchanges directly between the vascular compartments may be neglected.

The lack of usable data in the literature makes an accurate quantitative analysis of the system unfeasible. The various resistance and capacitance values are nonlinear functions of transmural pressures and the nature of these functions has not yet been explored.

It is possible, however, to calculate approximate values for the resistances for a normal case, using available published data for average pressures and flow rates. These average values are shown in Table 3-2.

As the information on simultaneous recording of various pressures as a function of time is lacking, the values of the capacitances were obtained experimentally by simulating the electrical model and approximately matching the arterial, capillary, and jugular bulb pressure curves from Bering.[6]

The values of the fluid capacitances so obtained by analog simulation range from 10^{-5} to 10^{-3} ml per mm Hg. A careful study of the volume-pressure relationship of a small blood vessel has not yet been done, and in particular there are no such data available for the blood vessels of the brain. The capacitance value for the aorta, as calculated from the volume-pressure relationship, is about 6.8×10^{-2} ml per mm Hg or about 1.7×10^{-2} ml per mm Hg per cm length of aorta. Some quantitative results on the distensibility of the small blood vessels of the frog were obtained by Jerrard and reported by Burton.[4] The volume distensibilities of the arteries, capillaries, and veins of the frog were estimated to be 0.4, 0.008, and 1.0 percent per mm of Hg, respectively.

Pulsation of CSF Pressure

There is some controversy in the literature as to the origin of the pulsation of CSF pressure. Bering[6] proposes that the arterial system

Table 3-2. Fluid-to-electrical-analog Conversion Factors and Units

FLUID		ELECTRICAL	
Pressure	1 mm Hg	Voltage	0.1 volt
Volume	1 ml	Charge	10^{-6} coulomb
Time	1 min	Time	0.06 sec
Flow	1 ml/min	Current	16.6×10^{-6} amp
Resistance	1 mm Hg/ml/min	Resistance	6×10^3 ohms
Compliance	1 ml/mm Hg	Capacitance	10^{-5} farads

via the choroid plexus is the major contributor to the pulsation. Hamit and associates[9] presented an entirely different view, namely, that the CSF pulse followed a venous rather than an arterial pattern; in addition, they have proposed that the static pressure of the CSF is maintained by the arterial blood pressure and that rapid dynamic changes in CSF pressure are effected mainly through the venous channels. Dunbar and others[10] found that the pulse waves in the CSF in the ventricles, cisterna magna, and lumbar subarachnoid space are almost identical to each other and to the arterial pulse wave. They propose that the venous pressure inside the skull and spine determines the mean CSF pressure, upon which arterial pulsations are superimposed.

In the model, transmission of pressure pulses from both arterial and venous (sinus) compartments occurs. The relative contribution of each to the CSF pulsation is determined by the rate of change of the vascular pressures and the capacitance values between the CSF and vascular compartments (artery and sinus). The average value of the CSF pressure (P_{csf}) lies between average values of the arterial pressure (P_a) and venous sinus pressure (P_s) and is a function of the various resistance values. As the capacitance values are different nonlinear functions of the transmural pressures, the average levels of P_a, P_{csf}, and P_s also influence the transmission of pulses.

Decreasing the transmural pressure of either vascular compartment tends to increase the capacitative flow from that compartment to the CSF. Raising the P_{csf} toward arterial pressure selectively increases transmission of the arterial pressure pulses to the CSF, while lowering the P_{csf} tends to favor transmission from the sinus.

It is inappropriate to attribute the CSF pulsations to either the arterial or the venous system alone, but rather to some function of both. The contribution of each component to the total is influenced by the average levels of P_{csf} with respect to P_a and P_s, the capacitance values, and the rate of change of P_a and P_s at any time.

Pathology of Hydrocephalus

In terms of the proposed model, the basic pathology of obstructive hydrocephalus is an abnormally high value of R_4, i.e., an increased resistance to CSF absorption. The average CSF pressure (P_{csf}) is related to the average values of capillary (P_{cap}) and sinus (P_s) pressures via the relationship

$$P_{csf} + P_s + \frac{R_4}{R_2 + R_4}(P_{cap} - P_s) \qquad (34)$$

As the value of R_4 is raised by an amount ΔR_4, the corresponding increase in the CSF pressure, ΔP_{csf}, is approximately given by the relationship

$$\Delta P_{csf} = \frac{\Delta R_4}{R_2 + R_4}(P_{cap} - P_s) \qquad (35)$$

assuming that $R_2 + R_4 + \Delta R_4 \approx R_2 + R_4$ because R_2 is much greater than R_4.

The pathology of hydrocephalus due to choroid plexus papilloma is an abnormally high rate of CSF production, i.e., the value of R_2 is reduced. This tends to increase the ratio $(R_4/R_2 + R_4)$ and thus increases the CSF pressure.

Treatment of Hydrocephalus

Several methods of treating hydrocephalus can also be represented on the model. Shunting procedures are represented by the addition of a new resistance path, R_7, from the CSF compartment to some other compartment at lower pressure. With a functioning shunt, the CSF outflow is initially increased to

$$\frac{(P_{csf} + \Delta P_{csf}) - P_s}{R_4 + \Delta R_4} + \frac{(P_{csf} + \Delta P_{csf}) - P_{oc}}{R_7} \qquad (36)$$

where P_{oc} is the pressure in the compartment to which the fluid is shunted. The effect of a shunting procedure is to increase the CSF outflow rate to compensate for the elevation of R_4 such that ΔP_{csf} is reduced to zero.

Choroid plexectomy demonstrates a rather different form of therapy. This technique constitutes the elevation of CSF formation resistance (R_2) to decrease the CSF formation rate from

$$\frac{P_{cap} - P_{csf}}{R_2} \quad \text{to} \quad \frac{P_{cap} - P_{csf}}{R_2 + R}.$$

CSF Pressure and Cerebral Blood Flow Rate

An important relationship which applies to many pathophysiological phenomena is the interrelationship of R_5, P_{csf}, and CBF.

An increase in P_{csf} increases the volume of the CSF compartment, which in turn tends to decrease the vascular volume, due to a fixed intracranial volume constraint. Assuming that the length of the vascular tree is relatively unchanged, any decrease in vascular volume is due to a reduction in the diameters of the vascular vessels along the whole vascular tree or at some localized points. Since the resistance to fluid flow is inversely proportional to the fourth power of the diameter of the vessel, a small reduction in the vessel diameter increases the fluid resistance by a significant amount and diminishes the cerebral blood flow.

Although the vascular volume change is distributed among all vascular compartments (R_1, R_3, R_5, and R_6), the greatest change in vascular resistance occurs at the venous segment R_5. The arterial segment, R_1, changes by only a small increment since the transmural pressure is very high in comparison to clinically observed alterations in P_{csf}. The change in R_3, the capillary resistance, is minimal; Burton[4] has shown that the capillary acts essentially as a rigid tube.

The elevation of R_5 out of proportion to the change in total resistance, when considered together with capillary resistance (R_3), causes the pressure drop ($P_{cap} - P_s$) to increase. This increase, being a determinant of P_{csf}, causes P_{csf} to rise. In control terminology, there is a positive feedback in the system. This interaction of R_5 and P_{csf} continues until a new stable point is reached with new average pressures and flow rates. Since the parameter values depend on transmural pressures, they also attain new values during this adaptation.

Elevation of R_5 due to local venous compression, without associated changes in other vascular compartment resistances (R_1, R_3 and R_6), is even more effective in raising P_{csf}, because the decrease in blood flow rate is less marked.

Mass Lesion

The effect of a mass lesion on the cerebral fluid system has two major components. The mechanisms involved with each component may be more clearly seen if they are considered individually. The first effect is the addition of the volume of the lesion to the system; the second effect is the compression of the neighboring vasculature.

The addition of the volume of the lesion to the system implies a simultaneous and equal decrease in the volume of some other compartment or compartments. This situation is analogous to adding, for example, a volume of fluid to the CSF compartment. In this case, the rate of volume growth of the lesion (analogous to the rate of inflow of fluid added to the CSF compartment) is the determining factor with respect to changes in P_{csf}. It may be considered in terms similar to those of the hydrocephalus discussion if the "CSF inflow rate" includes both the CSF formation rate and the rate of volume growth of the lesion. The sum of these two terms equals the CSF outflow rate.

$$\frac{P_{cap} - (P_{csf} + \Delta P_{csf})}{R_2} + \text{Rate of lesion volume growth}$$

$$= \frac{(P_{csf} + \Delta P_{csf}) - P_s}{R_4} \quad (37)$$

This relationship between rate of fluid addition to the CSF compartment and P_{csf} has been demonstrated in man by Foldes and Arrowood.[11]

The intracranial mass lesion growth rate may vary from trivial in neoplastic lesions to brisk in hematomas of arterial origin. Thus, the first factor, which is related to the rate of volume growth of the lesion, is probably not a major contributor to P_{csf} elevation in most clinical neoplastic mass lesions.

The second mechanism, compression of neighboring vascular structures with a resulting increase in vascular resistance, is probably the more important clinically. The distinction must be made between locally increased vascular resistance due to tissue distortion surrounding a local mass, and the generalized increase in vascular resistance due to increasing P_{csf} as discussed above. The extreme case of "local vascular compression" interacting with generalized vascular compression to raise P_{csf} occurs when the "local" component is also geographically widespread, as in pseudotumor or cerebral edema.

A further facet of this mechanism to be considered is the formation of new vessels within the lesion. These new vessels, which act as shunts across a part of the vascular tree, tend to offset the increase in the vascular resistance due to local compression. If the lesion is relatively avascular, the equivalent resistance of the cerebral blood flow (CBF) path will increase and P_{csf} will rise. If the lesion is so vascular that the equivalent resistance to the CBF path remains relatively unchanged, then no increase in P_{csf} will occur due to this mechanism.

References for Fluid Flow

1. Langhaar, H. L.: *Dimensional Analysis and Theory of Models.* John Wiley and Sons, 1960.
2. Merrill, E. W., and Pelletier, G. A.: *Viscosity of human blood: transition from newtonian to non-newtonian.* J. Appl. Physiol. 23:178, 1967.
3. Shapiro, A. H., Jaffin, M. Y., and Weinberg, S. L.: *Peristaltic pumping with long wave lengths at low Reynolds number.* J. Fluid Mech. 37:799, 1969.

4. Fung, Y. C., and Yih, C. S.: *Peristaltic transport.* Report of Dept. of the Aerospace and Mechanical Eng. Sciences, Univ. of Calif., San Diego, Sept. 1967.

5. Bugliarello, G., and Hsiao, G. C.: *A mathematical model of the flow in the axial plasmatic gaps of the smaller vessels.* Biorheology 1969.

6. Macagno, E. O., and Hung, T. K.: *Computational and experimental study of a captive annular eddy.* J. Fluid Mech. 28:(Pt. 1) 43, 1967.

7. Schicht, L.: Hydraulic Aspects of Arteriosclerosis and Arterial Repair, in Allgower, M. (ed.): *Progress in Surgery,* Vol. 5. Hafner, New York. 1967.

8. Womersley, J. R.: *An elastic tube theory of pulse transmission and oscillatory flow in mammalian arteries.* WADC, Tech. Report TR 56-614, Wright Air Development Center, Jan., 1957.

9. McDonall, D. A.: *Blood Flow in Arteries.* Williams & Wilkins, 1960.

10. Whirlow, D. K., and Rouleau, W. T.: *Periodic flow of a viscous liquid in a thick walled elastic tube.* Bull. Math. Biophys. 27:355, 1965.

11. Atabek, H. B.: End Effects, in Attinger, E. (ed.): *Pulsatile Blood Flow.* McGraw-Hill, New York, 1964.

12. Sarpkaya, S.: *Experimental determination of the critical Reynolds number for pulsating Poiseuille flows.* Trans. ASME, Paper No. 66-FE-5.1966.

References for A Model System

1. Livinston, R. B., Woodbury, D. M., and Patterson, J. L.: Fluid compartments of the brain; cerebral circulation, in Ruch, T. C., and Patton, H. D. (eds.): *Physiology and Biophysics.* W. B. Saunders, Philadelphia, 1965.

2. Rubin, R. C., Henderson, E. S., Ommaya, A. K., Walker, M. D., and Rall, D. P.: *The production of cerebrospinal fluid in man and its modification by acetazolamide.* J. Neurosurg. 25:430, 1966.

3. Shearer, J. L., Murphy, A. T., and Richardson, H. H.: *Introduction to System Dynamics.* Addison-Wesley, Reading, Mass., 1967.

4. Burton, A. C.: Hemodynamics and the physics of the circulation, in Ruch, T. C., and Patton, H. D. (eds.): *Physiology and Biophysics.* W. B. Saunders, Philadelphia, 1965.

5. Key, S. S., and Schmidt, C. F.: *The nitrous oxide method for the quantitative determination of cerebral blood flow in man: theory procedure and normal values,* J. Clin. Invest. 27:476, 1948.

6. Bering, E. A.: *Choroid plexus and arterial pulsation of cerebrospinal fluid. Demonstration of choroid plexuses as a cerebrospinal fluid pump.* Arch. Neurol. Psychiat. 73:165, 1955.

7. Burton, A. C.: *On the physical equilibrium of small blood vessels.* Amer. J. Physiol. 164:319, 1951.

8. Burton, A. C.: *Role of geometry, of size and shape in the microcirculation.* Fed. Proc. 25:1753, 1966.

9. Hamit, H. F., Beall, A. C., and DeBakey, M. E.: *Hemodynamic influences upon brain and cerebrospinal fluid pulsations and pressures.* J. Trauma 5:174, 1965.

10. Dunbar, H. S., Guthrie, T. C., and Karpell, B.: *A study of the cerebrospinal fluid pulse wave.* Arch. Neurol. 14:624, 1966.

11. Foldes, F. F., and Arrowood, J. G.: *Changes in cerebrospinal fluid pressure under the influence of continuous subarachnoid infusion of normal saline.* J. Clin. Invest. 27:346, 1948.

4
Adaptation

In a broad sense, any system in which a parameter is adjusted to counteract a degradation in performance brought about by a change in the environment of the system could be called adaptive. In control literature, the term "adaptive control" is used for a certain class of closed-loop systems which perform the three essential functions of identification, decision, and modification of the system.[1] That is, the controller may be called upon to compute or identify the characteristics of the system while it is in normal operation. The controller must then make a decision concerning the way in which the available parameters of the system should be adjusted so as to improve the operation with respect to a desired performance. Finally, certain signals or parameters must undergo modification to accomplish this goal.

The human operator in a control situation is an example of a very versatile adaptive controller, one that can adapt to many different kinds of changes in operating conditions. The purpose of this chapter is to try to suggest possible physiological mechanisms which may play a role in human-operator adaptive behavior.

Two mechanisms play an important part in manual control and in providing the control system with adaptive capability.[2] These are: (a) the postural control system, and (b) the voluntary control system. The postural system is a feedback system which functions to maintain the posture or position of the body and its limbs. It is essentially a position servoloop with position sensors, motor elements, and some data processing. The postural servosystem is controlled by higher centers which provide reference inputs and control the parameters of some of the elements of this system, as shown in Figure 4-1. The voluntary control system provides the mechanisms for executing skillful, precise movements. It too, of course, is composed of sensory, motor, and computational components. It is quite different from the postural control system in that the controller is thought to be an open-loop intermittent system which executes preprogrammed control movement. These movements are proprioceptively open-loop in the sense that the postural control system appears to be inactivated, at least partially, when voluntary control movements are being executed. The feedback is obtained from a variety of sensors, including joint position receptors and the visual system, but the information obtained from these feedback elements must be processed by higher centers before they can effect a movement. There appears to be some evidence that the voluntary control system acts like a sampled-data, input-synchronized control system. An important part of the control of the voluntary move-

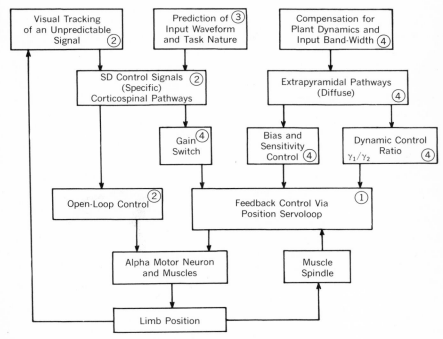

Figure 4-1. Simplified block diagram for movement control.

ments is the predictive system which is able to extract information from signals and responses to predict the future course of events. It is this predictive ability that is largely responsible for the input-adaptive properties of manual control systems.

The Muscle Unit

Muscle is unique in converting chemical energy directly into mechanical work. Limitation in the rate of this conversion acts as an apparent viscosity which plays a role in the overall dynamics of movement. In 1938, Hill[3] showed in a thermodynamic study of muscle that the force-velocity relationship could be approximated very closely by the equation

$$(v + b)(P + a) P_0 = \text{constant} \tag{1}$$

where a and b are constants that are dependent on the particular muscle, P is load, v is shortening velocity, and P_0 is isometric tension. When experiments were extended to lengthening, as well as shortening, muscle, Hill's equation failed. Katz[4] found a steep linear relationship between muscle opposition force and lengthening velocity.

For the purpose of building a model of this complex control system, some very simple approximations to a physiological model may be considered adequate. The force-velocity relationship will be approximated by straight lines, as shown in Figure 4-2. The slope of the straight line representing lengthening, B_1, is made many times greater than that for shortening, B_s, as seen in the solid lines of Figure 4-2. The following parameters based on measurements on human subjects adequately describe the force-velocity relationship: $P_0 = 100$ kg; $v_0 = 0.01$ m/sec; $B_s = 10^4$ kg-sec/m; $B_1 = 6 \times 10^4$ kg-sec/m. If a muscle is shortened at a velocity greater than that at which it is capable of shortening itself, it exerts no force. Thus, P is zero for velocities greater than v_0. When a muscle is stretched more rapidly than its critical stretching velocity, several things may happen. A phenomenon called "slipping" or "yielding" occurs first. Katz[4] found that contractile structures of a muscle may be damaged if its velocity of lengthening is increased rapidly while the muscle is active. Normally, the Golgi tendon organ (a tension-sensing device) reflexly causes the muscle to relax before this damage occurs.

It is expected that the model will seldom operate in the critical region. Thus, a compromise is used between complete relaxation and increased resistance caused by slipping. For velocities of lengthening

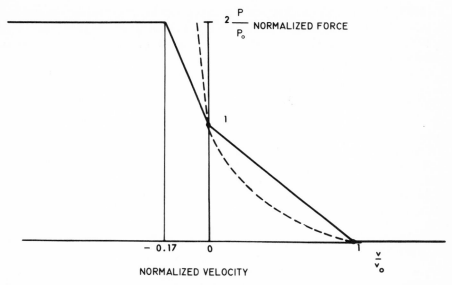

Figure 4-2. Comparison of mathematical model with experiment. Solid curve shows force-velocity relationship of mathematical model; dashed curve, force-velocity relationship of frog muscle as obtained experimentally. For frog muscle, $P_0 = 0.1$ kg, $v_0 = 4$ cm/sec.

which are greater than critical, the model saturates at $2P_0$, as shown in Figure 4-2. The resultant relationships are expressed in Eqs. (2) and (3):

$$P(\text{muscle}) = a(P_0 + Bsv), \quad 0 < v < v_0 \tag{2}$$

$$P(\text{muscle}) = a(P_0 + B_1v), \quad \frac{-v_0}{6} < v < 0 \tag{3}$$

where a indicates the level of stimulation; for maximal stimulation $a = 1$, for half-maximal stimulation $a = 0.5$. The asymmetrical characteristics become very important in smoothly terminating a rapid voluntary movement. Asymmetry increases the effective damping over that of a possible symmetrical relationship. Addition of the postural system (muscle spindle and afferent feedback) further increases man's ability of rapidly damping his motion.

The muscle spindle receptor is a differential length receptor found in parallel with contractile fibers of muscles of many species.[5,6] Its importance in human motor coordination is great. Its positional feedback characteristics cause the myotatic or stretch reflex. It may also send kinesthetic information to higher centers to help control complex motor coordination tasks. The spindle is connected in parallel with the muscle contractile fibers as shown in Figure 4-3, and its direct mechanical effect on the muscle is negligible. Thus, we can consider the length of the muscle, Xm, to be an input produced either through the alpha efferent nerve or through stretch by external forces. The afferent nerve carries information concerning the length of the nuclear bag to the central nervous system. Continuous signals representing a short-term average number of pulses per second are used in the model. The gamma efferent nerve excites the contractile element, or intrafusal fibers, of the spindle. This is another input that may bias the output of the nuclear bag or may, perhaps, act indirectly as an input that might control movement; this follow-up servoconfiguration has been suggested by Merton.[5] In the model we have simplified this input to a

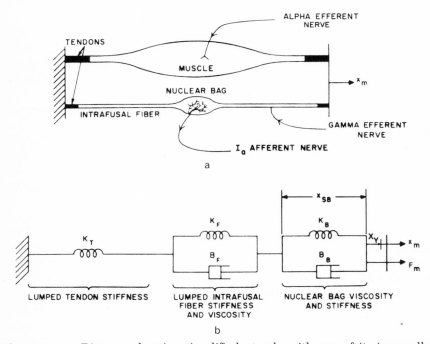

Figure 4-3. a. Diagram showing simplified muscle with one of its in-parallel spindle receptors separated for ease of illustration. b. Mechanical model of spindle receptor. Inputs are X_m (length of muscle) and X_r (artificial length caused by input to intrafusal fiber); output is X_{SB} (length of nuclear bag).

change in length of the intrafusal fiber, Xr, and have merely added it to Xm (Fig. 4-3b). The dynamics of this input may have to be considered more explicitly when we investigate coordinated movements through this input.

The response of the spindle receptor to a step input of stretch shows approximately 400 percent overshoot; this differential effect is called "phasic response."[2,5] After approximately 200 msec, the output settles down to its steady-state value; the steady-state response is called the "tonic response." There seems to be enough evidence to suggest that the phasic response is caused entirely by mechanical factors, and that the mechaincal-to-electrical transducer is a no-memory device. The transfer function for the mechanical model of Figure 4-3b is of the form

$$\frac{X_{SB}}{X_m} = H(s) = \frac{G_m \dfrac{(Ts + 1)}{1}}{\dfrac{(Ts + 1)}{2}\dfrac{(Ts + 1)}{2}}$$

where s denotes complex frequency.

The spindle system has a differentiating action over 0.5 to 3 Hz, an important part of the frequency range. Since it is in the feedback path of the proprioceptive stretch reflex, or postural servomechanism loop, this obviously acts as a damping element in movement. It can be shown that, in ordinary movements, the muscle apparent viscosity, which has similar dynamics, operates in a region of lesser interest, below 0.25 Hz. The muscle damping is the result of asymmetrical nonlinear energy conversion saturation and thus has, strictly speaking, no restricted frequency range. However, the spindle is an ideal damping element, and the gamma bias which sets its activity level and gain is well suited to modify the damping in an adaptive way.

Recent studies[5,6] have shown that the muscle spindle may be considered as a device signalling mechanical events by means of two different outputs (the primary and secondary afferent fibers), and that it is controlled by two additional inputs (the Y_1 and Y_2 efferent fibers). It may be concluded from the present evidence that, when a muscle is being stretched, the muscle spindle primary endings signal both the instantaneous length of the muscle and the velocity at which it is being stretched, while the secondary endings signal mainly the instantaneous length.[5] Also, the activity in the Y_1 fibers (causing contraction of nuclear-bag intrafusal fibers) will excite mostly the primary endings, whereas the activity in the Y_2 fibers (causing contraction of nuclear-chain fibers) will excite both the primary and the secondary endings. Although quantitative data and various details are lacking, it is conceivable that the double motor input to the spindle provides relatively independent control of the "bias" and of the "damping" of the servo-loop. This permits even more powerful control of overall dynamics and damping.

When a muscle contracts, it develops forces which are applied directly in series with Golgi tendon organs, causing them to discharge. It has been shown that the tendon organs have a much higher threshold and a much lower sensitivity to forces applied to a passive muscle than they do to active inputs.[7] A given tendon organ response to a unit step function increase in active muscle tension is

$$r(t) = K(1 + Be^{-bt} + Ce^{-ct}) \text{ pulses per sec}$$

where $t > 0$. The gain factor K, in the soleus muscle of an anesthetized cat, has a value of about 4 pps per gm (as contrasted with 0.04 pps per gm for passive forces). The exponential parameters are approximately: $B = 1$; $b = 2 \text{ sec}^{-1}$; $C = 2$; $c = 25 \text{ sec}^{-1}$. Thus, the signal which is sent to the spinal cord by a Golgi tendon organ is a filtered sample of the active forces that are being produced in the muscle. These afferent signals excite internuncial neurons in the spinal cord, which in turn inhibit the motor neurons that supply the same muscle. The specific role of the Golgi tendon organ in motor coordination and in spinal reflex activities has not yet been clearly demonstrated.

Supraspinal Mechanisms of Motor Control

Descending fibers from the higher centers end on the two types of motor neurons (alpha and gamma motoneurons) and on the inter-neurons of the spinal cord. Through these connections the descending influence from the brain interacts with the reflex pathways in determining the outflow from the motor neurons and the pattern of movement. The descending fibers arise at various levels, and include fibers from the cerebral cortex and certain nerve cell masses of the brainstem such as the red nucleus, vestibular nucleus, and reticular formation. Other parts of the brain, such as the cerebellum or basal ganglia, exert their influence by sending nerve fibers to the cells of origin of these descending tracts. The quantitative and even the qualitative information concerning the supraspinal control of the motor system is still too much in a stage of infancy to suggest any definitive model of this part of the motor system. The present knowledge of these mechanisms has been reviewed by Eldred and Buchwald.[8] The mechanisms chiefly responsible for postural control are distinct from those subserving discrete movements. Postural control seems to be mediated by fibers in the reticulospinal and vestibulospinal tracts, because ablation of these causes a lack of postural support, yet leaves discrete movement control intact.

Behavior Characteristics of the Motor Coordination System in Man

The behavioral characteristics of the human motor coordination system must be quantitatively defined in order to test the predictive ability of a physiological model. Studies on the human motor system before World War II were primarily oriented toward an understanding of the learning ability of a human operator. Very little, if any, attention was paid to the dynamic character of the operator's movements. During and after the war, however, with the advent of sophisticated military equipment, it became necessary to include the dynamic characteristics of the human operator in the design of these systems.

A complete identification of the dynamic behavior of the human motor system is an unsolvable problem from an engineering viewpoint with today's analytical tools. Very little is available in control theory literature on identification of complex nonlinear systems. Most often, an initial attempt is made to obtain some understanding of such a system using the tools of linear systems analysis, that is, the performance of the system is studied using simple inputs such as sinusoids, steps, and ramps, and the analysis is based on the assumption of quasilinearity. This has been completely discussed by Stark.[2]

Mathematical Models and Identification Problem

The identification of biological systems is an extremely difficult problem because of the complexity and randomness of these systems. This problem is further compounded due to the "black-box" nature of the system and inadequate measurement techniques. Thus, any mathematical model of these systems only approximates very crudely the behavior of the actual system for a certain class of inputs. However, a mathematical model is useful in organizing one's thoughts and in presenting the available information in a concise manner. A good model, naturally, should suggest new experiments that will modify the model to make it even better.

Navas and Stark[10] have reviewed the various models proposed for hand-tracking control systems. The model (Fig. 4-1) includes the visual feedback path and the integrator in the loop which makes the system a position servo the additional switch providing an intermittency or quasi-sampled-data action, a proprioceptive feedback path, the alternation between voluntary and reflex action for control, and some delay compensation. The time delays in this model have been estimated to be

$$\text{Visual latencies} = Td_1 = 40 \text{ msec}$$
$$\text{Conduction time} = Td_2 = 15 \text{ msec}$$
$$\text{Muscular contraction time} = Td_3 = 30 \text{ msec}$$

Although admittedly tentative and incomplete, the model of human movement has clarified in a semiquantitative way certain features of interest to anyone studying adaptation. A distinction is made between two systems: a lower-level diffuse reflex postural system the dynamic behavior of which is controlled via the muscle spindle, and a higher-level specific control system which is open-loop with respect to the proprioceptive feedback. The interaction between these two systems is complex and antagonistic in part. The voluntary system seems to be a discrete or discontinuous control which can switch off the opposing position servo for rapid movements.

Prediction of input wave form and task nature seems to be a higher control function controlling the movement via the direct specific pathway. Adaptation of this type should operate independently of the gamma system. When plant dynamics change, different damping is required, and this may well be a function of the diffuse postural system. It is known that mental set can change the stability margin of the postural system. Certain possible controls such as the gamma ratio would be ideal for this functional role. A proposed hypothesis for this adaptive mechanism is: adaptation to stick dynamics is via the postural loop; adaption to input wave form is via the direct proprioceptive open-loop path. This hypothesis may be tested experimentally by measuring the postural feedback system parameters and comparing with their values in the normal mode.

Adaptation in the human operator takes several forms. The adaptation to a predictable signal involves recognition of the parameters of the input wave form in a carefully preprogrammed sequential set of signals to drive the muscles. There is some evidence that proprioceptive sensory feedback is necessary in a physiologically generated form, i.e., by active movement, before these "engrams" or adaptive matched filters can be formed. On the other hand, general sensory pattern classification of, for example, visual patterns may well operate in this manner without requiring active motor participation by the subject. Adaptation to constraints on random signals has been clearly demonstrated.

The postural servo is an ideal element for adaptive compensation. The present anatomical as well as physiological evidence suggests that the higher centers of the nervous system can control the α, γ, and δ_2 motoneurons independently, and that this control may be exerted relatively independently for different muscles, at any rate for those as different in function as the agonist and antagonist muscles of a limb.[5] The postural servo system with double motor innervation to the spindle is ideally suited for an adaptive role.

Dynamical Responses of the Movement Coordination System of Patients with Parkinson's Syndrome

Parkinson's syndrome is a common affliction, especially of elderly persons, well known in the neurological literature for the past hundred years. The muscles and nerves of parkinsonian patients are generally not at all affected by the disease (until late secondary changes occur). Evidence for this comes from observation of patients who were able to walk normally when sleepwalking, as well as from the ability of certain brain operations to modify this syndrome. In the late 1920's it was demonstrated that interruption of the stretch reflex markedly reduced the rigidity which is a prominent sign of this syndrome. Steady-state and transient experiments described earlier, when performed on a series of parkinsonian patients, not only provided evidence concerning the nature of the defect of Parkinson's syndrome, but also are valuable in understanding the organization of the normal control system. Parkinson's syndrome is a more complex disorder than can possibly be explained with out present knowledge. For example, the phenomenon of tremor still eludes the neurophysiologist, although the presence of this oscillation is tantalizing. In addition to the presence of the rest tremor, the interaction of this tremor with volitional movements is unpredictable and erratic. The response of the patient is slowed and inaccurate. Surprisingly, after the movement is completed, the tremor dies down for approximately one-half second.

In Figure 4-4 the frequency responses obtained using unpredictable sinusoidal target signals are shown for three subjects: a normal person and two patients with mild and moderate degrees of parkinsonian rigidity, respectively. It can be seen that the band-width of the patients'

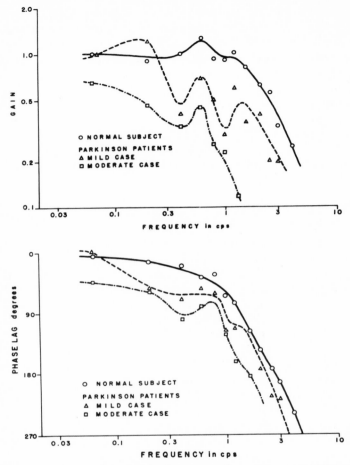

Figure 4-4. Effect of Parkinson syndrome on frequency response gain and
phase data.

movement response is severely restricted. This is consistent with the
mode of a higher-gain postural feedback system operating against
voluntary control of the muscles.

References

1. Gibson, J. E.: *Nonlinear Automatic Control.* McGraw-Hill, New York, 1963.

2. Stark, L.: Neurological feedback control systems, in Alt, F. (ed.): *Advances
   in Bioengineering and Instrumentation.* Plenum Press, New York, 1966.

3. Hill, A. V.: *The heat of shortening and the dynamic constants of muscle.* Proc. Roy. Soc. Ser. B 126:136, 1938.

4. Katz, B.: *The relation between force and speed in muscular contraction.* J. Physiol. 96:45, 1939.

5. Matthews, P. B. C.: *Muscle spindles and their motor control.* Physiol. Rev. 44:219, 1964.

6. Granit, R. (ed.): *Nobel Symposium I—Muscular Afferents and Motor Control.* John Wiley, New York, 1966.

7. Houk, J. C., Jr.: *Sensory transduction of muscle forces by Golgi tendon organs: experiments and models.* Ph.D. Thesis, Harvard University, Cambridge, Mass., 1966.

8. Eldred, E., and Buchwald, J.: *Central nervous system: motor mechanisms.* Ann. Rev. Physiol. 29:573, 1967.

9. Stark, L., Vossius, G., and Young, L. R.: *Predictive control of eye tracking movements.* IEEE Trans. Human Factors in Elect., 1962, pp. 52-57.

10. Navas, F., and Stark, L.: *Sampling or intermittency in hand control system dynamics.* Biophys. J. 8:252, 1968.

PART TWO

*Engineering in the Analysis
of Physiological Systems*

5

Bioengineering Techniques for the Nervous System

It is now a generally accepted principle of neurophysiology that the fundamental process of communication between nerve cells occurs by means of signals composed of electrical pulses. While it may seem surprising, from the apparent continuity of animal behavior, that the fundamental events of nervous activity are inherently discontinuous, this fact has been universally confirmed by numerous experimenters who now routinely monitor the seemingly endless flow of pulses transmitted from cell to cell in the brain, hoping to clarify the role these signals play in the processes of control and communication. The singular experimental difficulty is that the basic units of control and communication often do not lie at the level of the single nerve cell (neuron), but are derived from large complexes of cells organized together into aggregates that are loosely called "networks." In general, however, signals from only an insignificant fraction of an entire network can be observed at any one time, and experimental limitations often enable these signals to be observed only under restricted or artificial conditions. Confronted then by nature's most complex biological system, but constrained at the same time by the most restrictive experimental circumstances, the neurophysiologist continuously searches for whatever assistance can be furnished by available analytical methods. Among the most powerful and useful tools in this context are the techniques of mathematical modeling and simulation.

97

Neurophysiologists can now monitor the electrical signals associated with activity in individual cells of the brain. This is achieved by inserting into the brain wire electrodes insulated along their entire extent except at their tip, which is polished down to the extremely small dimensions of the neuron (i.e., the electrode tip may be only several microns in diameter). When in proximity to the surface of a single neuron, a microelectrode can detect the minute changes of the electrical field which result from activity in the cell, thus making it possible for pulses produced by the neuron to be amplified and recorded. Somewhat smaller electrodes, usually made of glass with a conducting fluid inside, can be made to penetrate the boundary membrane of the neuron and detect the electrical concomitants of a variety of internal processes in the cell, some of which will be described below. By using several microelectrodes of either of these types, signals from more than one cell can be observed simultaneously (Fig. 5-1).

The present chapter will deal with the behavior of networks whose neurons exhibit component processes which can be observed with an

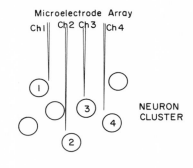

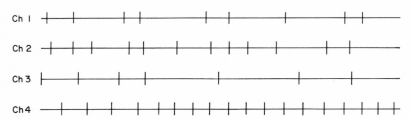

Figure 5-1. Top: Schematic diagram of neuron cluster (only the cell bodies are shown) together with microelectrode array which monitors pulse activity in selected cells of the cluster. Bottom: Typical group of pulse trains recorded simultaneously from neuron cluster. Each electrode has been positioned to monitor output of a single cell.

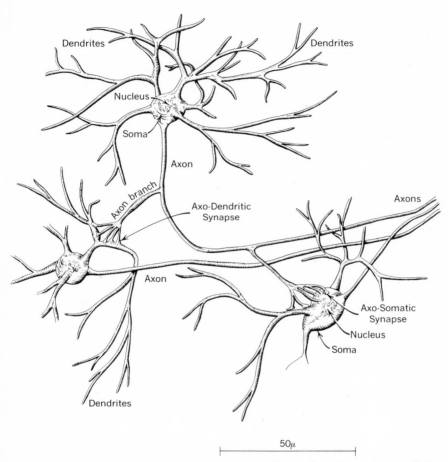

Figure 5-2. Semidiagrammatic view of neuron cluster showing various cellular
structures.[8]

electrode placed inside the cell. Molecular events will not be discussed,
although there is a considerable literature on this subject (see references).

Basic Internal Functional Processes of the Neuron

The nature of the processes to be studied can be understood by
referring to Figure 5-1, where a hypothetical, but typical, result of an
experiment is shown in which a group of electrodes placed in a brain

have been positioned in the proximity of a neuron cluster. In such an experiment, each electrode monitors the pulses from a single cell.

The mechanism by which the nerve impulse is propagated has been the object of intense investigation for several decades. It is now known that the mechanism involves extremely rapid exchanges of sodium and potassium ions across the boundary membrane of the axon, and that this exchange is dependent upon a momentary instability of the membrane induced by the appearance of an impulse at an adjacent area of membrane. This serial induction of instability, with the resulting exchange of ions, is associated with a rapid reversal in the transmembrane potential, which is registered by the nearby electrode as a pulse around one millisecond in duration. Our understanding of this phenomenon is due in large part to the work of Hodgkin and Huxley, who were also the first to develop a mathematical model for the process.[47] A number of other workers have treated this problem from a mathematical point of view, or have otherwise attempted to simulate the process by means of electrical circuits (see references).

If the electrode is allowed to penetrate the cell and record the transmembrane potential (i.e., the potential inside the cell relative to that of the surrounding tissues of the body, which is regarded as ground potential), the details of the nerve impulse are seen along with a variety of other events.

In the Channel 3 (Fig. 5-3) record, there is an upward deflection of the membrane potential, (marked by ▲), and each event is associated with a pulse in the Channel 1 record. This deflection is the result of the propagation of the pulse from cell #1 along its axon, and of its

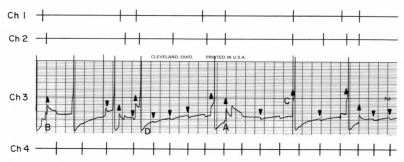

Figure 5-3. Cell cluster as in Figure 5-1, except that electrode for Ch 3 has been allowed to penetrate cell 3 and record the internal potential which exhibits certain intracellular events occurring between pulses. Note the coincidence between certain deflections in the potential and the occurrence of pulses in cells 1 and 4.

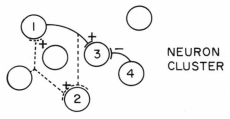

Figure 5-4. Neuron cluster as in Figure 5-1, but with axons and synaptic connections shown to indicate the connections inferred from the record in Figure 5-3. Dotted connections are those inferred from statistical analysis of the firing patterns of neurons 1 and 2.

reaching a synapse at some dendrite of cell #3. The nerve impulse causes the release of a chemical *transmitter substance,* which diffuses from the synaptic terminal to the membrane of cell 3 where it quickly, but momentarily, alters the properties of that membrane to permit a rush of charged ions to cross it and thereby change the transmembrane potential. This ionic exchange and the associated change in potential is called the "synaptic potential" (▲). Synaptic potentials are characterized by rather sudden changes in membrane potential followed by gradual returns to the baseline (A, Fig. 5-3). When arriving impulses follow in rapid succession, additive steplike changes in potential result (B, Fig. 5-3). This process is referred to as *summation.*

An examination of Figure 5-3 reveals that some of the synaptic potentials in Channel 3 are invariably associated with pulse events in one of the other cells in Channels 1 or 4. From this it can be concluded that there are synaptic connections between them, as in Figure 5-4. Some are characterized by abrupt changes in potential *away* from the threshold level. These too are synaptic potentials, induced by the arrival of pulses from Channel 4. During these events, the ionic fluxes are in the opposite direction; hence, the potential change is reversed. An impulse in Channel 4 causes the release of a chemical substance at the synapse which alters the membrane in a fundamentally different way from that of the substance producing the upward-directed synaptic potentials. These two types of events are distinguished by noting that those marked ↑ deflect the membrane potential toward the threshold value, whereas those marked ↓ shift the potential away from the threshold. Since the former tend to produce firing, they are called *excitatory postsynaptic potentials* (EPSP's); and since the latter tend to suppress firing, they are called *inhibitory postsynaptic potentials* (IPSP's) (Fig. 5-5).

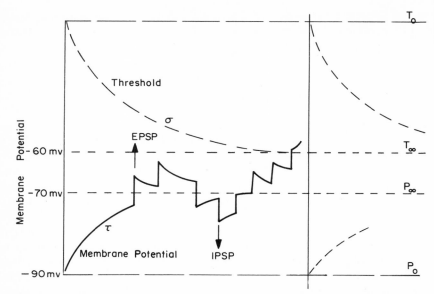

Figure 5-5. Schematic diagram summarizing the detailed assumptions common to a wide class of neuron models.

The typical neuron may have a very large number of excitatory and inhibitory impulses converging upon it from other cells. Each arriving impulse is associated with a synaptic potential which, when summated with the existing potential value, moves the membrane potential toward, or away from, the threshold level, depending on whether it is excitatory or inhibitory. The basic rule always applies: the postsynaptic cell fires whenever the net input moves its membrane potential to the threshold value. Only then is any information regarding the interior state of the cell transmitted to any other cell.

Integrated Models of the Basic Component Processes

A model will not reveal any new information about the component processes at the subneuronal level, but it can tell us a great deal about the behavior of networks consisting of neurons with the properties already described. The assumptions of the model are:

1. The nerve impulse can be approximated by a pulse or delta function.

2. Whenever neuron j of the network produces an impulse, it will produce a synaptic potential of amplitude S_{ij} at cell i if it has a connection with cell i.

3. With each cell there is associated an equilibrium or *resting potential* P_∞ to which the membrane potential eventually returns after an impulse or after synaptic input. The membrane potential always approaches the resting potential at a rate proportional to the difference between the current potential and P_∞. That is, the potential always decays exponentially toward P_∞ except when an impulse is produced, or when a synaptic potential is added. The time constant of decay, τ_i, is called the *membrane time constant,* and governs the decay rate of the membrane potential of the neuron at all times except during impulse generation (Fig. 5-5). Neurons normally have resting potentials of 70 to 90 millivolts, with the inside negative with respect to the outside. Time constants are in the range of 1 to 20 milliseconds.

4. An impulse is generated in cell i whenever the membrane potential in cell i goes from a value less than the current value of threshold to a value greater than threshold. After an impulse, the membrane potential is reset to its restoration value P_{i_0}. From assumption 3 it follows that after an impulse the membrane potential $P_i(t)$ decays exponentially from P_{i_0} to P_{i_∞} according to the equation

$$P_i(t) = (P_{i_\infty} - P_{i_0})(1 - e^{-t/\tau_i}) + P_{i_0} \qquad (1)$$

(when it is measured from the moment of the impulse), provided no other input arrives.

5. With every neuron i is associated a *threshold function* T_i which determines the critical firing level of a neuron in terms of the current value of the membrane potential at which an impulse will be generated. Present-day models based on neurophysiological evidence utilize several different assumptions regarding this variable. The two most often used, and probably the simplest, are:

a. Constant threshold.

b. Exponential threshold: After each impulse, the threshold potential is reset to some value T_{i_0} from which it decays exponentially to some asymptotic value T_{i_∞} with decay constant o.

$$T_i(t) = (T_{i_\infty} - T_{i_0})(1 - e^{-t/o_i}) + T_{i_0}$$

From assumption 4 it follows that an impulse will be produced in cell i whenever $[P_i(t) - T_i(t)]$ changes sign from negative to positive. These assumptions are known to be in error in certain areas:

1. In the model presented above, the synaptic potentials S_{ij} are assumed to be algebraically additive, an assumption known empirically to be incorrect.
2. The synaptic potentials S_{ij} are incorrectly assumed to be independent of the membrane potential $P_i(t)$ at the time of their production. There is good evidence to suppose that both excitatory and inhibitory synaptic potential amplitudes are linear functions of the membrane potential, and this could easily be included in any model. Similarly, the synaptic potentials are assumed to be instantaneous in their rise times. This is often not the case and several models incorporate modifications of this assumption.
3. The assumption that after each pulse the membrane potential and threshold are restored to fixed values effectively implies that the sequence of events leading to an impulse is no longer influential after the pulse is produced. This postulate of our model is not true for all cells.
4. The assumptions concerning the behavior of the threshold potential are quite arbitrary. They are based on the observation that immediately after a neuron has fired an impulse it is less "excitable" (a vague term) than it is at a later time. This relative inexcitability, called *refractoriness,* could arise from the fact that after an impulse the membrane potential is restored to a low value P_{io}, at which time large excitatory potentials, ordinarily capable of triggering a spike (i.e., capable of raising the membrane potential to the threshold level), cannot overcome the added distance to the threshold level. There is some evidence, however, that the relative refractoriness of the neuron is not due to this increase in distance of the membrane potential from the asymptotic threshold level alone, but actually results from an elevation of the threshold level after each impulse generation.

    There is also evidence that the threshold level at which an impulse is generated depends on the rate of change of membrane potential as well as on its absolute value. That is, an impulse will be generated at a lower level if the membrane potential is rising rapidly than it will if the rate of rise is low. This property is known as *accommodation* and is generally not included in models at present.

5. All of the preceding assumptions, however inaccurate in physiological terms, are completely deterministic from a mathematical point of view.

There are three principal techniques now used to study the behavior of network models, all of which in one way or another mechanize the solution of the describing network equations by a process known as *simulation*.

In many cases the neuron networks have been simulated using a digital computer program that incorporates the describing equations for all the neurons in the network and solves them numerically at each point in time (see references). The principal advantages of digital simulation are the ability to handle large networks of cells (up to a hundred or more) with complex interconnections, and the greater control it offers over the statistical parameters of the model.

In a number of cases involving a few cells, analog computer simulations have been successfully used (see references). Not only is the programming of small networks relatively easy, but the variables associated with each cell can be monitored continuously and checked for their similitude with real neurons.

A third method for simulating networks is to build special electrical analogs of neurons. A number of these have been successful, and some even include a number of component processes at levels below those that have been discussed here. Their principal advantage is the relatively low cost of their implementation.

Neuron Models in the Study of Input-Output Relations in Single Neurons

In the study of signals in the nervous system, one most frequently encounters situations in which a microelectrode samples only the output pulse activity in a single neuron, as seen, for example, in one of the channels shown in Figure 5-1. From this single pulse train, one wishes to draw inferences concerning the possible input-output relations for this cell. That is, by squeezing all the information out of that pulse train, we would like to determine (a) the intrinsic properties of that cell (such as the parameters embodied in the model), and (b) the type, number, and temporal history of the pulse trains reaching the observed neuron from other sources. In short, we wish to know, from the observations of output only, the connections of the cell, its input, and its transformational properties. However, because a large number of input signals can arrive at the cell without producing a cell firing (Fig. 5-3), and because many different input configurations can produce comparable patterns of cell firing, it is clear that the problem, as stated, is generally insoluble.

Nevertheless, with the aid of models, explicit hypotheses regarding the network in which the observed cell is imbedded can be subjected to simulation procedures, and the output of the real neuron can be compared with its counterpart in the hypothetical network. In constructing such a hypothetical network, all available anatomical and physiological information known to be pertinent to the neuron in question is used. This information is translated into appropriate terms for the model. For example, experimental data must be used in estimating the intrinsic parameters of the neuron, the number and type of synaptic connections, and the types of discharge pattern likely to be reaching the cell of interest from other cells in the network, etc. When the model parameters have been completely specified, the information can be run on a computer to produce a pulse train which is the model equivalent of the real pulse train.

We must be content with a probabilistic or statistical expression describing the behavior of the pulse train. That is, the output pulse trains can be described only by a probabilistic representation. Therefore, it is not reasonable to expect the output of a model of a neuronal network to duplicate exactly the temporal history of an experimentally recorded neuronal pulse train. Instead, we ask that the statistical measures applied to both the real cell and the model of that cell be essentially indistinguishable.

One of the most commonly used techniques for this purpose is to construct a frequency histogram of the interval between pulses, the *interspike interval histogram*. This is obtained by measuring the time t_j between pulses j and j + 1 in the spike train, for all j. The kth bin in the histogram contains the total number of intervals in the train for which

$$(k - 1)\,\Delta t \leq t_j < k\,\Delta$$

where t is the bin width of the histogram. It is useful to consider this histogram as an estimate of the probability distribution governing the length of intervals between neuron pulses. This distribution expresses the probability that a neuron will fire an impulse in the interval (t, t + Δt) when it is known that the neuron last fired at t = 0. Alternatively, we can view this distribution, $P_1(t)$, which we shall call the "interspike interval density," as expressing the probability that if a neuron fires at the present instant, the probability of its *next* firing in an interval Δt seconds in duration t seconds from now is equal to

$$P_1(t) \, \Delta t$$

The interspike interval density can be generalized to obtain a probability density not only of the next firing time, but of all past and future firing times. This density function is known as the "renewal density," $H(t)^*$, and expresses the probability that, in an interval of duration t, a neuron will fire an impulse (not necessarily the first) at time t when it is known to have fired at $t = 0$.[21] Since this impulse must be either the first, second, or n^{th} impulse since the pulse at $t = 0$, it is clear that $H(t)$ must be the sum of all the interval probability distributions $P_1(t)$, $P_2(t) \ldots$, $P_n(t)$, where $P_k(t)$ is the interval density of times between impulse s and impulse $s + k$.

Some typical interspike interval histograms and corresponding autocorrelation functions for neurons are shown in Figure 5-6, indicating some of the types of discharge patterns which arise naturally from neuronal networks.

A physiologically meaningful model can be proposed which includes the additional piece of information that large EPSP's are known to occur in these cells. This suggests the possibility that a periodically firing input source generates an excitatory postsynaptic potential very nearly equal in amplitude to the distance between the resting potential and the asymptotic threshold potential (Fig. 5-7). With random fluctuations in membrane potential ("noise" in the membrane), some of these, say 50 percent, would be expected to trigger postsynaptic spikes. There would then be a probability $p = 0.5$ that a given pulse would not appear, and a probability $p^2 = 0.25$ that two pulses in succession would not appear, etc. If the model is valid, the height of the successive peaks in the experimental histogram should decline in a geometric ratio, as in fact they do. This model is certainly plausible on physiological grounds and suggests that further experimentation, such as intracellular recording, would find a regularly firing cell, which generates large EPSP's, connected to these cells.

Naturally, only direct experimentation on the neurons in question can ever determine the precise mechanism by which these observed patterns are produced.

---

*When the neuronal pulse train is regarded as a sequence of delta functions, the renewal density is proportional to the conventionally defined autocorrelation function.[23, 24]

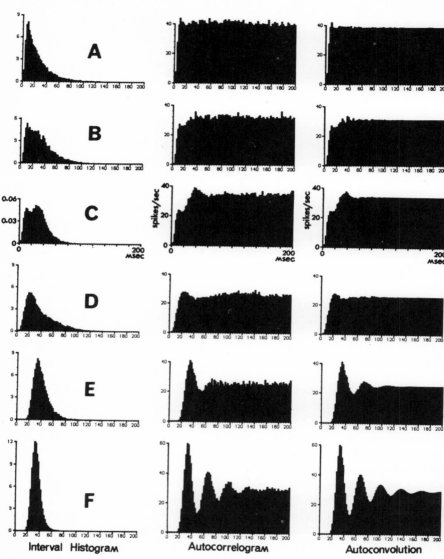

Figure 5-6. Left column: Interspike interval histograms from six neurons in the visual system of the cat. Middle column: Autocorrelogram from the same data records. Note the periodicity in the lowest record. Right column: Predicted form that the autocorrelogram should have if all the intervals of the pulse train are independent ("autoconvolution").[27]

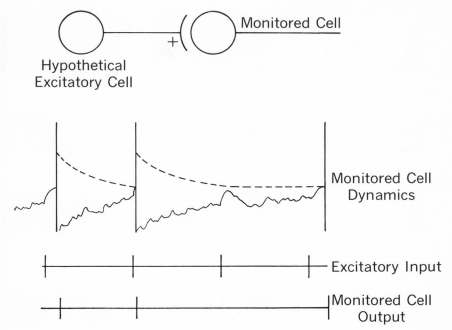

Figure 5-7. Highly periodic excitatory source has fixed probability of firing postsynaptic cell, but postsynaptic cell noise randomly delays cell firings with respect to the occurrence of the EPSP.

Neuronal Models in the Study of Network Structure

A technical difficulty is that the attempt to account for the output of a single neuron was based solely on the output from that neuron without any clues from the activity in other cells. When there is access to data from several neurons simultaneously, as in Figure 5-1, we can utilize the joint information in those signals to draw inferences about network structure. These inferences can then be incorporated into a network model as before, and the output of the model can be compared with the original data set.

For any two channels of pulse trains the *cross-channel waiting time* from pulse train A to pulse train B is determined by measuring the distribution of elapsed times from all impulses in A to the next impulse in B.[25] This distribution estimates the probability $w_1(t)$ that if cell A fires now, cell B will fire its *next* pulse in the time interval (t, t + Δt). This probability is given by

$$w_1(t)\,\Delta t$$

This distribution is analogous to the interspike interval histogram. The higher order waiting times, $w_i(t)$, are obtained by measuring the distribution of times from all pulses in A to the ith succeeding pulse in B. The sum of the waiting time up to order n

$$W(t) = \sum_{i=1}^{N} w_i(t)$$

is an estimate of the cross-correlation function between cells A and B. The sum should include not only the forward waiting times but also the backward waiting times. The cross-correlation function estimates the probability that if cell A fires an impulse at t = 0, there will be a pulse in cell B (not necessarily the first) at time t, where t can be positive or negative. If cells A and B are independent the cross-correlation function is constant, i.e., knowledge of cell A firing times is not useful in predicting the times of firing of cell B.

On applying the cross-correlation estimator to a sufficiently long record of cell activity, we find none of the cross correlations are flat, and therefore we must reject a hypothesis based on their independence. But their dependency must be accounted for. The Channel 4—Channel 3 cross-correlation function is interpreted as indicating that immediately after a cell 4 firing (at t = 0) there is a reduction in the probability of cell 3 firing. Conversely, there is an increased probability of cell 3 firing after cell 1 fires, as shown by the Channel 1—Channel 3 cross correlation. The reason is clear from the intracellular record of cell 3 shown in Figure 5-3. The reduced probability of cell 3 firing after cell 4 fires results from the fact that cell 4 *inhibits* cell 3, precisely as one would expect. The enhanced probability of cell 3 firing after cell 1 fires arises expectedly from the fact that cell 1 *excites* cell 3.

This does not explain the enhanced probability of cell 3 firing after cell 2 because cell 2 does not appear to produce synaptic potentials in cell 3. This can be explained, however, by computing the cross correlation between cells 1 and 2. The data indicate a high correlation between cells 1 and 2. This can be accounted for by assuming that cell 1 excites cell 2 or by assuming that cells 1 and 2 are both excited by a common source. Our references to the intracellular record of Figure 5-3 have been useful in interpreting the cross correlations, but clearly we could have been led to the same conclusions about intercell connections without this information. The important point is that correlational analysis of the four pulse trains in Figure 5-1 leads immediately to two possible network models that could generate the data (the possible connections are shown as dashed lines in Figure 5-4). Synaptic and cell parameters must then be adjusted to determine which of the two simulated models provides a more satisfactory fit of the autocorrelation and cross-correlation measures.

Neuronal Models in the Study of Complex Network Behavior

From an experimental point of view, the capability of neurophysiologists to observe a large number of neurons simultaneously and consistently within a given network appears, at present, to be out of the question. For the foreseeable future, then, we must be content to observe only the smallest fraction of an operating network, or must observe the entire behavioral consequences of that network at some level other than the neuronal level. Nowhere, it would seem, is there a greater need for large-scale simulation and modeling than in the study of what have been termed the "emergent properties" of neuronal networks, that is, those properties of neuron circuits which are derived from complex interactions and interconnections, rather than being intrinsic properties of the component neurons themselves. We shall consider a few examples of these.

The classic example of a neuron network is one organized on the pattern of "lateral inhibition." In such a network, each neuron makes inhibitory connections with a fraction of the population surrounding it. Such a pattern of organization was first observed among the sensory neurons in the compound eye of the horseshoe crab *Limulus*. The neuronal transducer elements of that eye are effectively spread geographically along a two-dimensional grid, one neuron at each node of the array.

Early work by Hartline and his collaborators[33,34] established that (a) every individually illuminated sensory cell responds with a discharge frequency proportional to the light intensity, and (b) each neuron in the grid inhibits its neighbors in proportion to its own firing rate, but with an intensity that diminishes linearly with the distance separating the cells. Using a network composed of model neurons with standard properties, and coupling these in a two-dimensional array according to the conventional type of inhibition described earlier (with the IPSP amplitude a function of distance between cells) make possible the simulation of the two-dimensional pattern of firing in the network as a whole, and permit examination of the behavior of the network as an optical data processor.

To model the excitatory effect of light on these cells, the asymptotic resting potential of each neuron is set to some value *greater* than the asymptoic threshold, and linearly increasing with light intensity. Then, as light intensities increase, the time to reach threshold is diminished and the firing rate increases monotonically.

One of the most important properties of this network which can be demonstrated by simulation is the behavior of the system under various conditions of illumination. If, for example, a square area of the network grid is uniformly illuminated and the surrounding areas are kept dark, the average firing frequency of the neurons in the lighted areas will, of course, be greater than that of the dark neurons. Cells in the middle of the illuminated zone achieve a firing rate which is a balance between the excitatory effect produced by the light and the inhibitory effect produced by the surrounding neurons. Neurons in the dark are not excited by light and exhibit a low firing rate. Cells on the light side of the light/dark boundary receive the same light stimulus as the rest of the illuminated cells, but their inhibition is less than that of the other illuminated cells, because some of their inhibitors lie on the dark side and are firing at a lower rate. As a result, their firing rate is higher than the average for the lighted zone. Conversely, the dark neurons near the light/dark boundary have a firing rate less than the average for the dark side. Thus, at the boundary, the local firing rate gradient is steeper than the overall average difference in firing rates for the two regions. The result is an accentuation of the boundary region from the point of view of firing rates of the sensory neurons, a phenomenon known as "edge enhancement." This property, which arises from the intrinsic organization of the network connections, serves to sharpen the optical boundaries. It has been pointed out[36] that such an arrangement acts effectively as a neuronal lens, "focussing" the image in an eye system where an actual lens is not present.

There are, of course, many other remarkable properties of lateral inhibitory networks.[32-36] Before closing this topic, it is important to note that the principle of lateral inhibition in the structure of neuron networks appears to be extremely widespread in many parts of many rather different nervous systems, and seems to be an organizational principle basic to a variety of otherwise dissimilar sensory systems.[32]

A second example of the use of network models is in the study of those organizational principles and mechanisms which underlie the production of patterned activity in groups of neurons. A classic example is the so-called respiratory networks of the vertebrate brain. These are areas in which neurons fire bursts of impulses at very regular intervals, which in turn drive neurons at other locations that directly control the muscles of breathing. One group of cells is active during the inspiratory phase of respiration, and alternates with an antagonist group which drives the muscles of expiration during the exhalation phase of the cycle (Fig. 5-8). Surprising as it may seem, it is not an easy matter to construct an artificial network of neurons that shows this kind of alternating and rhythmic behavior, and what electrical records we have of real neurons from these centers have not appreciably clarified the type of underlying mechanisms involved. Almost nothing is known about the connections either within a population or between the two populations of neurons. Although both receive sensory information from a number of areas, including stretch receptors in the lungs, the alternating and rhythmic behavior of these cells appears to be an intrinsic property of the network and is not dependent on sensory input, nor is it derived from any individual cells within the group.

As yet, there is no generally accepted model for these networks, but an example of a proposed network is shown in Figure 5-9. Two subnetworks are employed which correspond to the two alternating cell groups. Each subnetwork is inhibitory to the other, but is internally excitatory to itself. The neurons in each group have the standard property of model neurons with the exception of the behavior of the threshold variable. In order to terminate each burst, it is necessary to increase the effective threshold of each cell in proportion to the recent frequency of activity in that cell. This can be modeled by assuming that the threshold decay constant, o, of each cell in the network model is raised additively by a constant amount after each firing, following which o itself decays exponentially to its resting value. This tends to limit or terminate the repetitive firing of each neuron, reducing its mutually excitatory effect on neurons in its own group and halting its inhibition of the antagonist group, which, under some general excitatory input, then begins to fire. This basic pattern of activity is also important in other neuronal circuits.[37-39]

Finally, we cite an example of network simulation which attempts to bridge the tremendous gap between individual neuronal events and

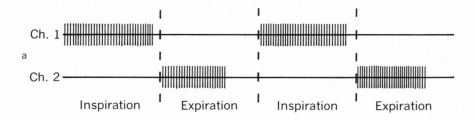

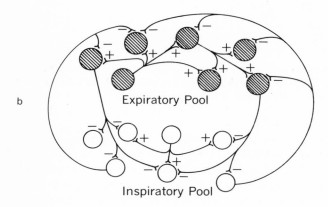

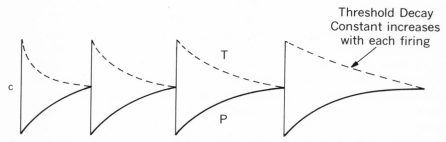

Figure 5-8. a. Typical two-neuron discharge pattern in respiratory centers of brain. b. Network connectivity for respiratory pattern model. c. Cell dynamics for component neurons of network synaptic potentials. Threshold decay constant is increased with each firing and itself decays exponentially. This tends to make neuron less and less excitable as burst continues, tending to terminate the discharge.

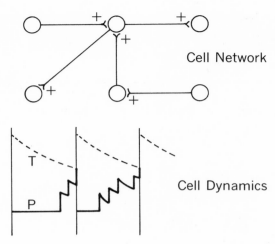

Figure 5-9. Random network connections and cell dynamics for component neurons in the network designed to simulate the gross activity of the electroencephalogram.

phenomena that appear to be resultant effects of thousands or more cells. One of the earliest of all attempts to exploit neuronal models was the work of Farley and Clark[42] which was directed toward the question of the origin of "brain waves." These electrical waves, termed technically the "electroencephalogram," can be recorded easily from the scalp of man or from the exposed cortex of the brain of many animals. Under some circumstances, these waves are strikingly periodic, and the most prominent of these rhythmic components, at around 10 cycles per second in man, is the well known "alpha rhythm" (Fig. 5-10).

The source of these waves was, and remains, unknown, but Farley and Clark attempted to determine whether rhythmic oscillations of this type could be a natural phenomenon in a large network of neurons in which no individual component had intrinsically rhythmic properties. In other words, they tested the concept that rhythmic behavior such as the alpha rhythm was an "emergent" network property. To answer this question, they used an extremely large network of model neurons whose properties were essentially equivalent to those outlined earlier. Their network consisted of a two-dimensional (36 × 36) grid of 1296 neurons whose connections were assigned randomly according to various probability distributions. Only excitatory connections were used, with each cell typically having about 10 inputs. The only difference between their neuron model and those described previously is

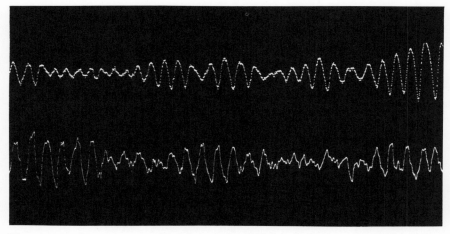

Figure 5-10. Lower trace: Actual scalp recording of electroencephalogram of
          a man. Note similarity of the 10/second oscillations in each
          record.

that their membrane potential was immediately reset to the asymptotic
value after each impulse.

The network was simulated on a large digital computer and quantitative measures of the network were employed. The simplest of these
was to count the total number of cells in the network which had fired
during some time Δt, and plot this "activity level" as a function of time.
Many network configurations were found in which more-or-less rhythmic waxing and waning of total activity occurred, suggesting that the
alpha rhythm might be an envelope of the total amount of neuronal
activity in the cortex. This study, of course, does not prove that this
activity is the source of the alpha rhythm, but does establish the fact
that complex wavelike patterns of activity could reverberate around
networks of rather simple elements such as we have described. Clearly,
the existence of such modes of behavior could not be predicted in
advance, nor could it even be observed in the real brain if it in fact
occurred.

Other Systems Approaches to
Neurophysiological Problems

In this brief review we have stressed the applications of mathematical techniques, probability theory, and computer simulation at a

single level of description, namely, the functional level of the neuron and groups of neurons. Many important phenomena, certainly those that form the underlying molecular basis for the phenomena we have discussed, lie at lower levels of organization and can be approached only at that level.

Many of these phenomena at lower levels have been subjected to study by the same tools described here. Among these we might mention various models of membrane function, and models of synaptic processes (see references). Conversely, some of the most effective uses of engineering theory, particularly of control and communication theory, have been applications to the overall input-output properties of large neurological servomechanisms. The classic application by Stark and his co-workers[55] of control systems theory to the study of the neurological networks regulating the width of the pupil of the eye was unquestionably the first real advance in the theoretical analysis of real cybernetic systems, and we might take this work as the essential starting point of modern engineering approaches to biological systems. A number of extremely interesting studies of neurophysiological systems of this type have been dealt with through analytical models (see references). These models, however, do not depend directly on the notion of a neuron, and hence cannot be immediately related to the phenomena we have considered here. For this reason, we have omitted them from the present discussion.

References

1. Brazier, M. A. B.: *The Electrical Activity of the Nervous System,* ed. 3. Williams and Wilkins, Baltimore, 1968.

2. Bullock, T. H., and Horridge, G. A.: *Structure and Function in the Nervous System of Invertebrates.* Freeman, San Francisco, 1965.

3. Burns, B. D.: *The Uncertain Nervous System.* Edward Arnold Ltd., London, 1968.

4. Hodgkin, A. L.: *The Conduction of the Nervous Impulse.* Charles C Thomas, Springfield, Ill., 1964.

5. Katz, B.: *Nerve, Muscle, and Synapse.* McGraw-Hill, New York, 1966.

6. Quarton, G. C., Melnechuk, T., Schmitt, F. O.: *The Neurosciences: A Study Program.* Rockefeller University Press, New York, 1967.

7. Rosenblith, W. A. (ed.): *Sensory Communication.* John Wiley, New York, 1961.

8. Stevens, C. F.: *Neurophysiology: A Primer.* John Wiley, New York, 1966.

9. Proc. IEEE, Special Issue on Studies of Neural Elements and Systems. 56, No. 6, 1968.

10. Deutsch, S.: *Models of the Nervous System.* John Wiley, New York, 1967.

11. Harmon, L. D., and Lewis, E. R.: *Neural modeling.* Physiol. Rev. 46:513, 1966.

12. Moore, G. P., Perkel, D. H., and Segundo, J. P.: *Statistical analysis and functional interpretation of neuronal spike data.* Ann. Rev. Physiol. 28:493, 1966.

13. Reiss, R. F. (ed.): *Neural Theory and Modeling.* Proc. 1962 Ojai Symposium. Stanford University Press, Palo Alto, Calif., 1964.

14. Geisler, C. D., and Goldberg, J. M.: *A stochastic model of the repetitive activity of neurons.* Biophys. J. 6:53, 1966.

15. Junge, D., and Moore, G. P.: *Interspike-interval fluctuations in aplysia pacemaker neurons.* Biophys. J. 6:411, 1966.

16. Perkel, D. H., Moore, G. P., and Segundo, J. P.: *Continuous-time simulation of ganglion nerve cells in aplysia.* Biomed. Sci. Instrum. 1:347, 1963.

17. Hiltz, F. F.: *Simulated membrane junctions and additional feedback characteristics for an artificial neuron.* IEEE Trans. Biomed. Engrg. BME-12, 1965, pp. 94–104.

18. Lewis, E. R.: *Using electronic circuits to model simple neuroelectric interactions.* Proc. IEEE 56:931, 1968.

19. Taylor, W. K.: Computers and the nervous system, in: *Models and Analogues in Biology.* Cambridge University Press, London, 1960, pp. 152–168.

20. Calvin, W. H.: *Evaluating membrane potential and spike patterns by experimenter-controlled computer displays.* Exp. Neurol. 21:512, 1968.

21. Cox, D. R.: *Renewal Theory,* Methuen, London, 1962.

22. Gerstein, G. L., and Mandelbrot, B.: *Random walk models for the spike activity of a single neuron.* Biophys. J. 4:41, 1964.

23. Gerstein, G. L., and Kiang, N. Y-S.: *An approach to the quantitative analysis of electrophysiological data from single neurons.* Biophys. J. 1:15, 1960.

24. Perkel, D. H., Gerstein, G. L., and Moore, G. P.: *Neuronal spike trains and stochastic point processes I. The single spike train.* Biophys. J. 7:391, 1967.

25. Perkel, D. H., Gerstein, G. L., and Moore, G. P.: *Neuronal spike trains and stochastic point processes II. Simultaneous spike trains.* Biophys. J. 1:419, 1967.

26. Rodieck, R. W., Kiang, N. Y-S., and Gerstein, G. L.: *Some quantitative*

methods for the study of spontaneous activity of single neurons. Biophys. J. 2:351, 1962.

27. Rodieck, R. W.: *Maintained activity of cat retinal ganglion cells.* J. Neurophysiol. 30:1043, 1967.

28. Bishop, P. O., Levick, W. R., and Williams, W. O.: *Statistical analysis of the dark discharge of lateral geniculate neurons.* J. Physiol. 170:598, 1964.

29. ten Hoopen, M.: *Multi-modal interval distributions.* Kybernetik 3:17, 1966.

30. ten Hoopen, M.: *Impulse sequences of thalamic neurons—an attempted theoretical interpretation.* Brain Res. 3:123, 1966.

31. Ogawa, T., Bishop, P. O., and Levick, W. R.: *Temporal characteristics of responses to photic stimulation by single ganglion cells in the unopened eye of the cat.* J. Neurophysiol. 29:1, 1966.

32. von Bekesy, G.: *Sensory Inhibition.* Princeton University Press, Princeton, N.J., 1967.

33. Hartline, H. K., Ratliff, F., and Miller, W. H.: Inhibitory interaction in the retina and its significance in vision, in *Nervous Inhibition.* Pergamon Press, New York, 1961, pp. 241–284.

34. Lange, D., Hartline, H. K., and Ratliff, F.: The dynamics of lateral inhibition in the compound eye of *Limulus.* II, in Bernhard, C. G. (ed.): *The Functional Organization of the Compound Eye.* Pergamon Press, New York, 1967, pp. 425–449.

35. Ratliff, F.: *Mach Bands: Quantitative Studies on Neuronal Networks in the Retina.* Holden Day, San Francisco, 1965.

36. Reichardt, W.: Theoretical aspects of neural inhibition in the lateral eye of *Limulus,* in Gerard R. W., and Duyff, J. W. (eds.): *Information Processing in the Nervous System.* Proc. Int. Union Physiol. Sci. XXII Int. Cong., Leiden, 1962, Vol. III, pp. 65–84.

37. Harmon, L. D.: *Neuromimes: action of a reciprocally inhibitory pair.* Science 146:1323, 1964.

38. Reiss, R. F.: *A theory and simulation of rhythmic behavior due to reciprocal inhibition in small nerve nets,* in *Proc. 1962 AFIPS Spring Joint Computer Conf.* National Press, Palo Alto, 1962, Vol. 21, pp. 171–194.

39. Wilson, D. M., and Waldron, I.: *Models for the generation of the motor output pattern in flying locusts.* Proc. IEEE 56:1058, 1968.

40. Anderson, P.: Rhythmic 10/sec. activity in the thalamus, in Purpura, D. P., and Yahr, M. D. (eds.): *The Thalamus.* Columbia University Press, New York, 1966, pp. 143–151.

41. Beurle, R. L.: *Properties of a mass of cells capable of regenerating pulses.* Trans. Roy. Soc. Ser. B 240:55, 1956.

42. Farley, B. G., and Clark, W. A.: Activity in networks of neuron-like elements, in Cherry, C. (ed.): *Information Theory.* Fourth London Symposium, Butterworths, London, 1961.

43. Farley, B. G.: A neural network model and the slow potentials of encephalography, in Stacy, R. W., and Waxman, B. (eds.): *Computers in Biomedical Research.* Academic Press, New York, 1965, Vol. I, Chap. 11, pp. 265–294.

44. Calvin, W. H., and Stevens, C. F.: *Synaptic noise and other sources of randomness in motoneuron interspike intervals.* J. Neurophysiol. 31:574, 1968.

45. Cole, K. S.: Theory, experiment and the nerve impulse, in Waterman, T. H., and Morowitz, H. J. (eds.): *Theoretical and Mathematical Biology.* Blaisdell, New York, 1965, pp. 136–171.

46. Fitz Hugh, R.: *Mathematical models of threshold phenomena in the nerve membrane.* Bull. Math. Biophys. 17:257, 1955.

47. Hodgkin, A. L., and Huxley, A. F.: *A quantitative description of membrane current and its application to conduction and excitation in nerve.* J. Physiol. 117:500, 1952.

48. Kruckenberg, P., and Sandweg, R.: *An analog model for acetylcholine release by motor nerve endings.* J. Theor. Biol. 19:327, 1968.

49. Martin, A. R.: *Quantal nature of synaptic transmission.* Physiol. Rev. 46:51, 1966.

50. Moore, J. W.: *Specifications for nerve membrane models.* Proc. IEEE 56:895, 1968.

51. Rall, W.: Theoretical significance of dendritic trees for neuronal input-output relations, in Reiss, R. F. (ed.): *Neural Theory and Modeling.* Stanford University Press, Palo Alto, 1964, pp. 73–97.

52. Rall, W.: *Electrophysiology of a dendritic neuron model.* Biophys. J. 2:145, 1962.

53. Stevens, C. F.: *Synaptic physiology.* Proc. IEEE 56:916, 1968.

54. Verveen, A. A., and Derksen, H. E.: *Fluctuation phenomena in nerve membrane.* Proc. IEEE 56:906, 1968.

55. Stark, L.: Neurological feedback control systems, in Alt, F. (ed.): *Advances in Bioengineering and Instrumentation.* Plenum Press, New York, 1966, Chap. IV.

56. Stark, L., Iida, M., and Willis, P. A.: *Dynamic characteristics of the motor coordination system in man.* Biophys. J. 1:279, 1961.

57. Thorson, J.: *Small-signal analysis of a visual reflex in the locust. I and II.* Kybernetik 3:14, 1966.

58. Young, L. R., and Stark, L.: *Variable feedback experiments testing a sampled data model for eye tracking movements.* IEEE Trans. of the Professional Technical Group on Human Factors in Electronics. HFE-4(1), 1963, pp. 38–51.

6

Peripheral Sensory Mechanisms

Every living being must interact with its environment. Such interactions involve a two-way flow of information, the sensing of relevant environmental features by the living organism, and its appropriate reaction to whatever the perceived features signify to it. The exploration of the environment is the function of the so-called *exteroceptors,* which are an important class of sensory receptors capable of responding to various forms of energy directed at the organism from the external world. Other classes of sensory receptors are the *interoceptors,* which are responding to various stimuli from within the animal, and *proprioceptors,* which sense the position of various parts of the body. Another receptor classification, probably even more useful for our purposes, is according to the form of energy which can best activate a particular sensor. Accordingly, there are a variety of *mechanoreceptors* whose initial stimulus is some sort of mechanical force. Such receptors subserve the senses of touch, pressure, hearing, and equilibrium, to mention only some of them. Other major categories include the *photoreceptors,* whose appropriate stimulus is light, *thermoreceptors,* responding to warmth and cold, *chemoreceptors,* which serve taste and smell, responding to chemical substances in solution, and the *electroreceptors* of certain fish, which are capable of detecting electric-field patterns in the surrounding water.

Sensory receptors have always been considered particularly fascinating subjects for investigation by bioengineers and biophysicists, probably because under optimal conditions they consistently outperform man-made systems in their sensitivity and information-handling capacity. For example, at the lowest limit of our hearing ability, only about 10^{-16} watt per cm^2 acoustic power is required for the detection of the presence of a sound. The pit viper needs only about 10^{-11} small calories of radiant heat for 0.1 second to be able to detect a change in his environment, presumably the presence of prey. A rod in the human eye can be excited by the absorption of a single light quantum. The sensitivity of the electric fish is such that it responds to a current change of approximately 10^{-5} μA per cm^2. The echo-locating ability of certain bats is such that they can avoid thin wires placed in their flight path. All these examples, and many more that could be cited, point to the high degree of "engineering" achievement by nature's evolutionary processes, which culminate in the sense organs as we know them. The more mobile an organism becomes, the more complex the environment in which it must survive, and the more highly specialized its sensory organs become. Such specificity permits the evolution of a given system to its ultimate refinement. Thus, we trade universality for sensitivity, and breadth for depth. Even within a given modality, the range of stimulus parameters which are appropriate for excitation of a sense organ is quite limited. Our visual system, for example, is only capable of discerning less than one octave (400 to 750 mμ) of the electromagnetic wave spectrum as light. Fortunately, even though vastly different in their sensitivities toward various forms of external energy, sensory receptors tend to function similarly on both psychological and physiological bases. To explain, the function of all sensory organs is to provide the brain with suitably coded information concerning the current state of the environment. One major constraint on the operation of all sense organs is the necessity to evolve a code which is acceptable to the receiving station (the brain), and which can be efficiently transmitted through the available links of communication between the sense organs and the brain, that is, through the sensory nerves. Since the mode of transmission in a single nerve fiber is essentially digital (in other words, a nerve impulse or spike is either present or absent), *all* information provided by a sense organ about the environment must be encoded in two dimensions, i.e., which fibers within the corresponding sensory nerve are active, and what is the instantaneous firing rate of any given fiber. This appears to be a very severe restriction at first glance, but one should consider that, along with those sensory modalities in which great information-handling capacity is required, go hand-in-hand a very large pool of nerve fibers capable of carrying digital messages. The optic nerve, for example, consists of more than one million fibers. The necessity of signalling their information through essentially similar channels and to essentially similar receiving stations forces the various sensory receptors into a fairly tight mold, and creates a frame of reference against which the various sensations are evaluated. For example, no matter what sensory modality is active, the nervous system is seeking information from its input channels which would enable it to determine the following characteristics of the stimuli:

(a) the quality of the stimulus, i.e., sweet or sour
(b) the intensity of the stimulus, i.e., bright or dim
(c) the spatial location and extent of the stimulus, i.e., where from and how big
(d) the displacement pattern of the stimulus in space and time, i.e., the identification of the auditory clues generated by a passing racing car

The characteristic coding patterns relating to these stimulus attributes tend to show similarities from one sensory modality to the other; thus, it is not surprising that the mechanisms, and indeed the basic structure, of the sensory receptors reveal a great deal of commonality. The deep similarities are usually overlooked, or at least not emphasized, because of the superficial differences among sense organs. However, these similarities are of such extent and importance that a fairly unified treatment of the various sensory receptors is possible. Such a treatment is presented in this chapter; in order to avoid abstract discussions, the auditory system is chosen to serve as a specific example of the principles and mechanisms that are treated. Often, a concept or behavior is introduced as it is manifested in the operation of the hearing organ, and its generality is mentioned subsequently. First, some psychological attributes of sensory stimuli, and of sound in particular, are treated, then the general plan of a sensory organ is discussed from the viewpoint of electrophysiology. Here, a relatively detailed analysis of electrical phenomena recordable from the inner ear and from the auditory nerve is included in the discussion. The truly distinguishing feature of any particular sense organ is its *accessory structure,* which makes possible the acceptance of a particular form of energy and its transduction by the organ. As an example of the possible complexity of such a structure, the middle ear-inner ear mechanohydraulic transducer is given a quantitative treatment. Finally, the central control of the incoming sensory information is examined by discussing feedback mechanisms of the auditory system.

Some Psychological Attributes of Sensory Stimulation

Sensitivity and Range

One of the most important problems concerning a given sensory modality is the determination of the minimum amount of energy which can be detected by an observer. Two indices are useful in this context, *absolute sensitivity* and *differential sensitivity,* or their reciprocals, *absolute threshold* and *differential threshold.* The absolute threshold of a sense organ is defined in terms of the magnitude of the stimulus which can be detected 50 percent of the time, while differential threshold is the magnitude of the stimulus *change* which can be detected with that frequency. It is essential to understand that the threshold concept is a statistical, as opposed to a deterministic, description. A stimulus of threshold strength goes unnoticed by a subject half the time it is present. A slightly weaker stimulus is detected less than 50 percent of the time, and one that is slightly stronger is noticed in more than 50 percent of the trials. A function which relates the probability of detection to the strength of the stimulus is called the *psychometric function.* The energy content of a stimulus is proportional to both the duration and the intensity of the stimulus. It is not surprising that detection threshold can be reached with either very intense brief stimuli or with weak long ones. A roughly reciprocal relationship exists between stimulus duration and intensity at threshold in most sense modalities, suggesting the presence of an integrating mechanism. In vision, this is expressed by the Bunsen-Roscoe law: $I \times T = C$, where I is light intensity, T is the duration of the light, and C is a constant luminous energy. In audition, if tones of sufficiently long duration are presented, then the threshold can be described by the sound pressure of the stimulus. Below approximately 500 msec, the threshold also depends on the duration on the stimulus tone. The time dependence can be extremely well fitted by the simple relation

$$(I - I_0)T = E = \text{constant}$$

where I and T are the intensity and duration of the tone, respectively, E is the threshold energy, and I_0 is a hypothetical minimum intensity which can be an adequate stimulus for the auditory system. Of course, both I_0 and E are frequency dependent. As T increases, I asymptotically approaches I_0, and is indistinguishable from it for all practical purposes when T exceeds approximately 500 msec.

For long-duration tones, the threshold is usually expressed in terms of sound pressure. Depending on whether threshold determination is made with the aid of an earphone or in a free sound field, somewhat different sound pressure values are required to elicit auditory sensation. The field measurements yield better thresholds (i.e., lower sound pressure elicits a response) than do the earphone measurements. The difference is primarily due to impedance mismatch under the closed-ear condition at low frequencies; at high frequencies, diffraction of sound waves due to the head also contributes to the discrepancy.

A typical threshold curve obtained by listening under an earphone is shown in Figure 6-1. This curve rises steeply at both high and low frequencies, the most sensitive region being in the vicinity of 3000 Hz. This bandpass filter characteristic is fairly typical for the various living creatures whose hearing was studied; the differences among species lie in the location of the most sensitive frequency and in the width of the useful frequency range. The shaded region in Figure 6-1 represents the auditory area within which human beings can experience

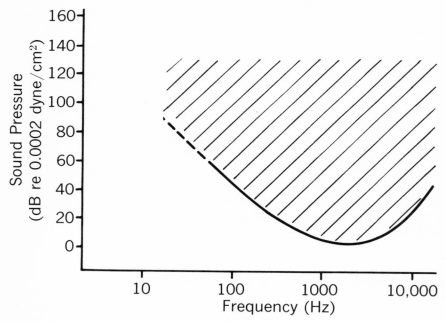

Figure 6-1. Threshold sound pressure for human beings at various frequencies. Ruled area indicates the intensity frequency domain where useful hearing is possible.

useful auditory experience. Above the sound pressure of approximately 130 dB (re 0.0002 dyne per cm²), relatively independent of frequency, hearing is accompanied by pain, and exposure to such intense sounds can result in irreversible damage to the auditory system. It should be mentioned that prolonged exposure to sounds in the 100 to 130 dB intensity range can also cause impairment.

It is most illuminating to consider that, in the region of maximum sensitivity, the excursion of the eardrum at threshold sound intensities is of the order of 10^{-10} cm. This minute displacement is only about ten times greater than the amplitude of movement of air molecules due to Brownian motion. It is clear then that the auditory system is sensitive almost to its practical limit; further improvement in sensitivity could actually be detrimental, since one might be forced to perceive faint sounds in a background of perpetual static. It was already noted that a single visual receptor cell can be activated by a single photon. Perception of light sensation can be mediated by the simultaneous absorption of single photons by five, and possibly only two, photo receptors in the eye. The lower limits of operation of our sensory systems thus involve incredibly small amounts of energy. Significantly, these extremely sensitive receptors, particularly the ear and the eye, also possess a very wide dynamic range of intensities within which they can function, and function well.

The great range of audible intensities is covered by the hearing mechanism with exceptional acuity. While at threshold the relative difference limen (differential sensitivity) reaches 6 decibels, it decreases hyperbolically to approach an asymptote corresponding to a fraction of a decibel. In general, approximately 30 dB above threshold, the relative difference limen falls below 1 decibel. The resolving power of the ear in the frequency dimension is similarly quite excellent. The relative difference limen is of the order of 4 percent at very low frequencies, and it decreases with the increase of frequency to the approximately constant value of 0.5 percent above 1000 Hz. The difference limen at any frequency is greater for fainter sounds.

Strength and Quality of the Sensation

It is important to make a clear distinction between the strength of a stimulus (a physical quantity) and the strength of the sensation evoked by this stimulus (a subjective magnitude). Thus, we relate sound intensity and loudness, light intensity and brightness, sucrose content and sweetness. These relationships can best be understood by treating our prototype, the auditory system. Sound intensity is measured in decibels relative to 10^{-16} watt per cm^2. Loudness is measured in *sones,* where 1 sone is the loudness of a 1000-Hz tone presented 40 dB above its threshold intensity. The sone scale is then established by asking listeners to estimate the loudness of tones at different frequencies and intensities in comparison to the loudness of the reference tone (1000 Hz at 40 dB). The loudness of a tone which thus subjectively appears to be n times that of the reference tone is designated as n sones. The relationship between intensity and loudness is not a straightforward one-to-one correspondence. As with the magnitude of other types of sensation, for hearing the magnitudes of sensation and of the stimulus are related by Stevens' power law. Accordingly, $S = k \times I^p$ where S is the sensation, I is the intensity, the constant k depends on the choice of units, and the exponent p is approximately 0.55 for loudness functions.

Just as loudness is the psychological correlate of intensity, pitch is the psychological correlate of frequency. Thus, pitch is a subjective measure, whereas frequency is a physical measure. Pitch, though directly related to frequency, is not uniquely determined by it. In fact, the intensity of a tone of constant frequency can significantly affect the pitch experience. In general, for low frequency tones the apparent pitch decreases with the increase of intensity, while the pitch of a high frequency tone increases with intensity. Between 1000 and 3000 Hz the pitch of a tone is relatively independent of its intensity. A scale for relating pitch to frequency can be constructed by choosing a reference tone (1000 Hz at 40 dB above its threshold), and then comparing the pitch of all other tones to that of the reference. The unit of pitch is the *mel,* which is defined such that the pitch of the 1000-Hz reference tone is designated as 1000 mels. Then, for example, the tone which has a subjective pitch half as large as the pitch of the reference tone would correspond to 500 mels (note that the frequency of the 500-mel tone would not, in general, be 500 Hz). The shape of the pitch-vs-frequency function is such that the pitch is usually expressed in a smaller number of mels than is the corresponding frequency in Hz.

The relationship between pitch and frequency is of course along a different continuum from the one between intensity and loudness. The former is similar to the correspondence between the wavelength of light and the perceived hue in vision, or between the relative NaCl-vs-HCl content and the observed saltiness in taste vs sourness in taste.

Principles of Sensory Receptor Function

Some General Ideas

Sensory receptor systems are transducers which convert various forms of physical energy into electrical processes that are compatible with the functioning of other portions of the nervous system. It has already been mentioned that information transfer between distant points in the nervous system is mediated by trains of electrical pulses traveling in nerve fibers. Thus, a sensory receptor must absorb the stimulus energy and emit a train of nerve impulses in response. A sense organ usually consists of hundreds or thousands of receptors, each receptor transducing whatever portion of the total incoming stimulus energy it absorbs. The distribution of the stimulus to the most appropriate group on receptors is the function of an *accessory structure,* which is frequently found in intimate contact with the receptors, or of which the receptors form a part. Such accessory structures are the optical apparatus of the eye and the acoustical apparatus of the inner ear. These structures function not merely as distributors of energy, but also as appropriate and indispensable environments for the transduction process. The accessory structures are probably the most interesting components of the sensory receptors from the engineer's viewpoint. Considerable attention will be given in a later section of this chapter to the cochlea, which forms the extremely complex accessory structure of the ear.

In simple sensory systems, a sensory neuron functions both as a receptor of physical stimulus and as the transmitter of pulsatile responses. Extensive research indicates that the general scheme of operation of such a primary sensory neuron involves an intermediate step between physical stimulation and digital message formation, this step being a graded electrical activity known as the *generator potential.* Generator potentials have been identified in the crayfish stretch receptor, in the Pacinian corpuscle, in the vertebrate olfactory epithelium, and in the muscle spindle. In all these simple sensory receptors, the ending of the first-order nerve fiber is differentially sensitive to a particular form of incident energy. Upon absorption of such energy, a local electrical process, the generator potential, is set up. This process is graded, that is, its magnitude is proportional to the strength of the stimulus, and its duration may or may not be as long as the stimulus, depending on the rate of adaptation. The generator potential is non-propagating in the sense that it is transmitted only by electrotonus, i.e., by passive electrical conduction. The generator potential is depolarizing, and it acts on an electrically excitable portion of the sensory neuron, producing all-or-none responses which are propagated centripetally on the axon of this first-order neuron. It should be emphasized that the simple receptor neuron produces two types of potential, a graded nonpropagated depolarization, and a sequence of propagated spikes. These potentials originate in different geographical locations within the neuron, the graded response in an electrically inexcitable portion of the neuronal membrane, and the spikes in an electrically excitable segment. Structural differences between the two types of membrane have not yet been identified. Clear separation of the two processes within the same sensory neuron can be proven by their differential sensitivity to certain drugs. The mode of operation of the primary sensory neuron can be summarized with the aid of Figure 6-2 adapted from Grundfest.[1]

It should be noted that this scheme fits Bodian's description[2] of the generalized neuron extremely well. Accordingly, different portions of a neuron are determined by function instead of by external morphology. The typical neuron can be described as having three main functional parts. These are the receptive pole or dendritic zone, the transmission apparatus—namely, the axon, and the distribution apparatus—the telodendria. The location of the cell body, which is concerned only

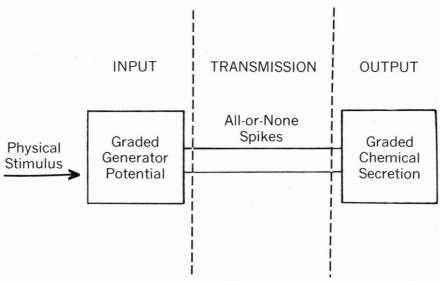

Figure 6-2. Schematic representation of a simple primary receptor cell.[1]

with metabolic functions, is immaterial. In sensory neurons the receptive apparatus is highly specialized to respond to specific physical stimuli. In response to these stimuli, a local graded generator potential is set up which depolarizes the initial segment of the axon where transmitted impulses are generated. It is of importance to note that the receptor pole does not sustain all-or-none discharges; these originate at some distance from the receptor membrane. The distal end on the neuron typically responds to the arrival of spike discharges by liberating chemical transmitters, which in turn activate secondary neurons. The transmission of information from one neuron to another is for the most part mediated by chemical transmitter agents. This mediation takes place at a specialized junction of close contact between neurons (or between neurons and muscle or gland cells), which region is termed the *synapse*. The chemical transmitter is secreted in the presynaptic region, diffuses through the narrow synaptic cleft (a clear demarcation of 150 to 250 Å between the pre- and postsynaptic membranes which is contiguous with the intercellular substance), and creates a graded depolarization or hyperpolarization of the postsynaptic membrane. Depending on whether the result of synaptic transmission is depolarizing or hyperpolarizing, one distinguishes excitatory and inhibitory junctions. Chemically operated synapses predominate in vertebrates and these are the only types that have been identified in the mammalian nervous system. However, it is now clear that chemical transmission is not the only possible means of synaptic activation. In electrical synapses, transmission is achieved by the flow of current through the synaptic cleft. Such a flow of current is unidirectional; thus, both chemical and electrical synapses transmit information only in one direction. A very important physiological distinction between the two types of synapse is the presence or absence of a time delay between the pre- and postsynaptic electrical events. In a chemically operated synapse, due to the diffusion time of the transmitter substance through the synaptic cleft, a well-defined time delay of 0.5 to 1.0 msec takes place between the arrival of the presynaptic impulse and the appearance of the postsynaptic potential. In an electrical synapse the transmission time is negligible.

Up to this point, only simple primary sensory cells and their synaptic connections have been discussed. The distinguishing feature of these cells is that they respond to external stimuli and sustain two kinds of electrical response, the graded generator potential and propagated spikes. These primary sensory neurons synapse at their distal end with secondary neurons. A more complex type of (so-called secondary) sensory receptor system is seen in the vertebrate special sense organs (the olfactory epithelium being the only exception). Here a specialized receptor cell receives the external stimulus, transforms it into a form that is compatible with the receptive properties of adjacent cells, and passes this information on to these cells. The distinctive feature of the receptor cell is that it does not possess an axonal segment, and thus it does not conduct all-or-none responses. There is no need for pulsatile transmission because the sensory cell communicates with neurons immediately adjacent to it. Since the receptive pole of these neighboring neurons is probably chemically excitable, the output of the sensory cell is, in all likelihood, the secretion of a chemical transmitter substance. Between the sensory cell and the following neuron, the information transfer would then take place via the already discussed synaptic mechanism. The secretory nature of the sense cell is of primary importance; these cells may or may not generate an electrical potential. If a potential does appear, it can be depolarizing, hyperpolarizing, or both, depending on the stimulus parameters. In any case, such a potential may be only a secondary process, essentially an electrical sign of secretory activity in the cell. If electrical activity does appear in the sensory cell but does not sustain pulsatile transmission, this potential should not be confused with the generator potential of the simple sensory cell. Davis[3] proposed that the term "generator potential" be reserved for the graded electrical activity that is directly responsible for the initiation of all-or-none neural responses. In contrast, the term "receptor potential" was recommended for describing the graded electrical activity which occurs in the sensory receptor cell. Thus, whatever graded potentials are recordable from the sensory cells of vertebrate special sense organs are properly described as receptor potentials. Such potentials have been obtained intracellularly from the taste buds, from retinal cones, and from the hair cells of the lateral line organ. The slow potentials recordable from the olfactory epithelium are true generator potentials, since the cells here are primary receptors. The microphonic potential of the cochlea is thought to be an aggregate external manifestation of receptor potentials originating in the hair cells.

A general scheme of secondary receptor systems (i.e., those in which the stimulus reception and pulsatile transmission take place in separate cells) is shown in Figure 6-3. It might be noted that the sensory cell in the above scheme is really a specialized neuron; thus, it would be more appropriate to label the following neuron as a second-order neuron. However, it is so common in the literature on auditory physiology to consider as first-order neurons those whose cell bodies lie within the spiral ganglion that in order to avoid conflicting nomenclature this designation is adopted here.

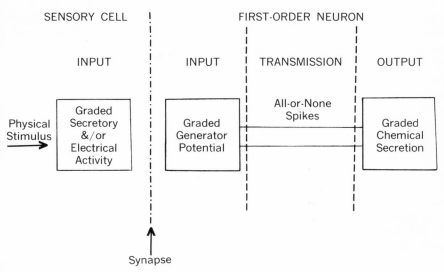

Figure 6-3. Schematic representation of a secondary receptor system.

The Electrophysiology of the Peripheral Auditory System

It is now appropriate to consider the application of the above general scheme to the auditory sense organ. The sensory cells of the receptor system are the external and internal hair cells whose adequate stimulus is probably mechanical deformation. In Figure 6-4 a schematic cross section of the organ of Corti and part of one turn of the cochlea are seen; the positions of the sensory cells can easily be visualized with the aid of this drawing. The hair cells are specialized sensory cells in the class of mechanoreceptors. They are innervated by the terminal branches of the 8th nerve fibers. The upper ends of the hair cells are firmly embedded in a stiff tissue layer, the reticular lamina. From this upper end of each cell, a group of spatially well-organized hairs (stereo cilia) project upward, perpendicular to the recticular lamina. The far ends of the majority of these hairs fit depressions in the bottom layer of the tectorial membrane. Thus, any relative sliding movement between the tectorial membrane and the reticular membrane results in the shearing and bending of the hairs. Such bending is assumed to distort the top portion of the hair cell body, and this distortion or deformation may act as the stimulus in the transduction process. Two

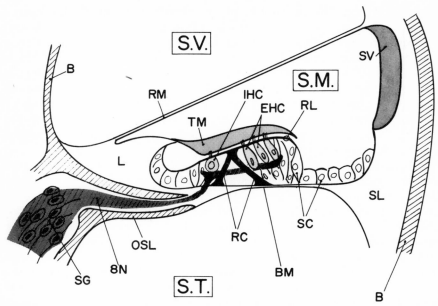

Figure 6-4. Schematic diagram of the organ of Corti. The 8th nerve fibers are indicated by the shaded bundle. Code: S.V.—scala vestibuli, S.M. —scala media, S.T.—scala tympani, RM—Reissner's membrane, TM—tectorial membrane, BM—basilar membrane, IHC—internal hair cell, EHC—external hair cell, RL—reticular lamina, SC— supporting cells, SL—spiral ligament, SV—stria vascularis, L— limbus, B—bone, SG—spiral ganglion, OSL—osseous spiral lamina, AN—auditory nerve, and RC—rods of Corti.

types of electrical phenomena have been identified with the hair cells, the cochlear microphonic (AC) and the summating potential (DC). Neither of these potentials has as yet been reliably recorded intracellularly from mammalian hair cells; what we measure is a gross aggregate response of a large number of cells producing vectorial summation of their outputs for microphonics and algebraic addition for summating potentials. These potentials are generally accepted as receptor potentials, although it now seems likely that at least a portion of the summating potential is merely a DC distortion component. Davis' view[3] is that the microphonic potential acts directly on the synaptic junction at the base of the hair cell, causing the liberation of chemical transmitters which then diffuse through the synaptic cleft. While chemical transmission appears to be more probable, Davis does not rule out the direct electrical stimulation of the dendrites of the following neuron, i.e., the possibility of an electrical synapse. Histological evidence is fairly strong that the junction between the hair cell and the first-order neuron is a chemically operated synapse. Such evidence consists of the demonstration of a well-developed synaptic cleft, a large concentration of vesicles in the base of the hair cell, and the appearance of typical presynaptic structures such as synaptic bars. The chemical transmitter agent has not yet been identified. Although the receptor potential (cochlear microphonic) could act as an important link in the chain of events which lead to the liberation of a chemical transmitter, it does not need to do so; the receptor potential could merely be an electrical sign of the secretory activity of the cell. The receptor potential could conceivably originate in the modulation of DC current flow forced through the hair cells by two biological batteries. One of these is the intracellular negativity (−80 mV) of the hair cells themselves; the other is the positive polarization of the scala media (+80 mV), the so-called endocochlear potential generated by the stria vascularis. The modulation of the current could occur due to the changing hair cell resistance caused by mechanical deformation. The receptor pole of the first-order auditory neuron consists of the long nonmyelinated dendritic branches which make extensive synaptic contacts with the bases of hair cells. Dendritic processes are not usually electrically excitable, which means that they do not support all-or-none propagated activity. Instead, in response to the arrival of chemical transmitter agents, a local depolarization, the generator potential, is set up. This potential is conducted by electrotonus to that region of the fiber which is electrically excitable and where pulsatile responses are initiated by the generator potential. This fiber region is generally assumed to be associated with the appearance of myelination; thus, in the auditory organ, it would correspond to the neighborhood of the habenula perforata; that is, spikes probably appear first where the nerve fibers pass through the basilar membrane and enter into the osseous spiral lamina. The generator potentials of the auditory system have not yet been identified experimentally.

If microphonic magnitude is measured at various points along the cochlear partition using stimuli of varying frequency but constant sound pressure, "electrical tuning curves" are obtained for the cochlea (Fig. 6-5). These curves show that microphonic potential is generated at various cochlear locations, in amounts that are harmonious with the shape of the traveling wave envelope, which is to be discussed in a later section.

The receptor potentials of the cochlea are related in magnitude to the vibratory amplitude of the cochlear partition. Since there is a

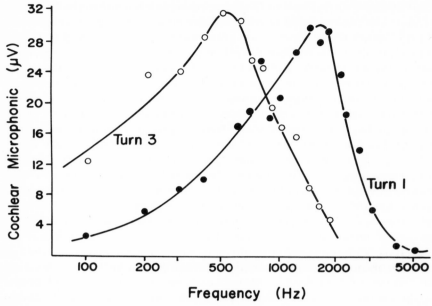

Figure 6-5. Cochlear microphonic magnitude as the function of frequency at a constant intensity, recorded at two sites within the guinea pig cochlea.[18]

mechanical frequency analysis in the cochlea (see following section), this localization of certain frequencies to certain regions is translated into corresponding localization of the receptor potentials. Because these potentials are presumed to mediate neural responses, it is not surprising that fibers originating in a region where CM and SP are maximal would be maximally stimulated. It is clear, therefore, that a crude spatial analysis takes place in the cochlea, and thus in the auditory nerve, in which certain fibers are maximally responsive to certain frequency components in the input signal. This spatial coding carries pitch information which is then determined by which neurons fire at any time.

In addition, the time pattern of the sound stimulus is reflected in the receptor potentials, and consequently in the firing pattern in time of a given neuron. This time coding is, of course, also capable of carrying information to the higher auditory centers regarding various characteristics of the stimulus. Thus, we see that all auditory information is carried in this dual code: which fiber responds and how?

If an electrode is placed in the 8th nerve trunk, then even without direct sound stimulation some spontaneous neural activity is seen. The average spontaneous rate of firing varies from fiber to fiber in the range of from 1 every few seconds up to 200 per second. By using tonal sound stimuli, one can significantly alter the spontaneous activity of the fiber. With appropriate threshold criteria (such as a given increase in firing rate or synchronization of activity), one can see for any fiber how stimulus intensity and frequency relate in providing threshold response. When plotting intensity as a function of frequency at threshold for any given fiber, one obtains characteristic tuning curves. These curves are asymmetrical, being steep toward high frequencies and relatively shallow on the low-frequency side. This asymmetry and the existence of a favored frequency region are again manifestations of the spatial frequency analysis which takes place in the cochlea. These neural curves possess a very definite "best" or characteristic frequency (CF). Above the threshold curve, any combination of frequency and intensity will elicit a response (within the so-called response area), whereas below the curve no combination will.

The characteristic frequency, not surprisingly, is related to the spatial location of the fiber within the nerve trunk. Fibers with high CF are on the periphery, while those with low CF are in the center of the trunk. This corresponds to the fact that peripheral fibers innervate the more basal, higher frequency, regions of the cochlea, and vice versa. It has been observed that fibers with high CF are relatively more numerous than are those with low CF. This fact may be a manifestation of the relative importance of spatial coding at various frequencies. As will be seen shortly, the auditory system possesses the ability to encode pitch information at low frequencies without a need for spatial relationships; however, it appears that at high frequencies the spatial code is the only available means of signalling the pitch of a sound.

The temporal pattern of neural impulses in the 8th nerve is customarily studied by statistical means. When recording from a given fiber the responses to a series of identical clicks, one observes that there is considerable variability in the response. The most commonly used statistical method for smoothing out the variability is the use of post-stimulus time (PST) histograms. These depict graphically the average latency of spikes from the time of stimulus delivery. PST histograms are usually obtained on-line with the aid of special-purpose small digital computers. The method is to count the number of spikes within various time intervals after the delivery of the stimulus, and to average the counts for repeated stimulus presentations.

At low frequencies, individual fibers tend to synchronize their activity with certain multiples of the period of the eliciting stimulus.

The synchronization time is largely determined by the refractory period of the fiber. As Wever[4] has postulated, such synchronization results because there is a group of fibers responding to each cycle of a low-frequency stimulus. Not all fibers in the group fire for all cycles, but there is sufficient staggering of the firing times so that the group gives a well-synchronized burst during every period. Such "volleying," according to Wever, is utilized by the central auditory system to identify pitch, by simple counting, at low frequencies.

We can now give a rather simplified summary of the coding properties of the peripheral auditory system. It is clear that information pertaining to the frequency and intensity pattern of the stimulus is encoded into a complex place-time representation. At high frequencies, pitch information is adequately represented by the geographical location of the responding fiber. At low frequencies (below 200 to 400 Hz), placement cannot account for our pitch acuity. Instead, the nervous system must rely upon the rate of neural firing bursts to convey pitch information. In the frequency region between 300 and 4000 Hz, both mechanisms operate with varying degrees of importance. Intensity of stimulation appears to be coded in the number of active fibers per unit time. In case of complex time-varying auditory stimulus patterns, the interaction of place and time clues, varying from instant to instant, carries the related information.

Mechanical Frequency Analysis in the Ear

We have made repeated references to the accessory structures in the sense organs. These structures are most intimately tuned to optimally receive certain types of physical energy and to deliver these, with whatever modification is necessary, to the sensory cells which transduce the energy. Accessory structures can be very simple, like the lamellae of the Pacinian corpuscle. This onion-like elastic capsule transforms the incoming mechanical force to cause a distortional strain in the enclosed nerve ending, which responds with a generator potential. In contrast is the awesome complexity of the mechanical structure of the inner ear. Both the Pacinian corpuscle and the cochlear hair cells are mechanoreceptor sensory cells; however, while the accessory structure of the former needs simply to deliver the stimulus, that of the latter must perform a complex pretransduction analysis of the incoming signal. Such an analysis in the inner ear involves the frequency-selective distribution of the incoming sound energy over a spatial array of sensory receptor cells. The essential result of this analysis is that certain groups of sense cells are differentially stimulated by sounds of differing input frequency.

Before the functioning of the inner ear is discussed, the operation of the middle ear is briefly treated, since this mechanism is also necessary in order that air-borne sound energy should be effectively delivered to the sensory cells.

Aerial sounds reaching the ear are funneled to the eardrum, which is set into vibratory motion. This motion is transmitted by the three middle-ear bones (ossicles) to the fluid-filled inner ear.

The middle ear is necessary to transmit sounds from air to the inner ear, because sound transmission between media of different acoustic properties is accompanied by considerable loss of energy. Specifically, the transmission loss is determined by the ratio of the acoustic impedances of the two media. Acoustic impedance is defined as the ratio of sound pressure to velocity of propagation; it is determined by the density and elasticity of the material in which the sound propagates. The specific acoustic resistance of air is 41.5 g-sec^{-1}-cm^{-2}, wheras that of sea water is 161,000 g-sec^{-1}-cm^{-2}.

The ratio of these two resistances (R_2/R_1) is 3880; thus, the coefficient of transmission (T) is about 0.001, as obtained from

$$T = (4R_2/R_1)/(1 + R_2/R_1)^2 \tag{1}$$

Since the fluids that fill the cochlea have approximately the same acoustic properties as sea water, it is clear that, without the benefit of an impedance-matching device, transmission from air to the inner ear can be accomplished only with a loss of approximately 30 dB. The middle ear serves as an acoustic transformer in partially reducing the impedance disparity between air and the inner ear fluid. The middle ear provides a pressure amplification of approximately 18.2; consequently, it affords an impedance transformation of 333 times, and produces a transmission coefficient of approximately 0.3.

The physical shape of the basilar membrane is crucial for the development of appropriate vibratory patterns within the cochlea. The membrane, when uncoiled, resembles a narrow wedge about 40 microns wide at the basal end and 500 microns at the apical end. As a consequence of its physical dimensions, there is a compliance gradient along the membrane; it is quite stiff at the base and rather compliant at the apex.

As the stapes is set in vibration by the more peripheral ossicles, it exerts varying pressure on the perilymph in the scala vestibuli. The pressure is communicated to the perilymph in the scala tympani via the cochlear partition. Since the cochlear walls are rigid and the cochlear fluids are incompressible, a pressure increase at the oval window (due to the inward movement of the stapes) is of necessity accompanied by a bending of the cochlear partition toward the scala tympani and also an outward bulge of the round window membrane. An outward motion of the stapes would set up movements in the opposite direction. Clearly, the vibrations of the stapes are communicated to the cochlear partition by way of the cochlear fluid. Very slow pressure changes can equalize through the helicotrema, which acts as a high-pass filter. The stiff basal portion of the cochlear partition vibrates in phase and "drives" the rest of the membrane. It was shown by a series of ingenious experiments of von Békésy[5] that the resulting motion pattern of the partition can be best described as a traveling wave. By observing the movements of the basilar membrane in cadaver ears and in mechanical scale models, von Békésy had noted that, when the stapes is set in vibration, a traveling wave progresses down the partition toward the apex. The amplitude of vibrations changes from point to point, but it reaches a definite maximum. This maximum occurs at different points along the partition, depending on the stimulating frequency. High

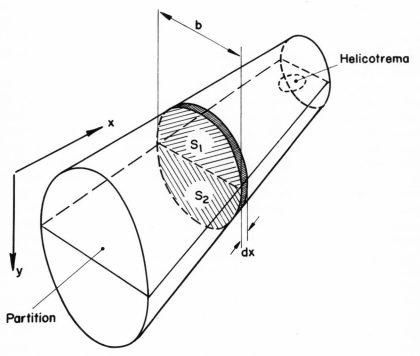

Figure 6-6. Simplified configuration of the cochlea.

frequencies produce maxima close to the stapes, and the lower the frequency becomes, the further the maximum moves toward the apex. Of course, at any point the membrane sustains only transversal motion, whereas a wave front travels from the base toward the apex. The wave length, and thus the velocity of propagation, decrease with distance as a consequence of increasing membrane elasticity. Since the velocity decreases, beyond the maximum vibratory regions the wave rapidly extinguishes, and the membrane is at rest far apically from the maximum.

The cochlear partition is thus seen to operate as a mechanical frequency analyzer in that, for different driving frequencies, different points along the basilar membrane vibrate maximally. The envelope of vibrations is markedly asymmetrical, shallow toward the base and steep toward the apex. High tones are localized near the stapes, while low tones activate a larger portion of the membrane, but mostly the apical region.

Numerous attempts have been made to derive a mathematical model for the motion of the basilar membrane on the basis of the hydrodynamical properties of the cochlea.

The actual configuration is so complex that, even with drastic simplifications of geometry and boundary conditions, solutions for the partial differential equations that describe the system cannot be obtained in closed form. The derivation should proceed with the aid of Figure 6-6. The following definitions are required:

1. The density of the perilymph is δ.
2. The particle velocities of fluid in the x direction are u_1 and u_2, respectively, in the two scalae.
3. The velocity of the partition in the y direction is v.
4. The pressures of the two scalae are P_1 and P_2, respectively.
5. The width of the cochlear partition is b.

The equations of continuity for the two scalae are

$$\frac{S_1 \, dx}{\delta} \frac{\partial \delta}{\partial t} + \frac{\partial(S_1 u_1)}{\partial x} \, dx + v \times b \times dx = 0 \tag{2}$$

$$\frac{S_2 \, dx}{\delta} \frac{\partial \delta}{\partial t} + \frac{\partial(S_2 u_2)}{\partial x} \, dx - v \times b \times dx = 0 \tag{3}$$

Newton's laws of fluid motion give

$$\frac{-\partial(S_1 p_1)}{\partial x} = \delta S_1 \frac{\partial u_1}{\partial t} \tag{4}$$

$$\frac{-\partial(S_2 p_2)}{\partial x} = \delta S_2 \frac{\partial u_2}{\partial t} \tag{5}$$

Furthermore, if the velocity of sound in the liquid is c then

$$\frac{\partial \delta}{\partial t} = \frac{1}{c^2} \frac{\partial p}{\partial t} \tag{6}$$

Combining the above group of equations yields

$$\frac{S_1}{\delta c^2} \frac{\partial p_1}{\partial t} - \frac{1}{\delta} \int_0^t \frac{\partial^2 (S_1 p_1)}{\partial x^2} \, d\tau + v \times b = 0 \tag{7}$$

$$\frac{S_2}{\delta c^2} \frac{\partial p_2}{\partial t} - \frac{1}{\delta} \int_0^t \frac{\partial^2 (S_2 p_2)}{\partial x^2} \, d\tau - v \times b = 0 \tag{8}$$

If the impedance of the membrane per unit area is denoted by $Z(x)$, and if it is assumed that $S_1 = S_2$ and that $P = P_1 - P_2$, then the displacement equation of the partition assumes the form

$$\frac{\partial^2 (S \times Z \times y)}{\partial x^2} - \left[\frac{1}{c^2} + \frac{2 \times b \times \delta}{S \times Z} \right] \frac{\partial^2 (S \times Z \times y)}{\partial t^2} = 0 \tag{9}$$

Solution of the above equation corresponds to a wave traveling along the cochlear partition from base to apex. The amplitude increases up to a point beyond which it rapidly decreases. The point of maximum vibratory amplitude depends on the stimulus frequency. Thus, it is evident that a qualitative correspondence exists between the behavior of this model and the operation of the cochlea.

Thus far we have discussed cochlear mechanics when stimulation of the inner ear is accomplished via the eardrum-ossicular chain route. Another possibility of stimulation is to set the skull in vibration. It has been shown that such bone-conducted stimulus also originates a traveling wave on the cochlear partition which is the exact counterpart of the ordinarily established traveling wave. Bone conduction, because of the variety of paths and mechanisms through which vibratory energy can be communicated to the cochlea, is a much more complex process than air conduction. At frequencies below 800 Hz, the skull, when stimulated with a vibrator, moves in phase as a rigid body. Since the ossicular chain is only loosely coupled to the skull, it lags behind the motion of the skull, and thus there is a relative displacement between the stapes and the oval window. This relative stapes motion can then stimulate the cochlea in the same way that air-conducted sounds do. This type of elicitation of hearing sensation is called inertial bone conduction. At higher frequencies, the skull vibrates in segments and there are compression waves traveling in the bone. These waves compress and expand the cochlear capsule. Because of the volume difference between the scala tympani and the scala vestibuli, a pressure differential develops across the cochlear partition which sets this partition into vibratory motion. Regular traveling waves then develop. This mode of elicitation is called compressional bone conduction.

Feedback Mechanisms

Feedback plays an extremely important role in the maintenance of appropriate physiological conditions in living systems, and also in the goal-seeking activities of an organism. Feedback regulation of sensory function is apparently quite common, and its importance is just becoming evident. One type of feedback mechanism, exemplified by the well-known pupillary or fixation reflexes of the eye and by the less well-known acoustic reflex of the ear, serves gross orientation, tuning, and volume control purposes. Other mechanisms operate on the cellular level, probably to centrally control the overall sensitivity or "set point" of the sensory receptor. We will get acquainted with these systems by studying them in connection with the hearing organ.

Two major feedback circuits can be identified which directly influence the operation of the peripheral auditory system. One of these, the acoustic reflex, affects the transmission characteristics of the middle ear, while the other system, comprising the efferent auditory paths, affects the transmission characteristics of the hair cell-neuron junction within the organ of Corti.

The Acoustic Reflex. When the two middle-ear muscles (stapedius and tensor tympani) contract, they pull on the ossicular chain in essentially opposite directions. Co-contraction of the muscles results in very little movement, but in a significant increase in lateral tension of the bony chain. This tension causes a partial immobilization of the chain, with a consequent alteration in its transmission properties. The two middle-ear muscles can be activated by various stimuli, including loud sounds, cutaneous stimulation, and associative internal stimuli such as chewing and vocalization. Finally, a perpetually changing muscle tonus can be observed, even without any stimulation. When the reflex is elicited by sounds, it is described as the acoustic reflex. In Figure 6-7, a block diagram of the acoustically elicited middle-ear muscle reflex is presented. The ascending (forward) path of the clearly present feedback loop consists of the way stations of the afferent auditory system up to the superior olivary complex. From this brainstem center, connections are made to the motor nuclei of the two muscles. The descending (feedback) tract consists of these motor nuclei, the path of motor

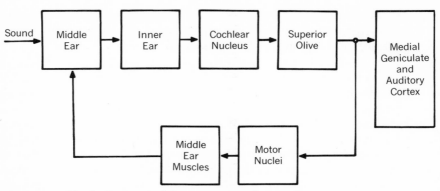

Figure 6-7. Block diagram of the acoustic reflex pathway.

innervation, and the muscles themselves. It is observed that the acoustic reflex is bilateral, that is, the muscles contract in both middle ears. This bilateral nature of the reflex can be used to good advantage in measuring various properties of the mechanism. The primary effect of the acoustic reflex is a reduction of sound transmission efficiency of the ossicular chain, primarily at low frequencies. In general, the reflex is effective in regulating tones whose frequency is less than 1000 Hz. In humans, reflex activity is usually measured by determining changes (due to alterations in properties of the middle ear) in the acoustic input impedance of the ear. Such measurements can be accomplished with appropriate impedance bridges. Clearly, as the contracting muscles influence the transmission properties of the middle ear, a concurrent change must occur in the acoustic driving point impedance of the middle ear. This change is measured, and is used as an indirect index of reflex activity. It has been shown that, for the low frequencies of greatest interest, the acoustic impedance measured at the eardrum can be used to obtain the approximate transmission function of the middle ear. Thus, impedance measurements can give excellent information about changes in middle-ear mechanics resulting from reflex action.

The acoustic reflex is a relatively high-intensity phenomenon. Threshold for reflex activity, as determined by the impedance method, ranges between 60 and 90 dB relative to audiometric threshold for the eliciting sound. Above threshold, reflex activity increases proportionately with stimulus strength within an approximately 30-dB range; for higher intensities the reflex "saturates," in that no further increase in activity is seen. Upon eliciting the reflex with sudden-onset sound stimuli, a well-defined latency can be noted. At the onset of the stimulus, the latency (ranging between 25 and 150 msec) is inversely related to stimulus strength. At the cessation of the stimulus, the relaxation latency appears to be independent of the previous stimulation, and an average of 75 to 100 msec is reported.

In studying the onset and cessation dynamics of the reflex upon suddenly initiated and terminated (step-function) stimuli, one sees a marked asymmetry between the on and off response, in that the former is always significantly more rapid than the latter. In addition, it is seen that, with increasing stimulus intensity, the on response rise time decreases, while the off time response remains essentially unchanged. The change in response speed at the onset clearly indicates that the reflex possesses a magnitude-dependent nonlinearity. Muscle relaxation, however, appears to be an essentially linear process.

It was mentioned that the tonus of the stapedius and tensor tympani muscles is constantly changing. This resting activity is minimum during sleep and maximum during arousal and alertness. It has been proposed that the change in muscle tonus might serve two functions. First, constantly changing muscle tonus introduces sufficient variations in the impedance of the ear to smooth out undesirably sharp resonances. Second, it is speculated that, in analogy with vision, when stabilized retinal images are shown to fade out, the perpetually changing muscle tonus might serve to provide a changing background to audition, and thus facilitate the maintenance of auditory attention.

Other theories have also been formulated to explain the role of the middle-ear muscle reflex. Only two of these will be briefly reviewed here. Since the acoustic reflex is elicited only by relatively intense sounds, and since as the result of muscle activity the transmitted sound to the inner ear is attenuated, it is an attractive hypothesis to assume that the acoustic reflex acts as a protective mechanism, not unlike an automatic volume control. Undoubtedly, the reflex behaves in such a regulatory fashion, but it is questionable whether there are such loud sounds in nature that the evolution of such a complex regulatory system would be required. The lack of a clear-cut answer to this question seems to relegate the protective role of the reflex to a position of secondary importance. As an alternative to the protection theory, Simmons[6] has offered the following ideas. Since middle-ear muscle contractions occur as a prelude to and during chewing, vocalization, and head movements, contractions associated with such motor events appear to serve the purpose of reducing self-generated sounds of the organism so that the signal-to-noise ratio of auditory reception can be improved. It will be remembered that muscle contractions produce significant low-frequency attenuation. It is the low-frequency band in which internal sounds have their major energy components; thus, the reflex is quite effective in attenuating them. Such motor activity-related muscle action can clearly possess considerable survival value, and should therefore be favored by the evolutionary process.

The Efferent Auditory System. The crossed olivocochlear bundle (OCB) of nerve fibers is the primary path through which gating of auditory information is accomplished by higher auditory centers. As such, this bundle comprises the primary feedback path in the auditory system. Direct electrical stimulation of the OCB results in the partial inhibition of sound-generated activity of the auditory nerve. All experiments indicate that the efferent system has only an inhibitory function, in that its activity can only depress the afferent neural responses.

It has been demonstrated that the efferent fibers of the auditory nerve definitely respond to sound presented to the opposite ear. On the basis of this observation, then, it is clear that the efferent system is normally activated by sound, and that such activity generally reduces afferent neural signalling. At the present time we have no definite notion regarding the dimension of the auditory experience which is modified by the efferent action, or how this modification might take place. It is hypothesized that the efferent fibers inhibit the afferent activity postsynaptically (in reference to the hair cells) by acting directly upon the extensive dendritic tree of the afferent system just below the hair cells. The very extensive efferent nerve endings which are completely intertwined with the afferent dendrites would make such interaction quite plausible. Another possibility should not be discounted, however: it was recently demonstrated that some DC intracochlear potentials, apparently generated by the hair cells, can be measured during OCB activity; these potentials presumably reflect postsynaptic activity of the olivocochlear fibers. These potentials hyperpolarize the hair cells, and thus act as presynaptic inhibitory agents for the afferent system.

References

1. Grundfest, H.: *Electrical inexcitability of synapses and some consequences in the central nervous system.* Physiol. Rev. 37:337, 1957.

2. Bodian, D.: *The generalized vertebrate neuron.* Science 137:323, 1962.

3. Davis, H.: *Some principles of sensory receptor action.* Physiol. Rev. 41:391, 1961.

4. Wever, E. G.: *Theory of Hearing.* John Wiley, New York, 1949; *Electrical potentials of the cochlea.* Physiol. Rev. 46:102, 1966.

5. von Békésy, G.: *Experiments in Hearing.* McGraw-Hill, New York, 1960.

6. Simmons, F. B.: *Perceptual theories of middle ear muscle function.* Ann. Otol. 73:724, 1964.

7. Alpern, M., Lawrence, M., and Wolsk, D.: *Sensory Processes.* Brooks-Cole, Belmont, Calif., 1967.

8. Bullock, T. H.: *Initiation of nerve impulses in receptor and central neurons.* Rev. Mod. Physiol. 31:504, 1959.

9. Case, J.: *Sensory Mechanisms.* Macmillan New York, 1966.

10. Corso, J.: *The Experimental Psychology of Sensory Behavior.* Holt, Rinehart and Winston, New York, 1967.

11. Granit, R.: *Receptors and Sensory Perception.* Yale Univ. Press, New Haven, 1955.

12. Grundfest, H.: Excitation by hyperpolarizing potentials. A general theory of receptor activities, in Florey, E. (ed.): *Nervous Inhibition.* Pergamon Press, London, 1961.

13. Mountcastle, V. B.: Physiology of sensory receptors: introduction to sensory processes, in Mountcastle, V. (ed.): *Medical Physiology* C. V. Mosby, St. Louis, 1968, p. 1345.

14. Candiollo, L., Filogamo, G., and Rossi, G.: *The morphology and function of auditory input control.* Beltone Inst. for Hearing Res. Transl. No. 20, 1967.

15. Dallos, P.: *Dynamics of the acoustic reflex: phenomenological aspects.* J. Acoust. Soc. Amer. 36:2175, 1964.

16. Davis, H.: *Biophysics and physiology of the inner ear.* Physiol. Rev. 37:1, 1957.

17. Fex, J.: *Efferent inhibition in the cochlea related to hair cell dc activity: study of postsynaptic activity of the crossed olivo-cochlear fibers in the cat.* J. Acoust. Soc. Amer. 41:666, 1967.

18. Goldstein, M. H., Jr.: The auditory periphery, in Mountcastle, V. (ed.): *Medical Physiology.* C. V. Mosby, St. Louis, 1968, p. 1465.

19. Kiang, N. Y., Watanabe, T., Thomas, E. C., and Clark, L. F.: *Discharge Patterns in Single Fibers in the Cat's Auditory Nerve.* MIT Res. Mon. No. 35, MIT Press, Cambridge, 1965.

20. Laszlo, C.: *A Model Oriented Study of the Cochlear Microphonic Response in the Auditory System.* Unpublished M.S. Thesis, McGill University, Montreal, 1966.

21. Rasmussen, G., and Windle, W. (eds.): *Neural Mechanisms of the Auditory and Vestibular Systems,* Charles C Thomas, Springfield, Ill., 1960.

22. Spoendlin, H.: *The Organization of the Cochlear Receptor.* Krager, Basel, 1966.

23. Stevens, S. S., and Davis, H.: *Hearing: Its Psychology and Physiology.* John Wiley, New York, 1938.

24. Wansdronk, C.: *On the Mechanisms of Hearing.* Philips Res. Rep. Suppl. No. 1, 1962.

25. Wever, E. G., and Lawrence, M.: *Physiological Acoustics.* Princeton Univ. Press, Princeton, N.J., 1954.

26. Whitfield, I. C.: *The Auditory Pathway.* Williams and Wilkins, Baltimore, 1967.

27. Zwislocki, J.: Analysis of some auditory characteristics, in Luce, R., Bush, R., and Galanter, E. (eds.): *Handbook of Mathematical Psychology.* John Wiley, New York, 1965, Vol. III, p. 1.

7
Models of
Endocrine Systems

The application of systems theory to problems of endocrine systems may serve three general purposes. First, a model may serve to collect a variety of information concerning the system under study, and thus to perform a sort of compact library function. Second, the model may serve as a complex hypothesis concerning the behavior of the system, so that deductions made from the model may predict results of experiments as yet not performed on the system, or alternatively, of experiments not used in the construction of the model. The actual performance of such experiments and the comparison of their results with those predicted by the model will then lead either to validation of the model or to its revision to conform with the new experimental results. Third, models may be used in teaching the physiology of the system being modeled. Depending on the depth of the knowledge to be imparted, models serving either collecting or predicting functions may be used for teaching, so that this use places no additional restriction upon the way in which the model is formulated. It seems clear that the second function imposes more rigid constraints upon the model than does the first, and that a model which is useful for predicting must in general be useful for collecting. The present discussion will therefore center upon those special features of endocrine systems which must be considered in the formulation of models that can function predictively.

An endocrine system is any system in which the secretion of a hormone plays a dominant role. In order to be secreted, a hormone must be biosynthesized from some precursor molecule. This function is normally carried out in a single endocrine gland, but may take place in the blood itself, e.g., angiotensin II. The synthesized hormone is then released into the circulation. Here it passes to its target tissue, where it controls one or more specific functions. In the process of such passage, the hormone will be distributed in the circulation and at times in other body compartments. It may be bound to some protein component within the plasma, or enter the red blood cell, and it may be presented to other tissues where its metabolism, degradation, and excretion will begin. Such tissues may or may not be target tissues for the hormone. The biosynthesis of the hormone in the endocrine glands is, in general, part of a control system which, in turn, is part of a large homeostatic system.[1] Often, but not always, mechanisms controlling biosynthesis involve components of the central nervous system, where certain stimuli and information about existing availability of hormone (negative feedback) are integrated by a control system into some demand function which ultimately signals the endocrine gland to stimulate or inhibit its biosynthetic and secretory processes. At times such controller functions may be performed within the endocrine gland itself, e.g., in the parathyroid gland, but the situation in most cases is as described. Consequently, models of endocrine systems will usually consist of some combination of models of the following four more-or-less discrete functions:

1. The control functions, in which integration of the demand for hormone and its supply results in the signal sent to the endocrine gland in question.
2. The process within the endocrine gland itself, in which the hormone is synthesized and secreted.
3. The processes of distribution, metabolism, and excretion.
4. The action of a hormone upon its target tissues.

In the following discussion, these four processes will be considered separately and in reverse order from that above.

The Action of Hormones

Although the precise mode of action of most hormones has not been defined at the present time, it is generally held that, in most cases that have been studied, a hormone acts either by inducing a change in transport across some biological membrane or by changing one or more of the rate constants of some complex biochemical reaction. An appropriate model of the action of the hormone then becomes an appropriate model of the permeability processes of the membrane or of the interactions which form the biochemical process. In the case of permeability of biological membranes, most heuristically valuable models have been constructed from considerations of conservation of mass and of chemical species. So-called phenomenological models have been constructed based on such considerations and on the principles of irreversible thermodynamics.[2] Although such models can describe transport phenomena, they provide little or no understanding of the details of these phenomena, or of the precise chemical kinetics involved, upon which, presumably, the hormone must exert its action. If one extends such a model to include such processes, one encounters highly complicated mathematical relationships with prominent nonlinearities, and these in turn limit or prevent analytical solution. Such complexities are not, of course, a complete barrier to simulation, particularly since the advent of digital simulation. However, at the present time, no satisfactory and complete model of the action of the hormone upon a transport process has been advanced, since as Heckmann has observed, "The details of the kinetics of particle diffusion through biological membranes have not yet been clarified for even a single case."[3] Heckmann has demonstrated that kinetic data alone cannot prove or disprove the basic carrier-mediated diffusion model, which forms one of the central problems in transport today. He has advanced eloquent reasons for proposing that interpretation of the physical meaning of the kinetic constants, by means of careful experimentation, must precede such proof. In his view, such interpretation will permit formulation of a mathematical model of the *process* rather than merely of the events.

By analogy to the action of a hormone on a transport process, hormonal action on a complex biochemical reaction must be viewed in terms of a model of such a reaction. Fortunately, in this case such models are farther along the way to satisfactory development. Hess[4] has summarized some of the basic principles of biochemical regulation involved in the modeling of interacting reactions, and has noted especially the oscillatory behavior which may be generated by biochemical nets. Chance (see reference 4) has elegantly utilized the phase plane approach to depict such interacting oscillations. Models of the pathways by which certain hormones may influence metabolic processes, together with considerable kinetic data, are available for a variety of hormones, even though none of these can be considered complete. For example, the pathways by which glucocorticoid hormones act on metabolic processes, and the kinetics of action, are summarized by

Weber.[5] Most of the ingredients necessary for formulation of the model are given, even though the model itself has not been formulated except in terms of some of its structure. A *caveat* is necessary here, lest the reader consider that all that is necessary to make a dramatic contribution in this field is a degree of mathematical sophistication. Should he examine the problem of glucocorticoid action in detail, he would find a variety of parallel pathways of action, the relative importance of any one of which has not been assessed, together with a plethora of conflicting data. Thus, one cannot overemphasize the need for thorough familiarity with the biological problem and with the various experimental attacks that have been made upon it, if the model to be constructed is to avoid fatal flaws. There are other limitations in our understanding of hormonal action on metabolic processes that have prevented the formulation of detailed models of hormone action. For example, it is well known that the catecholamine epinephrine which acts on muscle to break down glycogen into glucose does so by activation of an enzyme, adenyl cyclase, which in turn increases the concentration of cyclic 3',5'-AMP. This substance is an allosteric cofactor in a kinase reaction, and leads to the activation of phosphorylase, which eventually breaks down glycogen. Most of the steps in this chain of reactions have been clarified, and the kinetic constants have been determined. However, the original enzyme, adenyl cyclase, though present in broken-cell preparations, has proved refractory to further purification.[6] Such a fact means that any model which can be formulated will of necessity have some limitation in detail. However, models have not been used in this area to formulate complex hypotheses. In particular, the present author feels that, in such a difficult experimental situation, the possible role of a model in leading to a critical experiment should not be overlooked. One special case of hormone action on a coupled biochemical reaction should be noted, in which one hormone controls, through its action, the biosynthesis of a second hormone. This case is discussed below.

Distribution and Metabolism

Models of hormone distribution, metabolism, and excretion have generally been formulated through the utilization of the compartmental hypothesis that, within a given theoretical compartment, there is instantaneous and complete mixing of the species under study. This hypothesis, together with the laws of conservation of mass and of chemical species, allows the formulation of mathematical models of these processes, leaving only the parameters of rate to be determined. This area has been by far the most active in the formulation of endocrine models.[7] Such models have been put to a variety of uses. Tait and his coworkers[8,9] have used compartmental models to calculate secretion rates, disappearance rates, and metabolic turnover rates for hormones, and to develop two different experimental methods for the estimation of these factors. As indicated,[7] the methods have been applied subsequently to a variety of other hormones in addition to aldosterone. These references constitute one of the clearest examples of the ultimate utility of mathematical models of endocrine systems, since they permit the confident calculation of essentially nonmeasurable variables in man. Compartmental models have also been used to examine the controlling function of hormone metabolism. For example, the work of Yates[10] has indicated that hepatic metabolism of cortisol plays a dominant role in the control of peripheral blood concentration of that hormone.

Hormone Biosynthesis

The complex chemical reactions leading to biosynthesis of a hormone usually take place within a single cell, but may involve a variety of subcellular elements. The modeling of such synthetic processes may involve features of the modeling of linked biochemical reaction, as well as features of movements of substances between a variety of compartments, and thus of transport across subcellular membranes. Many of the features of such a complex process are considered for the case of the adrenal cortex by McKerns and others.[11] The primary biosynthetic process of the adrenal gland, the synthesis of cortisol, is controlled by the pituitary hormone ACTH. Thus, biosynthesis of cortisol may also be viewed as the action of another hormone on a metabolic process. One hypothesis concerning the mechanism of action of ACTH indicates some of the utility of mathematical modeling in this area of endeavor. Koritz[12] has postulated that ACTH may act to facilitate movement of pregnenolone across the mitochondrial membrane. Koritz and Hall[13] have shown previously that this steroid can inhibit its own synthesis from the general precursor of steroids, cholesterol. Koritz postulated that the movement of pregnenolone resulting from the action of ACTH could release this inhibition, and thus increase the rate of biosynthesis of cortisol. A direct test of this hypothesis would involve the measurement of concentrations of pregnenolone within mitochondria. Methodological limitations preclude such a test at this time. Urquhart and Li[14] have accumulated considerable data regarding the response of the isolated perfused adrenal gland of the dog to a variety of dynamic forms of input of ACTH. Recently, Urquhart, Krall, and Li[15] have constructed an analog computer model of adrenal biosynthesis of cortisol which has incorporated the hypothesis of Koritz and Hall, and have wondered whether such a model could be consistent with their previously obtained dynamic data. Their positive finding does not, of course, prove the Koritz hypothesis, although it does indicate that at the present time it is at least plausible. On the other hand, a negative finding could have served to exclude this hypothesis. Further exploration of this technique will be necessary before its full utility will be defined. The potentially even more powerful combination of prediction from such models and direct testing of the biochemical processes still awaits exploration.

As indicated above, the biosynthesis of a hormone within an endocrine gland is one step in a process of signal flow extending from the control of the process to the action of the hormone and perhaps beyond. Thus, the biosynthesis of a hormone forms one stage of a communication process which, like any other process, must involve both information loss (increase in entropy) and noise. It is therefore reasonable to ask how much information can be transferred across any component of the system, and what the channel capacity is (bits per unit time). Li and Urquhart[16] have recently explored these questions for the adrenal synthesis of cortisol in response to ACTH, and have demonstrated that adrenal cortical secretion can be viewed as proceeding at eight different levels, so that the adrenal can be viewed as a 3-bit converter. The channel capacity in the dynamic case turned out to be of the order of 0.5 bit per minute. This type of analysis has two major implications: First, if the output of such a system is to be used as a means of making deductions about stages in the process which occur before it, then there is a maximum limitation of the information which can be deduced about preceding systems, because of the limited capacity of the biosynthetic system itself. Second, it has been traditional in the past to think about hormonal systems as operating in either the "on" or the "off" mode. Data such as those summarized above suggest that at least the endocrine gland itself is capable of sending much more information, unless such information disappears in the processes of distribution and transport of the hormone to its target tissues. It seems that such analyses of endocrine systems will be increasingly important in the future.

Control

In the case of most hormones, a demand signal is formulated through central neural mechanisms which in turn will control the biosynthesis of the hormone in question. In the case of the pituitary gland, this demand signal is thought to be, in general, in the form of the release by the central nervous system of polypeptide neurohormones from axonal terminals in the hypothalamus.[17] In the case of most other endocrine glands, such as the adrenals, the demand signal is relayed from the brain by means of a pituitary hormone like ACTH. In either case, factors such as distribution of hormone, blood flow to the interposed or final endocrine gland, and disappearance of the hormone from the blood must be considered. Such considerations will affect the capacity of the information channel in question to transmit its signal and, through the introduction of noise inherent in the information transfer process, may limit the variety of that signal as well, as has been indicated above. For example, if pituitary release of ACTH functioned only as either "on" or "off," so that the pituitary gland itself could be viewed as a 1-bit converter, then the 3-bit adrenal gland would be driven in only two modes. Additional variation might be introduced by variations in disappearance of ACTH from the plasma, or variations in presentation rate of ACTH to the adrenal mediated through changes in adrenal blood flow, though such changes would introduce a degree of confusion into the system. It has been shown for the overall adrenal control system[18] that this is not the case, and that the entire system may be viewed as a 3-bit converter, since eight levels of adrenal output can be observed in response to physiological stimuli.

The demand signal formulated by the central nervous system is in general the resultant of some combination of four classes of signals reaching some integrating point. These include (a) some function of the peripheral circulating concentration of the hormone being controlled (in general, classical negative feedback has been the point of view, although it has been shown for the case of cortisol[19] that the feedback must be nonlinear); (b) functions of other hormonal concentrations presented to the active sites in the central nervous system, thus permitting control of secretion of one hormone by another (for example, cortisol inhibits the secretion of vasopressin[20]; (c) signals of central nervous origin which drive the system in response to changes, for example, in the emotional state, or which entrain it in a rhythm such as the classical circadian rhythm[21]; and (d) central nervous derivatives of peripheral signals. These often indirectly reflect the functions of the hormone being controlled; for example, the signals related to blood pressure and volume are transmitted to the hypothalamus to control release of ACTH.[1] In general, the major emphasis has been placed on the first and fourth classes of signals in most models.

In most endocrine models that have been developed, there has been heavy reliance on modeling of components of the system involving hormone distribution and metabolism, with portions of the controller being modeled by an essentially compartmental approach, even though it is clear that flow of information rather than of mass is being considered. It has thus been convenient to treat the interaction of feedback and driving elements in the central nervous system as if there were some sort of "conservation of signal," although such a principle has never been defined and would not be generally valid. The actual construction of a model from such principles depends upon the introduction of certain simplifying assumptions which, in general, have produced linear models with ordinary differential equations. The basic assumption has often been one of linear proportional control.[22] Recently, DiStefano and Stear[23] have advanced a nonlinear model for control of thyroid hormone secretion and have indicated certain experiments which might serve to identify its parameters. Although the models are dynamic, they are solved only for the steady-state cases. This has been true of most other examples as well.

Recently, increased attention has been directed toward the modeling of the central nervous system controller elements, particularly for the case of the control of secretion of cortisol. Yates and his coworkers[24] have developed a model of adrenal cortical control based on an extension into the central nervous system of the principles of modeling of continuous systems. A very large number of assumptions have been made in the construction of this model. Dynamic behavior of the peripheral blood concentration of cortisol, upon which the model is based, can be inferred from the model advanced, or (at least with regard to most of the data) is equally consistent with a previous model with a different structure.[25] Although these models perform the library function, they may have limited usefulness in predicting the results of critical experiments, especially if these are designed to identify structures within the controller elements.

Kalman[26] has stressed the need for construction of so-called "minimal models" based on the least number of assumptions and the closest correspondence to experimental data. If such models can be constructed, a lack of correspondence between prediction and observation may be attributed to one of a very few potentially false assumptions, and these may be subjected to experimental tests. This sort of approach has been used[18,27] in the development of an equivalent model of the adrenal cortical control system, which has utilized some of the concepts of Boolean algebra and automata theory in the development of the model. The key principle involved is that, since the central neural elements must have very fast dynamics relative to the slow dynamics of hormone distribution, secretion, and metabolism, these earlier elements may be considered to be instantaneous, and have thus been modeled as static. The resultant model of central neural behavior with reference to the overall adrenal control system presents the central nervous system as a sort of switching circuit. Until it is possible to measure central neural dynamics of the processes involved in control of ACTH release, consideration of the time constants involved indicates that no evidence concerning these central neural dynamics can be derived even from the earliest stage at which physical-chemical measurement is possible, that of adrenal cortical secretion per se. It is interesting that in this model it was necessary to attribute only eight levels to the outputs of the adrenal gland itself or of any subsystem preceding it. Thus, the adrenal is modeled as a deterministic 3-bit converter which has the same information capacity as that proposed by Li and Urquhart. Ashby[28] has pointed out through his "law of requisite variety" that, if the output of a system has a given cardinality, then a subsystem controlling the output must have at least that cardinality. Thus, if there are eight levels at the output, there must be at least eight levels of information present throughout the system at each stage. However, if the hypotheses about the system are to be kept at a minimum, the cardinality of the output may be considered the maximum cardinality which can be inferred at any preceding stage in the system. That is, although a system leading into the 3-bit converter representing the adrenal gland may be continuous, it is impossible to infer from the output of the adrenal more than eight levels of input. Since this and Ashby's principle give identical upper and lower bounds for the cardinality of the system elements, taken together they offer one means of defining part of the mathematical structure of a minimal model if something is known about the biological noise in the components of the system on which direct measurements are performed. This approach has been useful in the case of the adrenal-cortical models[18] in leading to the prediction of a previously undescribed pathway of stimulation of ACTH release which was later verified[29] by direct experimentation. This experimental evidence led in turn to substantial revision of the model structure,[30] so that it appears that this method will permit the close working between model and experiment that is desirable if the model is to be used for prediction. Current work in this area is aimed at the extension of the modeling approach into the realm of finite-state dynamic models along the lines previously suggested.[26] It should be noted that this form of model can be coupled with continuous models of hormone biosynthesis, secretion, and metabolism for digital or hybrid simulation. At the present time, there has been no attempt to incorporate the action of the hormone into any such models except in an indirect way,[18] but this failure results primarily from a paucity of experimental data. Such data are now being obtained in a number of laboratories, and the near future should see the incorporation into single models of the four functions discussed at the outset of this chapter.

References

1. Gann, D. S.: *Systems analysis in the study of homeostasis, with special reference to cortisol secretion.* Amer. J. Surg. 114:95, 1967.

2. Katchalsky, A., and Curran, P. F.: *Nonequilibrium Thermodynamics in Biophysics.* Harvard Univ. Press, Cambridge, Mass., 1965.

3. Heckmann, K.: Die Permeabilität biologischer Membranen, in Karlson, P. (ed.): *Mechanisms of Hormone Action.* Academic Press, New York, 1965, p. 41.

4. Hess, B.: Biochemical regulations, in Mesarović, M. D. (ed.): *Systems Theory and Biology.* Springer-Verlag, New York, 1968, p. 88.

5. Weber, G.: Action of glucocorticoid hormone at the molecular level, in McKerns, K. W. (ed.): *Functions of the Adrenal Cortex.* Appleton-Century-Crofts, New York, 1968, Vol. 2, p. 1059.

6. Robison, G. A., Butcher, R. W., and Sutherland, E. W.: *Cyclic AMP.* Ann. Rev. Biochem. 37:149, 1968.

7. Pincus, G., Nakao, T., and Tait, J. F. (eds.): *Steroid Dynamics.* Academic Press, New York, 1966.

8. Tait, J. F., Tait, S. A. S., Little, B., and Laumas, K. R.: *The disappearance of 7-H^3-d-aldosterone in the plasma of normal subjects.* J. Clin. Invest. 40:72, 1961.

9. Tait, J. F., Little, B., Tait, S. A. S., and Flood, C.: *The metabolic clearance rate of aldosterone in pregnant and nonpregnant subjects estimated by both single-injection and constant-infusion methods.* J. Clin. Invest. 41:2093, 1962.

10. Yates, F. E.: *Contributions of the liver to steady-state performance and transient responses of the adrenal cortical system.* Fed. Proc. 24:723, 1965.

11. McKerns, K. W. (ed.): *Functions of the Adrenal Cortex*. Appleton-Century-Crofts, New York, 1968.

12. Koritz, S. B.: On the regulation of pregnenolone synthesis, in McKerns, K. W. (ed.): *Functions of the Adrenal Cortex*. Appleton-Century-Crofts, New York, 1968, Vol. 1, p. 27.

13. Koritz, S. B., and Hall, P. F.: *End-product inhibition of the conversion of cholesterol to pregnenolone in an adrenal extract*. Biochemistry (Wash.) 3:1298, 1964.

14. Urquhart, J., and Li, C. C.: *The dynamics of adrenocortical secretion*. Amer. J. Physiol. 214:73, 1968.

15. Urquhart, J., Krall, R. L., and Li, C. C.: *Analysis of the Koritz-Hall hypothesis for the regulation of steroidogenesis by ACTH*. Endocrinology 83:390, 1968.

16. Li, C. C., and Urquhart, J.: *Information transfer implicit in the adrenocortical secretory response to corticotropin*. Proc. 21st. Ann. Conf. Engrg. Med. Biol. 10:29.5, 1968.

17. Ganong, W. F.: Neuroendocrine integrating mechanisms, in Martini, L., and Ganong, W. F. (eds.): *Neuroendocrinology*. Academic Press, New York, 1966, Vol. 1, p. 1.

18. Gann, D. S., Ostrander, L. E., and Schoeffler, J. D.: A finite state model for the control of adrenal cortical steroid secretion, in Mesarović, M. D. (ed.): *Systems Theory and Biology*. Springer-Verlag, New York, 1968, p. 185.

19. Gann, D. S.: *Role of feedback inhibition in the corticosteroid response to hemorrhage in the dog*. Proc. 48th Mtg. Endocr. Soc., 1966, p. 63.

20. Dingman, J. F., Finkenstaedt, J. T., Laidlaw, J. C., Renold, A. E., Jenkins, D., Merrill, J. P., and Thorn, G. W.: *Influence of intravenously administered adrenal steroids on sodium and water excretion in normal and Addisonian subjects*. Metabolism 7:608, 1958.

21. Bartter, F. C., Delea, C. S., and Halberg, F.: *A map of blood and urinary changes related to circadian variations in adrenal cortical function in normal subjects*. Ann. N.Y. Acad. Sci. 98:969, 1962.

22. Roston, S.: *Mathematical representation of some endocrinological systems*. Bull. Math. Biophys. 21:271, 1959.

23. DiStefano, J. J., and Stear, E. B.: *On identification of hypothalamohypophysial control and feedback relationships with the thyroid gland*. J. Theor. Biol. 19:29, 1968.

24. Yates, F. E., Brennan, R. D., Urquhart, J., Dallman, M. F., Li, C. C., and Halpern, W.: A continuous system model of adrenocortical function, in Mesarović, M. D. (ed.): *Systems Theory and Biology*. Springer-Verlag, New York, 1968, p. 141.

25. Yates, F. E.: Physiological control of adrenal cortical hormone secretion, in Eisenstein, A. B. (ed.): *The Adrenal Cortex*. Little, Brown and Co., Boston, 1967, p. 133.

26. Kalman, R. E.: New developments in systems theory relevant to biology, in Mesarović, M. D. (ed.): *Systems Theory and Biology*. Springer-Verlag, New York, 1968, p. 222.

27. Schoeffler, J. D., Ostrander, L. E., and Gann, D. S.: Identification of Boolean mathematical models, in Mesarović, M. D. (ed.): *Systems Theory and Biology*. Springer-Verlag, New York, 1968, p. 201.

28. Ashby, W. R.: *An Introduction to Cybernetics*. Chapman and Hall, London, 1961.

29. Gann, D. S.: *Parameters of the stimulus initiating the adrenal cortical response to hemorrhage*. Ann. N.Y. Acad. Sci. In press, 1969.

30. Gann, D. S., and Ostrander, L. E.: *Extensions of a Boolean model of the adrenocortical response to hemorrhage*. Proc. 20th Ann. Conf. Engrg. Med. Biol. 9:8.2, 1967.

8

Compartments and Body Fluids

To function properly, cells must be maintained in an environment that has the right chemical composition. In addition, adequate performance of the circulatory system can occur only if body fluid volumes are kept within close limits. Maintenance of body solute and fluid content within the proper ranges is accomplished through complex feedback control systems which regulate interchange of these materials with the external environment.

It is necessary both to control total body content of water and solute and to have a proper distribution of water and solute among the various body compartments. This distribution is governed by physical and chemical forces rather than by physiological feedback mechanisms. We will consider first the mechanisms that determine movement of solute and water among the various body fluid compartments, and then the control of their exchange with the external environment.

Body Fluid Compartments

Water is considered to be distributed among several different "compartments" within the body. Nominally, these compartments have definite anatomical boundaries. *Intracellular* water is contained within the cells and separated from the extracellular compartment by cell membranes. The capillary wall divides the *extracellular* compartment

161

into an interstitial volume, containing the fluid which bathes the various cells of the body, and the plasma volume, which is contained within the circulatory system and may be considered as the interstitial fluid for the blood cells. The *transcellular* compartment is separated from the plasma compartment by the capillary wall, the interstitial fluid, and another layer of cells that has the ability to secrete various solutes, and thus to alter the composition of the transcellular fluid from that of the remainder of the extracellular fluid. Examples of transcellular fluid are the intestinal secretions, cerebrospinal fluid, intraocular fluid, and fluid contained within various body cavities such as the chest and abdominal cavities.

Measurement of Compartment Size

Compartment volume is estimated by use of the indicator dilution technique. This method depends upon the assumption that a particular material, when added in tracer quantities, will distribute uniformly throughout the desired compartment only. A known amount of the indicator is added to the compartment. After sufficient time for equilibration, a sample is withdrawn and the concentration of the tracer material is measured. Compartment volume is calculated from the following equation:

Volume of distribution
$$= (\text{Amount administered} - \text{Amount excreted})$$
$$/(\text{Equilibrium concentration}) \quad (1)$$

The requirements for a suitable tracer substance are:

1. It must be uniformly distributed in the desired compartment only.
2. It must not itself alter compartmental volumes.
3. It must be nontoxic.
4. Preferably, it should not be metabolized or synthesized.
5. Its concentration must be accurately measurable.

Properties other than the chemical concentration of a tracer can be measured. Radioisotopes are frequently used to measure compartment volumes. For example, the total water content of the body may be measured using tritiated water as the trace material. The tritium is given intravenously, and after a two-hour equilibration period a blood sample is obtained and assayed for radioactivity. For careful work, the amount of tritium lost in the urine and in expired air during the equilibration period must also be measured. The total body water is then calculated from the equation

$$\text{TBW} = \frac{(\text{CPM}_{inj} - \text{CPM}_{exc})}{\text{CPM/ml plasma water}} \tag{2}$$

where CPM is radioactivity (counts per minute), CPM_{inj} is the amount of radioactivity injected, and CPM_{exc} is the amount of radioactivity excreted in the urine and expired air.

Methods for measuring plasma volume depend on the fact that plasma proteins do not readily cross the capillary wall and thus tend to be retained in the plasma compartment. As a tracer, one can use an isotopically labeled plasma protein, e.g., radioiodinated albumin. As an alternative, one may use a substance which binds to albumin when injected into the bloodstream (Evans blue or T-1824). Another way of estimating plasma volume is to measure blood volume using labeled erythrocytes as the tracer, and then to calculate plasma volume from the relationship

$$\text{Plasma volume} = \text{Blood volume (100} - \text{Hematocrit)} \tag{3}$$

To serve as an indicator for the extracellular volume, a substance must readily cross the capillary wall but not enter the cells. Unfortunately, all of the substances used to measure extracellular volume enter the cells to some degree. In addition, these tracers do not mix easily with certain components of extracellular water, such as those in bone and connective tissue. Values obtained for the extracellular fluid volume thus depend on which tracer is used, and range from about 16 to 30 percent of total body weight.

One cannot measure the volume of intracellular water directly, but it can be calculated as the difference between total body water and extracellular water. Therefore, the measurement of intracellular water suffers from the errors inherent in determination of the extracellular space described above. Values for the various body water compartments in a 70-kg man are given in Table 8-1.

We have pointed out that the measured volume of the body fluid compartments depends on the specific tracer substance which is used. Thus, we cannot think routinely of a compartment as a space with well-defined physical boundaries. Instead, a compartmental system is described by the kinetic behavior of the particular material being tested. We can define a compartment as a pool or amount of material which acts kinetically as though it were well mixed. In some cases a compartment may be identical to a physiological space with definite

Table 8-1. Representative Values for Volumes of the Body Water Compartments in a 70-kg Man

Compartment	Volume, L
Intracellular	25
Plasma	3
Interstitial	12
Transcellular	1
Total body water	41

anatomical boundaries, e.g., the amount of some compound contained within the vascular system, but such a distinction frequently will not be possible. As an extension of this concept, the distinction between compartments can be made on the basis of chemical form rather than separation by a physical barrier. For instance, iodine may be present in the body as inorganic iodide or as organic iodine, contained in thyroid hormone. Interchange between the two forms of iodine can occur only through the processes of thyroid hormone synthesis and metabolism, which serve as a significant barrier to such interchange. Thus, inorganic iodide and hormonal iodine may be regarded as existing in two separate compartments, although both forms may occur together in many tissues of the body.

Water can move freely among the various body fluid compartments. It thus distributes so that its chemical potential is the same throughout the body. The relative amounts of water in the various body fluid compartments are determined by the balance between the hydrostatic and osmotic pressures acting on water movement.

In contrast to water, most solutes exhibit marked concentration differences among various fluid compartments. This is illustrated in

Table 8-2.   Ionic Composition of the Body Fluids

	Plasma Water* mEq/L	Interstitial Fluid mEq/L	Intracellular Fluid mEq/L
Na^+	148	141	10
K^+	4	4	150
Ca^{++}	4	4	—
Cl^-	109	115	15
HCO_3^-	28	29	10
SO_4^{--}	1	1	20

*Values are calculated from the concentrations of the ions in plasma, using the assumption that plasma is 93 percent water.

Table 8-2, which gives concentrations of certain ions in plasma, interstitial fluid, and intracellular fluid. Large concentration differences, e.g., those shown for sodium and potassium ions, must be maintained across cell membranes for phenomena such as membrane potentials to exist. However, the process of thermal diffusion tends to destroy these gradients. Two factors act to maintain these gradients:

1. The selective permeability characteristics of the cell membrane restrict diffusion of certain ions and solutes.
2. Active transport of ions and solutes against electrochemical potential gradients counteracts the diffusion process.

Diffusion Across Biological Membranes

Diffusion is the process by which matter is transported from one part of a system to another as a result of random molecular motions. In an isotropic medium, the rate of transfer of the diffusing substance per unit area of a section is proportional to the concentration gradient measured normal to the section (Fick's law of diffusion):

$$F = -D\, \partial C/\partial x \tag{4}$$

where F is the rate of transfer per unit area of section, C the concentration of the diffusing substance, x the distance measured in the direction of diffusion, and D the diffusion coefficient. When dealing with dilute solutions, D is usually considered to be a constant; however, in special cases such as polymeric solutions, D may vary markedly with concentration. The units of the diffusion coefficient are $length^2/time$.

Analytical solutions for the diffusion equation are available for many systems having different geometries.[1] Of special interest is the application of Eq. (4) in determination of the ease with which different solutes penetrate biological membranes. Consider the case of a cell with a constant internal volume, V, suspended in a solution containing a penetrating solute whose concentration is C_0 (Fig. 8-1). If the volume of bathing solution is large compared to the volume of cell contents, C_0 can be considered constant. Let the area for diffusion of the solute through the cell membrane be A, and let the thickness of membrane (in the direction normal to diffusion) be x. If the diffusion coefficient of the solute is much greater in the cell cytoplasm and external medium than in the cell membrane, the total resistance to diffusion may be considered to occur in the membrane. Application of Eq. (4) to this system results in

$$F = DA\, (\bar{C}_0 - \bar{C}_i)/x = V\, dc_i/dt \tag{5}$$

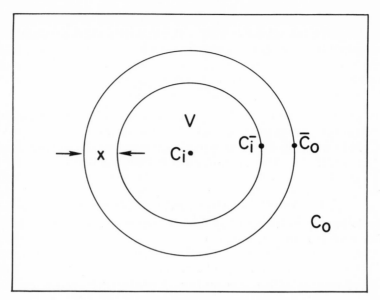

Figure 8-1. Symbols used in the application of the diffusion equation to movement of solute across the membrane of a cell suspended in a large volume of fluid.

where C_i is the concentration of solute in the cell fluid, $\bar{C}_0$ and $\bar{C}_i$ are concentrations of solute in the lipid constituents of the cell membrane at the outer and inner membrane surfaces, respectively, and t is time. The right-hand term in Eq. (5) represents the rate of accumulation of solute in the cell fluid, which must be equal to the net rate of diffusion into the cell. $\bar{C}_0$ and $\bar{C}_i$ are related to the concentrations C_0 and C_i by the oil/water partition coefficient, which is defined as the ratio of the solubility of the solute in oil to that in water at the same temperature. If we designate this coefficient as k, $\bar{C}_0 = kC_0$ and $\bar{C}_i = kC_i$. Then Eq. (5) may be written

$$V \, dC_i/dt = \frac{DkA}{x} (C_0 - C_i) = PA \, (C_0 - C_i) \qquad (6)$$

The quantity $P = Dk/x$ is defined as the permeability of the membrane to the solute and has the units of length/time.

With the initial condition that $C_i = C_i'$ when $t = 0$, Eq. (6) may be solved to give

$$\ln \frac{C_0 - C_i}{C_0 - C_i'} = -\frac{PAT}{V} \qquad \text{or} \qquad \frac{C_0 - C_i}{C_0 - C_i'} = e^{-\frac{PAT}{V}} \qquad (7)$$

From Eq. (7) it may be seen that the concentration of solute in the cell will approach the concentration in the external solution with a time constant of V/AP. Also from Eq. (7)

$$P = -\frac{V}{tA} \ln \frac{C_0 - C_i}{C_0 - C_i'}$$ (8)

Therefore, if a cell or group of cells with initial intracellular solute concentration C_i' is placed in a large volume of solution of concentration C_0, and intracellular concentration is periodically sampled, a plot of $\ln (C_0 - C_i)/(C_0 - C_i')$ vs t will have a slope of $-PA/V$. If the surface area-to-volume ratio of the cell is known, P may be determined directly. In cases in which the ratio A/V is not known, the permeability is taken as being equal to PA/V, with its units then being 1/time. Thus, when taking values for permeability from the literature, one must always be cognizant of the units used.

Permeability Characteristics of Biological Membranes

Cell Membranes

Movement of a material across the cell membrane as described by Eq. (6) implies that the solute dissolves in the substance of the membrane and moves through this substance by simple diffusion. Cell membranes consist of a bimolecular lipid layer covered on both sides by a protein coat. It has been shown that, when proteins are formed into artificial membranes, they do not offer significant resistance to free diffusion. It therefore appears that the diffusional barrier presented by membranes must be due to the double layer of lipid molecules. If a solute were to cross the membrane by dissolving in the lipid film and diffusing through it, energy would be required for the solute to leave the water phase and enter the lipid phase. The energy requirement would be higher for molecules possessing polar groups than for non-polar compounds, since roughly 5 kcal per mole is required to break hydrogen bonds between polar compounds and water. On this basis, one might expect permeability to be related to the degree of oil solubility of the solute. In general, permeability of the cell membrane to solutes increases as the lipid solubility of the solute increases. However, solutes with small molecular volumes tend to penetrate membranes much more rapidly than do large molecules having the same lipid solubility. The membrane thus acts as if it had "pores" through which low-molecular-weight solutes could pass without having to dissolve in the lipid film. Although such pores have not been identified anatomically, the cell membrane acts functionally as though it had a pore size slightly larger than the diameter of a urea molecule. Molecules smaller than urea, such as water, oxygen, carbon dioxide, and ammonia, diffuse through the membrane readily, while larger molecules apparently must cross the membrane in some other manner.

Permeability to ions is different from what would be predicted on the basis of ionic size, for two reasons. First, hydration of ions increases their effective diameter. Since the degree of hydration depends on surface charge, a small univalent ion such as Na^+ will bind more water and thus have a larger hydrated diameter than a larger univalent ion such as K^+. Second, the presence of a charge on an ion decreases its chance of entry into cells. The hydration sphere does not electrically neutralize the ion. The residual ionic charge still acts as a hindrance to penetration, as does the increased particle size resulting from hydration.

Active transport or metabolic pumping of ions against an electrochemical potential gradient (see below) may reduce the apparent permeability of the cell to ions. Thus, although sodium ions can diffuse across cell membranes, active transport of sodium out of the cells tends to keep its intracellular concentration low and to limit sodium ion almost entirely to the extracellular compartment.

It must be noted that the permeability to a given solute may vary among different cells by as much as a millionfold. Also, the permeability of a cell to a particular solute may vary markedly, depending upon physiological or pathological conditions such as the presence or absence of certain hormones.

Capillary Membranes

The capillary membrane is freely permeable to water and to all of the solutes in blood, with the exception of the plasma proteins. The proteins consist mostly of albumin (molecular weight 69,000), present in a concentration of about 5 g per 100 ml plasma, and globulins (m.w. 40,000 to 800,000) with a concentration of about 2 g per 100 ml. Because of their high molecular weights, the plasma proteins "leak" only slowly from the plasma compartment into the interstitial compartment, from which they are returned to the blood via the lymphatic circulation. Since the plasma proteins are the only major constituent of plasma to which the capillary is impermeable, their osmotic effect (the *colloid osmotic pressure*) is extremely important in controlling movement of water between the plasma and the interstitial compartments.

Osmosis and Ultrafiltration

Just as a solute will diffuse under a solute concentration gradient, water will move across membranes under a chemical potential gradient for water. The chemical potential of the solvent in a solution is increased by raising the temperature or pressure on the system, and decreased by increasing the concentration of solute particles in the solution. Since body temperature is usually maintained within a very narrow range, we need not consider effects of temperature change.

Pure water will have a higher chemical potential than the water in an aqueous solution. If the water and solute solutions are separated by a membrane permeable to water alone, water will move from the region where its potential is high (pure water) to the region where its potential is low (the solution). Obviously, water will also flow between two solutions of different concentration. The movement of pure solvent from the lesser to the greater concentration of solute, when two solutions are separated by a membrane which selectively prevents the passage of solute particles, is called *osmosis*. By applying a hydrostatic pressure to the solution, it would be possible to raise the chemical potential of water in the solution to that of pure water, so that there would be no net flow of water across the membrane. The hydrostatic pressure that must be applied to a solution to restore the chemical potential of the solvent in the solution to that of pure solvent at the same temperature is called the *osmotic pressure* of the solution. Osmotic pressure depends only upon the total number of particles per unit volume, rather than upon their size, molecular weight, or chemical composition. It is therefore a colligative property, as are vapor pressure, freezing point depression, and boiling point elevation. The osmotic concentration of a solution containing only one solute is expressed in terms of osmolarity:

Osmolarity = Molarity × Number
$\qquad$ of particles formed on dissociation of one molecule $\quad$ (9)

The number of particles obtained from one molecule depends on its chemical structure and degree of ionization. The effects of both of these factors are reflected in the osmotic or cryoscopic coefficient, defined as

$$G = \frac{\Delta t_f}{\Delta T_f} \qquad (10)$$

where Δt_f is the freezing point depression produced by a given molar concentration of the solute in question, and ΔT_f is the lowering of the freezing point by the same molar concentration of an ideal nonelectrolyte. The osmotic pressure of a solution may be calculated from the relation

$$\pi = (C \times G)RT \tag{11}$$

where π = osmotic pressure, atm
      C = molar concentration, mol/L
      R = gas constant, 0.08206 atm − L/(g-mol-°K)
      T = absolute temperature, °K
For a dilute nonelectrolyte solution, Eq. (11) reduces to

$$\pi = CRT \tag{12}$$

a form analogous to the perfect gas law.

The osmotic pressure of a solution may be determined by measuring any one of its colligative properties. The most convenient and frequently used method for biological solutions is measurement of freezing point depression.

Two solutions whose osmotic pressures are equal are said to be isosmotic to each other. If solution "A" has a higher osmotic pressure than solution "B," "A" is hyperosmotic to "B," and "B" is hypoosmotic to "A."

In the above discussion, we have considered the membrane to be permeable only to water. Actually, as we have noted above, biological membranes are also highly permeable to certain other substances such as urea. Solutes which readily penetrate the membrane will rapidly equalize their concentrations on both sides of the membrane by their own diffusion, without causing any net water movement. Only those solutes that do not penetrate will exert an osmotic effect and cause net movement of water across the membrane. The osmotic pressure due to nonpenetrating solutes is therefore called the *effective osmotic pressure*. The effective osmotic pressure thus depends upon the permeability of the specific membrane under consideration, while the total osmotic pressure does not.

Two solutions which have the same effective osmotic pressure with respect to a given membrane (for example, the cell membrane) are said to be isotonic. The total osmotic pressure of the body fluids is around 300 mOsm per liter. Since sodium ions do not readily penetrate cell membranes, a 300 milliosmolar NaCl solution (approximately 150 millimolar) is said to be isotonic to normal body fluids.

Again assume that a concentrated and a dilute solution are separated by a semipermeable membrane. By applying a sufficiently great hydrostatic pressure to the concentrated solution, it is possible to make the chemical potential of water in the concentrated solution greater than that in the dilute solution. Then water will be forced through the membrane against an osmotic gradient. This process, known as *ultra-filtration,* occurs in the capillaries.

Transcapillary Water Movement

The physical principles which govern water movement between body fluid compartments are well illustrated by an analysis of trans-capillary water movement. Figure 8-2 shows how hydrostatic pressure and colloid osmotic pressure might vary along a capillary. At the upstream end of the capillary, the hydrostatic pressure difference tending to force fluid out of the capillary (filtration pressure) exceeds the protein colloid osmotic pressure difference tending to draw fluid back into the capillaries (reabsorption). At the downstream end of the capillary, the situation is reversed, with the net force favoring reabsorption. However, some of the fluid that is filtered is not reabsorbed by the capillary, but instead is removed from the interstitial space via the lymphatic system.

A mathematical analysis of transcapillary fluid exchange has been presented by Wiederhielm.[2] A water balance for the interstitial compartment must include the net rate of inflow of water from the capillaries by diffusion and by leakage through "pores" (bulk flow), and the rate of removal of water by lymph flow. The net transcapillary diffusion rate for a capillary element of length dL is given by the relationship

$$\text{Diffusion rate} = K(P - P_T - \pi_{PL} + \pi_T)\, 2\pi r\, dL \tag{13}$$

where P is the hydrostatic pressure in the capillary, P_T the hydrostatic pressure in the interstitial space, π_{PL} the plasma colloid osmotic pressure, π_T the colloid osmotic pressure of tissue fluid, and r the capillary radius. P, K, and r are functions of distance along the capillary. As a simplification, the whole capillary may be divided into arterial and venous portions, which show filtration and reabsorption, respectively. The equations which result when Eq. (13) is applied to each portion of the capillary are

$$\text{Arterial segment:} \quad F = K_A A_A (P_A - P_T - \pi_{PL} + \pi_T) \tag{14}$$

$$\text{Venous segment:} \quad R = K_V A_V (P_V - P_T - \pi_{PL} + \pi_T) \tag{15}$$

where F is the rate of filtration, R is the rate of reabsorption, and A is area. The subscripts A and V refer to the arterial and venous segments, respectively.

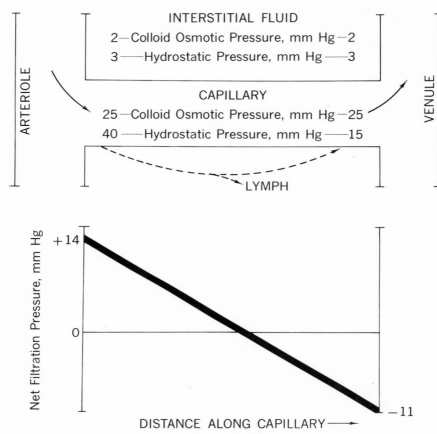

Figure 8-2. Variation of the net force for filtration or reabsorption along the length of a capillary. The dotted line indicates recirculation of fluid through the interstitial space, with part of the filtered fluid being reabsorbed by the capillary, and the remainder being returned to the circulation via the lymphatic system.

Plasma leakage by bulk flow is considered proportional to the hydrostatic pressure gradient across the venous capillary wall:

$$\dot{V}_{plasma} = K_1(P_V - P_T) \tag{16}$$

Tissue pressure is given by the relationship

$$P_T = K_4 \Delta V$$

where ΔV is the change in interstitial space volume from the volume at which tissue pressure would be zero, and K_4 is the elastance of the interstitial space.

The rate at which lymph leaves the interstitial compartment is assumed to be proportional to tissue pressure:

$$\dot{V}_{lymph} = K_2 P_T \tag{17}$$

The instantaneous rate of accumulation of interstitial fluid is thus given by

$$\dot{V} = F + R + \dot{V}_{plasma} - \dot{V}_{lymph} \tag{18}$$

and the change in interstitial volume over the time t is obtained by integration of Eq. (18):

$$\Delta V = \int_0^t (F + R + \dot{V}_{plasma} - \dot{V}_{lymph})\, dt \tag{19}$$

To determine the effects of changes in intracapillary pressure or plasma colloid osmotic pressure on interstitial volume from Eq. (19), the tissue fluid colloid osmotic pressure must be known. Over a limited range of protein concentrations, this pressure may be considered to be directly proportional to the interstitial protein concentration:

$$\pi t = K_3 C_{prot.\ int.} \tag{20}$$

Interstitial fluid protein concentration at time (t) is given by

$$C_{prot.\ int.} = \frac{(Q_{prot.})_0 + \int_0^t \dot{Q}_{prot.}\, dt}{V}$$

where $(Q_{prot.})_0$ is the interstitial protein content at time $t = 0$, V is interstitial volume, and the rate of accumulation of protein in the interstitial space is

$$\dot{Q}_{prot.} = \dot{V}_{plasma} C_{prot.\ pl.} - \dot{V}_{lymph} C_{prot.\ lymph} \tag{21}$$

where $C_{prot.\ pl.}$ and $C_{prot.\ lymph}$ are the concentrations of protein in plasma and lymph, respectively.

An analog computer program representing this set of equations is given in Figure 8-3. Estimated normal values for some of the model parameters are given in Table 8-3. The model is useful in studying the mechanisms by which capillary fluid exchange is altered in a variety of physiological or pathological conditions. These include alterations in capillary pressure, changes in plasma protein concentration, increased permeability of capillaries to protein (inflammation of the blood vessels), and obstruction of the lymphatic vessels.

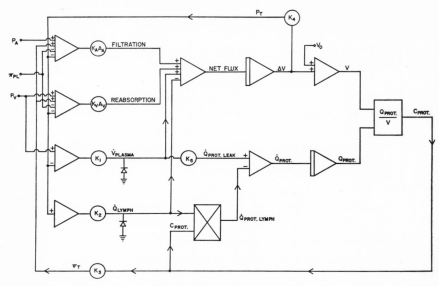

Figure 8-3. Analog computer program for simulating transcapillary water movement.[2]

Body Compartments and Fluid Balance

Movement of solutes across capillary walls occurs by diffusion, with the solute moving from a region of high concentration to one of low concentration. Transport of many solutes across cell membranes also occurs by simple diffusion. However, some substances such as glucose, amino acids, and certain ions (Na+, for example) cross cell membranes against electrochemical potential gradients, and thus are

Table 8-3.   Assumed Normal Values for Parameters Used in the Computer Program of Figure 3*

Parameter	Value
Arterial capillary pressure (P_A)	35 mm Hg
Venous capillary pressure	15 mm Hg
Plasma colloid osmotic pressure (π_{PL})	25 mm Hg
K_A/K_V	0.6
A_A/A_V	0.25
$K_A A_A/K_V A_V$	0.16
Plasma leakage	5% of total exchange
Interstitial space compliance	60%/mm Hg

* Values taken from ref. 2.

said to be actively transported. Active transport systems have the following characteristics:

1. The transport against a potential gradient requires expenditure of metabolic energy. Metabolic inhibitors prevent the release of this energy and thus prevent active transport. Decreasing the temperature, by slowing these metabolic processes, also decreases the rate of active transport.
2. The net flux of the transported solute does not continue to increase as its concentration difference is increased, but instead approaches a constant value. This phenomenon is called saturation.
3. The net transport of an actively transported solute may be decreased by the presence of another solute. In that case, the two solutes are said to show competition for the transport system.

Many of the characteristics of active transport systems can be explained by the carrier hypothesis, although no carrier molecules have been positively identified. According to this hypothesis, the transported molecule or ion unites with a carrier molecule on the low concentration side of the membrane. This creates a concentration gradient for the carrier-solute complex, which diffuses to the other side of the membrane. At this point the affinity of carrier for solute decreases, so that the solute is released into the more concentrated solution. In this manner the transported material moves against a concentration gradient. Production of the change in affinity of the carrier for the solute at the two sides of the membrane is presumably the energy-requiring part of the process.

It will be shown below that both simple diffusion and active transport play important roles in controlling the exchange of water and solute with the environment.

Body water content and the osmolarity of the body fluids are normally held within narrow limits by elaborate neuroendocrine control systems. These control systems are considered to act primarily upon the extracellular compartment for the following reason: The cell membrane can be assumed to be effectively permeable only to water because of its limited permeability to many solutes and the effect of active transport processes which tend to maintain transmembrane concentration gradients. In other words, there is no significant net flux of solute across the membrane under physiological conditions, so that the solute content of the intracellular compartment is essentially constant. Since water readily crosses the cell membrane to maintain osmotic equilib-

**Table 8-4.   Average Daily Intake and
Loss of Water in an Adult**

Route	Ml/24 hr
INTAKE	
Solid and semisolid food	1200
Liquid food and drink	1000
Oxidation of foodstuffs	300
LOSS	
Skin (insensible loss*)	500
Lungs	350
Feces	150
Urine	1500

*Sensible loss (perspiration) may vary from 0 to 2 liters per hour.

rium, cell water content will vary as a function of extracellular osmolarity.

It is frequently stated that extracellular fluid volume is determined mainly by the extracellular sodium content, whereas osmolarity is determined by extracellular water content. This is a considerable oversimplification. There is actually much overlap in the control systems which regulate volume and osmotic concentration.

Control of Extracellular Fluid Osmolarity

The osmolarity of extracellular fluid is regulated primarily through changes in water intake or excretion. Routes for water intake and loss with average values for a 70-kg man are shown in Table 8-4. Liquid intake and renal water excretion are the only two routes that can be manipulated to any extent in so far as control of water balance is concerned. A water overload is rapidly compensated for by the excretion of a large amount of very dilute urine. A water deficit results in a decrease in urine output and the production of a very concentrated urine. However, urinary water excretion cannot go below 300 to 500 ml per day (the obligatory urine volume) because of the necessity for excreting excess salts and the waste products of metabolism. Therefore, the primary response to water deficit is an attempt to increase water intake through the thirst mechanism.

Renal Excretion of Water and Salt

As shown in Figure 8-4, the kidney is divided into an outer cortex and an inner medulla. The medulla is divided into a number of triangular-shaped structures, the renal pyramids, which have their base at the junction of the cortex and medulla. The functional unit of the kidney is the nephron or renal tubule. There are approximately one million nephrons in a human kidney. The nephron consists of several parts which have both anatomical and functional differences (Fig. 8-5). The glomerulus is a capillary bed which serves as the filtering unit of the kidney. The proximal and distal convoluted tubules lie entirely in the cortex. The section connecting them, the U-shaped loop of Henle, has its long axis parallel to that of the renal pyramids. Loops of Henle may lie either entirely in the cortex (cortical nephron) or extend also into the medulla (juxtamedullary nephron). Each distal tubule empties into a collecting tubule, which then unites with other collecting tubules to form a collecting duct. The collecting ducts pass through the medulla in close proximity to the loops of Henle. The urine from all of the collecting ducts enters the renal pelvis and leaves the kidney through the ureter.

The first step in urine formation is ultrafiltration of about 20 percent of the plasma flowing through the glomerular capillaries. Thus, the filtrate entering the nephron has essentially the same composition as plasma, except that it does not contain the plasma proteins and some

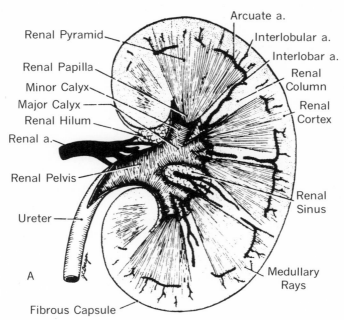

Figure 8-4. A coronal section of the kidney. (From Woodburne, R.T.: *Essentials of Human Anatomy,* ed. 3. Oxford Univ. Press, New York, 1965.)

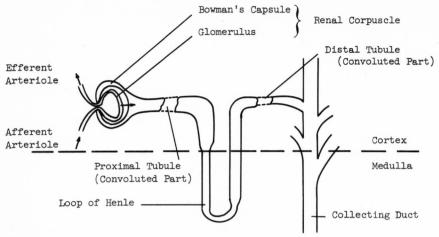

Figure 8-5. Anatomic divisions of the nephron.

of the plasma lipids. Some materials that are filtered pass through the renal tubules without change and are excreted at the same rate as they are filtered. Other solutes are reabsorbed from the tubular fluid into blood in the peritubular capillaries and are thus retained in the body (Fig. 8-6). Still other solutes are secreted into the tubular urine so that their excretion rate exceeds their filtration rate. In some cases, a material may be filtered, secreted, and reabsorbed so that its excretion rate represents the net effect of all three processes.

The major events occurring along the renal tubule concerned with the control of extracellular volume and osmolarity are:

1. Sodium is actively reabsorbed against a concentration gradient in the proximal tubule. Chloride is passively reabsorbed, following the positively charged sodium ions. Water is reabsorbed due to the osmotic pressure gradient created by sodium chloride reabsorption. Since the proximal tubule is always highly permeable to water, reabsorption in the proximal tubule is isosmotic. About 80 percent of the filtered water and sodium chloride is reabsorbed in the proximal tubule, so that only 20 percent enters the loop of Henle.
2. The loops of Henle play an essential role in the urinary concentrating mechanism, and thus in the renal control of body water content. The hairpin-like loops of Henle act as a countercurrent multiplier system to produce an osmotic gradient along the renal pyramids, with the highest osmolarity at the tip of the pyramid.

The manner in which this is accomplished is as follows (Figs. 8-7 and 8-8):

a. There is a sodium pump in the ascending limb of Henle's loop which pumps sodium from the lumen to the interstitium against a concentration gradient.

b. This sodium transport increases the sodium concentration and osmolarity of the interstitium.

c. Isosmotic proximal tubular fluid enters the descending limb, which is permeable to water and sodium, so that fluid passing down the limb can equilibrate with the interstitial fluid. The fluid thus reaches the tip of Henle's loop at a higher osmolarity than it had on entering.

d. The fluid entering the ascending limb is now more concentrated than it was initially. The pumps in the ascending limb can therefore develop a higher interstitial concentration, even though they are still developing the same tubular fluid-to-interstitial fluid concentration gradient.

e. As fluid passes up the ascending limb, its osmolarity decreases because sodium is being pumped out. This causes a concentration gradient along the medulla. Fluid entering the distal tubule is slightly hypotonic.

f. The ascending limb of Henle's loop is relatively impermeable to water. If this were not so, water would follow the sodium that was pumped out, and there would be no increase in interstitial concentration.

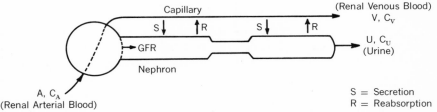

Figure 8-6. Secretion and reabsorption by the renal tubule. A filtered material may be reabsorbed from the tubular urine to blood in peritubular capillaries and returned to the systemic circulation via the renal vein. If a material is reabsorbed, its rate of excretion is less than its rate of filtration in the glomerulus. A material may be secreted from the peritubular capillaries into tubular urine by the renal tubular cells. In that case, more of the material is excreted in the urine than is filtered. Some materials may undergo both reabsorption and secretion at different points in the nephron.

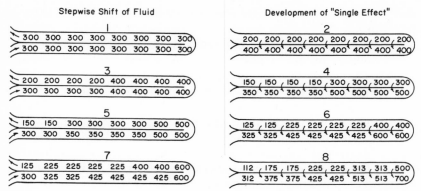

Figure 8-7. The principle of countercurrent multiplication of concentration applied to the loop of Henle. It is assumed that at any level along the loop a gradient of 200 mOsm per liter can be established between ascending and descending limbs by active transport of sodium ions. (From Pitts, R. F.: *Physiology of the Kidney and Body Fluids.* Year Book Medical Publishers, Chicago, 1963.)

The overall effect of the countercurrent multiplier is to produce a very high osmotic pressure at the tip of the renal pyramids without requiring the tubular cells to pump against a high sodium concentration gradient. The fluid entering the distal tubule from the ascending limb of Henle's loop is always hypotonic. If the distal tubular and collecting duct walls are not permeable to water, no water will be reabsorbed in these segments and the final urine will be hypotonic. On the other hand, if the walls of these segments are highly water permeable, distal tubular fluid will equilibrate osmotically with the peritubular blood, and thus the fluid entering the collecting ducts will be isotonic. As this fluid passes down the collecting duct, it will lose water to the more concentrated interstitial fluid. The final urine leaving the collecting ducts will have equilibrated with the very highly concentrated interstitial fluid at the tip of the pyramid, and thus will have an osmotic concentration several times that of plasma. The water reabsorbed from collecting duct fluid will be returned to the circulation via the vasa recta, a network of blood vessels which extend down into the medulla.

It is evident that, by varying the water permeability of the distal nephron (distal tubule and collecting duct), it is possible to alter greatly the urine concentration and the rate of urinary water excretion. Distal nephron water permeability is controlled by the plasma concentration of antidiuretic hormone (ADH). If ADH is present, the renal epithelial cells are highly water permeable, and a very concentrated urine is produced. In the absence of ADH, these cells are impermeable to water.

Then the dilute tubular fluid leaving Henle's loop remains hypotonic in the distal nephron, and a dilute urine is formed.

The action of ADH in controlling renal tubular water permeability is part of a complex neuroendocrine feedback control system which regulates body fluid osmolarity within limits of 283 ± 11 mOsm per L. The complete system consists of appropriate sensors or receptors, a central integrative mechanism, a neurosecretory mechanism that produces antidiuretic hormone, and the renal effector mechanism that governs the excretion of water.

ADH is an octopeptide which is formed in the cells of the supraoptic and paraventricular nuclei of the hypothalamus. From there it is transported to the posterior lobe of the pituitary gland by protoplasmic flow in nerve fibers that make up the supraopticohypophyseal tracts. In response to stimulation of various receptors, nerve impulses in these tracts cause liberation of ADH into the circulation.

At least two separate receptor systems seem to be involved. Osmoreceptors located within the zones of distribution of the internal carotid

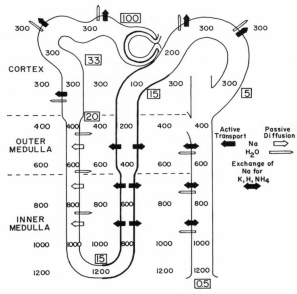

Figure 8-8. Exchanges of water and ions in the nephron during elaboration of a hypertonic urine. Concentrations of tubular urine and peritubular fluid are in milliosmoles per liter; boxed numerals indicate the estimated percent of glomerular filtrate remaining within the tubule at each level. (From Pitts, R.F.: *Physiology of the Kidney and Body Fluids.* Year Book Medical Publishers, Chicago, 1963.)

arteries, probably within the hypothalamus, sense the osmotic pressure of plasma. The injection into carotid loops of sufficient hypertonic sodium chloride solution to raise the osmotic pressure of carotid blood by only 2 percent results in prompt release of ADH. The time constant for ADH release in response to an increase in plasma osmolarity is about 4 minutes. A decrease in plasma osmolarity, such as accompanies water ingestion, inhibits ADH release, with a resultant increase in water excretion. There is a time delay, however, between the cessation of ADH secretion and the beginning of water diuresis, since the ADH already present in the system must first be destroyed. ADH is broken down in the liver and also apparently by the distal tubular cells upon which it acts. The time constant for the disappearance of ADH is in the order of 10 minutes.

Stretch receptors, apparently located in the left atrium with their afferent neurons passing centrally in the vagus nerve, also influence the secretion of ADH. Atrial distention inhibits ADH secretion and results in water diuresis, while decreased distention stimulates ADH secretion resulting in antidiuresis and thirst.

It should be noted that ADH may also increase the urea permeability of the distal nephron. Urea constitutes a substantial part of the total urinary solute excretion. If the distal nephron is permeable to urea, some urea will diffuse from the collecting duct lumen into medullary interstitial fluid. Some of this reabsorbed urea diffuses into the loops of Henle and is recirculated through the distal nephron. The remainder is returned to the body urea pool by the medullary blood vessels. The fact that urea may equilibrate between the medullary interstitial fluid and the collecting duct fluid allows the excretion of a urine having a high urea concentration without the need for active transport of urea.

Control of Extracellular Volume

Extracellular volume is regulated primarily by manipulation of renal sodium excretion. Thus, a decrease in extracellular volume is compensated for by increased sodium retention and an increase in the sodium content of the extracellular compartment. The salt retains water in the compartment, resulting in an increase in volume.

Possible mechanisms for regulation of sodium reabsorption include: (a) changes in hemodynamics resulting in variation in GFR, and (b) the effect of adrenal hormones on renal sodium transport mechanisms.

It has been suggested that changes in glomerular filtration rate (GFR) are the means by which rapid changes in sodium reabsorption are brought about. Sodium is reabsorbed from tubular urine in the proximal tubule, loop of Henle, and the distal tubule. About four-fifths of the filtered sodium is reabsorbed in the proximal tubule. If this ratio remains constant, an increase in filtration rate will result in delivery of more sodium to the distal parts of the nephron. This extra load will exceed the limited reabsorptive capacity of the distal tubule, so that more sodium will be excreted. Conversely, if filtration rate is decreased, the distal tubule will be able to reabsorb a greater fraction of the sodium which enters it, so that urinary sodium excretion will be decreased. The changes in GFR can be brought about by alterations in renal hemodynamics mediated by the sympathetic nervous system.

GFR changes probably do play a role in control of sodium reabsorption. However, the changes in GFR that would be required are small and not easily detectable, so that the importance of this mechanism is difficult to determine.

The role of adrenal hormones in sodium reabsorption is somewhat more clearly defined. It has long been known that adrenalectomy is followed by sodium loss in the urine, and that this loss can be reversed by the administration of adrenocortical hormones. Although all of the adrenocortical hormones possess some salt-retaining action, aldosterone is the most potent in this respect. Aldosterone appears to act primarily in the distal tubule. It therefore affects the renal tubular reabsorption of a small but highly significant fraction of filtered sodium.

The primary controller of aldosterone secretion is the hormone angiotensin, which stimulates aldosterone secretion by the zona glomerulosa cells of the adrenal cortex. Angiotensin is formed from a serum substrate (angiotensinogen) through the action of an enzyme, renin, which is formed in the renal juxtaglomerular cells. Although renin release is known to be increased by decreases in extracellular volume and sodium content, the exact stimulus for renin secretion is not known. One theory postulates the presence of intrarenal receptors which cause increased renin secretion in response to decreased pressure in the renal vasculature or renal tubules. Another theory postulates that the tubular fluid sodium concentration, or the sodium load delivered to a certain area of the renal tubule (the macula densa), controls renin release. Either of these theories by itself can explain most but not all of the experimental observations which have been obtained.

Extrarenal mechanisms are also involved in the control of aldosterone secretion. The rate of aldosterone output depends upon volume receptors which presumably are sensitive to extracellular volume, to some fraction of that volume such as plasma or interstitial fluid volume, to some derivative of volume such as intravascular or interstitial pressure or distention, or to cardiac output. Aldosterone secretion is increased by hemorrhage, erect standing, severe restriction of salt intake, positive-pressure breathing, and trapping of blood in the extremities

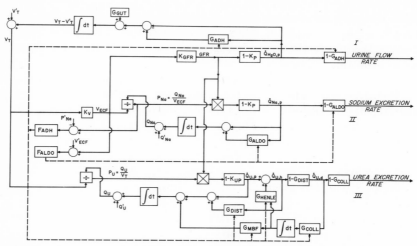

Figure 8-9. Block diagram showing some of the components of the system for control of extracellular volume and osmolarity. (Redrawn and modified from ref. 3.)

F_{ADH} = transfer function for ADH secretion

F_{ALDO} = transfer function for aldosterone secretion

GFR = glomerular filtration rate

G_{ADH} = transfer function for water reabsorption in the distal nephron (distal tubule and collecting duct)

G_{ALDO} = transfer function for sodium reabsorption in the distal tubule

G_{COLL} = transfer function for urea reabsorption in the collecting duct

G_{DIST} = transfer function for urea reabsorption in the distal convoluted tubule

G_{GUT} = transfer function for the rate of absorption of water from the intestines

G_{HENLE} = transfer function for recirculation of urea by Henle's loop

G_{MBF} = transfer function for removal of urea from interstitial fluid by medullary blood flow

K_{GFR} = transfer function relating GFR to extracellular volume

K_P = reabsorption coefficient for water and Na^+ in the proximal tubule

K_{UP} = reabsorption coefficient for urea in the proximal tubule

K_V = (extracellular fluid volume)/(total body fluid volume)

P_{Na} = plasma sodium concentration

P_U = plasma urea concentration

Q_{Na} = amount of sodium ion in extracellular fluid

Q_U = amount of urea in total body fluid

by the use of venous tourniquets. There are data to suggest that volume receptors reside in the large vessels of the chest, in the atria, in the intracranial area, in the arterial side of the circulation, and perhaps in the renal circulation itself.

Signals from the volume receptors pass to a central integrating mechanism, most likely located in the diencephalon. The integrating center initiates the secretion of aldosterone by one or more of the following efferent pathways:

1. Release of antidiuretic hormone (ADH) from the posterior pituitary. The ADH may act directly on the adrenal glands to cause aldosterone secretion. An alternate action of ADH may be to cause release of adrenocorticotrophic hormone (ACTH) from the anterior pituitary. ACTH would then stimulate aldosterone secretion.
2. The diencephalon itself may release a hormone, adrenoglomerulotropin, which stimulates the secretion of aldosterone.

The renal response to aldosterone appears to be rather slow, since an intravenous injection of aldosterone begins to affect sodium excretion only after a latency period of about an hour. Therefore, the control of sodium reabsorption by alterations in the rate of aldosterone secretion seems best suited for the day-to-day balancing of sodium intake and output, whereas control of sodium reabsorption by alteration in the rate of glomerular filtration seems better suited for compensation of acute volume depletion or expansion.

Many of the features of the control systems for extracellular volume and osmolarity, and for the interactions between the two systems, are

---

$\dot{Q}_{H_2O,p}$ = rate at which water leaves the proximal tubule in tubular urine

$\dot{Q}_{Na,p}$ = rate at which sodium ion leaves the proximal tubule in tubular urine

$\dot{Q}_{U,d}$ = rate at which urea leaves the distal convoluted tubule in tubular urine

$\dot{Q}_{U,h}$ = rate at which urea leaves the loop of Henle in tubular urine

$\dot{Q}_{U,p}$ = rate at which urea leaves the proximal tubule in tubular urine

V_{ECF} = extracellular fluid volume

V_T = total body fluid volume

The prime symbol (′) indicates an initial value.

demonstrated in the block diagram of Figure 8-9. The material balances for water are shown in the top third of the figure (Panel I). The rate of glomerular filtration is taken to be a function of extracellular volume. A constant fraction, K_p, of the glomerular filtrate is reabsorbed in the proximal tubule, with the remaining water, $\dot{Q}_{H_2O,p}$, passing to the distal tubule. The fraction of this fluid that is reabsorbed in the distal nephron (distal tubule and collecting duct) is given by the transfer function G_{ADH}, which is a function of ADH concentration. The net rate of water excretion by the kidney and the rate of water absorption from the gut (G_{GUT}) are added, and the result is integrated to give the accumulated change in total body water content at any given time.

The middle section (Panel II) of Figure 8-9 shows the material balances for sodium ion. Plasma sodium concentration, obtained by dividing extracellular sodium content by extracellular volume, is multiplied by GFR to obtain the rate at which sodium is filtered. The material balances for sodium over the renal tubule are the same form as those for water. The fraction of the total sodium entering the distal tubule that is reabsorbed, G_{ALDO}, is a function of plasma aldosterone concentration.

The renal excretion of urea is considered in the bottom section of Figure 8-9 (Panel III). Plasma urea concentration is obtained by dividing total body urea content by total body water, since urea readily distributes throughout total body water. The fractional reabsorptions of urea in the distal convoluted tubule and the collecting duct are given by G_{DIST} and G_{COLL}, respectively. Both of these transfer functions depend upon the plasma concentration of ADH, since this hormone alters the permeability of the distal nephron to urea as well as to water. A portion of this urea which is reabsorbed from the collecting duct is recirculated through the loop of Henle, while the remainder is returned to the body urea pool via the medullary blood vessels. The secretion rate for ADH (F_{ADH}) is determined by osmoreceptors which compare the plasma osmotic pressure with a reference value. In the system presented, plasma sodium concentration is used as a measure of plasma osmotic concentration, since in most cases these two values run parallel to each other. The rate of aldosterone secretion, F_{ALDO}, is determined by receptors which sense extracellular volume. The effects of a sudden water load on the system shown in Figure 8-9 have been studied using an analog computer.[3] In Figure 8-10 the effects predicted by the model for a water load of 1000 ml are compared with experimental data obtained for a normal subject. The predicted and experimental results show reasonably good agreement. Plasma osmolarity drops during the first

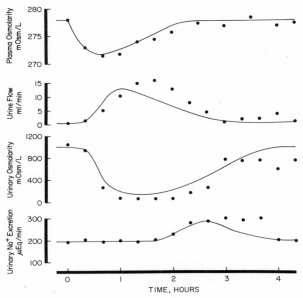

Figure 8-10. Comparison of predicted and experimentally determined responses to a rapid oral water load of 1000 ml. Responses of the
model shown in Figure 9 are shown as solid lines, while the
plotted points were determined experimentally. (Redrawn from
ref. 3.)

30 minutes while the water is being absorbed from the gastrointestinal
tract. As plasma osmolarity drops, urine flow rate increases markedly,
with a corresponding decrease in urine osmolarity. Within about three
hours, the water load has been excreted and plasma osmolarity has
returned to normal.

Although it is possible with this model to simulate reasonably well
the effects of acute changes in body fluid balance, many components
involved in longer term regulation have been omitted. For example,
the model does not include the thirst mechanism and control mechanisms for salt intake. A model of the thirst mechanism, plus additional
information on the regulation of body water content, have been presented by Reeve and Kulhanek.[4] However, a complete analysis of the
regulation of body fluid balance and osmotic concentration will require
the development of more quantitative data on the system parameters.

The Artificial Kidney

Many devices have been designed to perform the function of the kidney in patients with decreased renal function. The kidney designed by nature is very different from those which engineers have fabricated to do the same tasks. A brief discussion of artificial kidneys (hemo-dialyzers) is included here to illustrate the different approaches, and because artificial kidney design is currently an important area of bio-engineering effort.

In essence, the artificial kidney serves to return body electrolyte and water balance to normal, and to remove accumulated toxic products of metabolism. However, instead of using a number of specific transport mechanisms with intricate feedback control loops, the hemo-dialyzer operates on the principle of equilibration (or near equilibration) of blood passing through the dialyzer with a dialysis solution which has the desired solute composition. The equilibration is achieved by recirculating the blood from the patient through the machine in such a manner that the blood is separated from the dialysis fluid by only a thin semipermeable membrane. For example, in renal failure, potassium ion concentration tends to increase because of inadequate renal excretion. If it is desired to remove potassium from the patient's blood, the dialysis fluid is formulated with a low potassium concentration so that potassium will diffuse from blood into the dialysis fluid. In a patient with decreased urine formation, excess body water may be removed by making the dialysis bath hypertonic to plasma, so that there is a net filtration of water from plasma to the bath. Water may also be extracted from the blood by increasing the hydrostatic pressure in the blood flow channel, thus increasing filtration pressure. The examples cited here depend upon an adequate diffusion gradient, which means that either a very large dialysis bath must be used or the bath fluid must be changed very frequently.

The ideal hemodialyzer would have the following properties[5]:

1. The dialyzer must be efficient in the removal of nitrogenous and toxic products of metabolism. A measure of relative efficiency which is useful for the comparison of various hemodialysis systems is the dialysance

$$D = \frac{A - V}{A - B} \times BFR \qquad (22)$$

where D is the dialysance of a specific solute, A is the arterial or inflow concentration of that solute, V is the venous or outflow concentration, B is the bath concentration, and BFR is the blood flow rate through the dialyzer.

2. The dialyzer must be capable of removing water from a patient who is excreting insufficient urine.
3. The dialyzer should have a low internal volume (in the blood-conducting passages), eliminating the need for priming with blood, minimizing the possibility of fluid shifts during dialysis, and reducing blood loss at the end of dialysis.
4. It is desirable for a dialyzer to have a low internal resistance to flow, if blood is to be propelled through the system by arterial pressure alone. A blood pump can be inserted in the circuit, but this introduces additional complications.
5. The dialyzing membrane and the connecting tubing should be presterilized, simple to assemble, and inexpensive enough to be disposable.
6. The dialyzer should be reliable, safe, and require little attention during use.
7. It should be inexpensive and have low maintenance costs.

Although none of the dialyzers currently in use meets all of these requirements, several designs have proved to be reasonably satisfactory in chronic hemodialysis programs. One of the most widely used is the Kolff disposable coil kidney, for which a flow diagram is shown in Figure 8-11. Blood is pumped from an arm artery by a roller pump "E" through a twin coil dialyzer "A" at the rate of 200 to 400 ml per minute. The internal structure of the dialyzer coil and its membrane supports are shown in Figure 8-12. Dialysis fluid is recirculated past the dialyzer membrane at the rate of 10 to 30 liters per minute. A representative formula for the dialyzing fluid is given in Table 8-5. An average dialysis might take six hours, with the bath being changed once or twice during that time.

In a modification of the above system, the recirculating single-pass dialyzer, a small amount of dialysate fluid is continually removed from the system and replaced with fresh dialysate.

Numerous other types of hemodialyzers are available, including several which operate in parallel counterflow arrangements in which the blood flows between flat sheets of cellophane and the dialyzing fluid passes along the outside of the membrane (Fig. 8-13). Dialyzers of this type have the advantage that the blood flow channels offer a very low resistance to flow, thus eliminating the requirement for a blood pump. Ultrafiltration in these units may be accomplished by placing the dialysate side of the membrane under negative pressure.

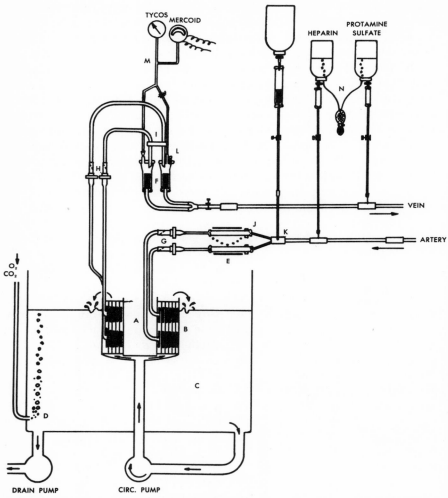

Figure 8-11. Flow diagram for the Kolff disposable twin coil kidney. (Courtesy of Travenol Laboratories.)

For small installations, dialysate may be recirculated through each dialyzer from individual supply tanks. In large hemodialysis centers, all dialyzers may receive dialysate fluid from a central supply. In that case the fluid is discarded after a single pass through the dialyzer.

Ideally, an artificial kidney would be easily portable so that dialysis could be done continuously. Extensive efforts have been made to optimize the geometrical arrangement, and to improve the efficiency of artificial kidneys in efforts to reduce both their size and their cost.[6] However, reducing the size of the mass exchanger is only a small part of the problem. Any artificial kidney which operates on the principle of dialysis alone will require large volumes of dialysis fluid to maintain adequate transmembrane concentration gradients. The amount of dialysate required could be markedly reduced if solutes could be selectively adsorbed from the blood, simulating the active transport processes which occur in the natural kidney. However, finding suitable adsorbents is very difficult because of the variety of materials which must be removed from the blood, and because of the fact that not all of these

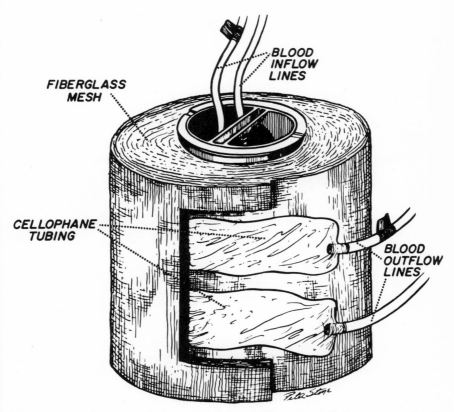

Figure 8-12. The internal structure of the Kolff twin coil dialyzing unit. (From ref. 5.)

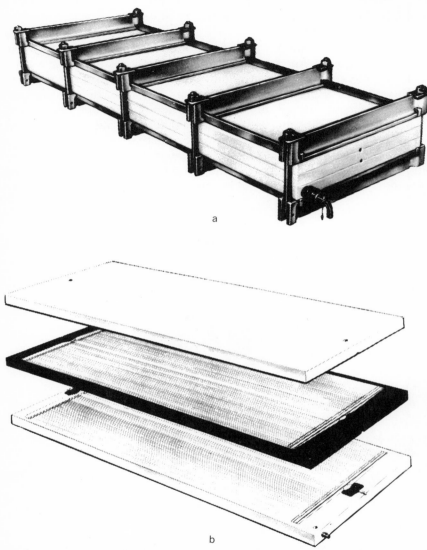

a

b

Figure 8-13. a. Two-layer Sweden-Kiil hemodialyzer. b. Exploded view showing the grooved plastic boards with two sheets of cellophane
membrane between each set of boards. Dialysate fluid flows
through the grooves of the boards outside the cellophane. Blood
flows between the cellophane sheets in a direction opposite to
that of dialysate flow. The boards are pressed together with the
external clamps, sealing the cellophane with rubber gaskets.
(Courtesy of Seattle Artificial Kidney Supply Company.)

Table 8-5. Composition of Dialysis Fluid

Component	Gm/100 L
NaCl	570
NaHCO$_3$	300
KCl	30
CaCl$_2$	28
MgCl$_2$	7.5
Invert sugar	400

substances have been identified. Attempts using materials such as activated carbon or processes such as electrodialysis have met with only slight success.

References

1. Crank, J.: *The Mathematics of Diffusion.* Oxford Univ. Press, London, 1956.

2. Wiederhielm, C. A.: *Dynamics of transcapillary fluid exchange.* J. Gen. Physiol. 52:29, 1968.

3. Koshikawa, S., and Suzuki, K.: *Study of osmo-regulation as a feedback system.* Med. Biol. Engrg. 6:149, 1968.

4. Reeve, E. B., and Kulhanek, L.: Regulation of body water content: a preliminary analysis, in Reeve, E. B., and Guyton, A. C.: *Physical Bases of Circulatory Transport: Regulation and Exchange.* W. B. Saunders, Philadelphia, 1967, pp. 151–177.

5. Freeman, R. B., Maher, J. F., and Schreiner, G. E.: *Hemodialysis for chronic renal failure. I. Technical considerations.* Ann. Intern. Med. 62:519, 1965.

6. Wolf, L., Jr., and Zaltzman, S.: *Optimum geometry for artificial kidney dialyzers.* Chem. Engrg. Prog. Symposium Ser. 64:104, 1968.

Suggestions for Further Reading

7. American Physiological Society: *Symposium on neural control of body salt and water.* Fed. Proc. 27:1127, 1968.

8. Dedrick, R. L., Bischoff, K. B., and Leonard, E. F. (eds.): *The artificial kidney.* Chem. Engrg. Prog. Symposium Ser. 64:1, 1968.

9. Dick, D. A. T.: *Cell Water.* Butterworths, Washington, 1966.

10. Jacquez, J. A., Carnahan, B., and Abbrecht, P.: *A model of the renal cortex and medulla.* Math. Biosci. 1:227, 1967.

11. Levine, S. N.: *A model for renal-electrolyte regulation.* J. Theor. Biol. 11:242, 1966.

12. New York Heart Association: *Biological Interfaces: Flows and Exchanges.* J. Gen. Physiol. 52:1, 1968.

13. Pitts, R. F.: *Physiology of the Kidney and Body Fluids,* ed. 2. Year Book Medical Publishers, Chicago, 1968.

14. Reeve, E. B., and Guyton, A. C. (eds.): *Physical Basis of Circulatory Transport: Regulation and Exchange.* W. B. Saunders, Philadelphia, 1967.

15. Stein, W. D.: *The Movement of Molecules across Cell Membranes.* Academic Press, New York, 1967.

PART THREE
Instrumentation

9

Examples of Transducers and Amplifiers Used in Research

One of the greatest problems encountered in biomedical engineering is that of making quantitative measurements of a living organism. In some instances, instruments developed primarily for use in industrial process control have been adapted for biological use; however, in most cases, these do not accomplish the task as efficiently as those designed especially for the problem at hand.

It is the purpose of this chapter to describe, in broad terms, instrumentation systems for use in biomedical engineering research with particular emphasis on problems associated with the input transducers.

General Characteristics of Instrumentation Systems

All instrument systems are composed of three basic elements: (a) a transducer, (b) a signal-processing unit, and (c) an output device. Figure 9-1 illustrates schematically such an instrument system. These instrument systems may be further classified according to (a) type of input, (b) sensitivity and dynamic range, and (c) effect of instrument on system being measured.

The input to the instrument may be a mechanical pressure, acceleration, electrical current, acoustic energy, or some form of visible or penetrating radiation. A characteristic of utmost importance in biological instrumentation is the effect on the system of taking the

Figure 9-1. Schematic representation of instrument system.

desired measurement. This may be termed the impedance of the instrument and depends, for its exact value, on whether the mechanical parameters, electrical parameters, or chemical parameters of the biological system are being measured. For example, if one is to determine the air flow velocity associated with a respiratory problem, care must be taken to assure that the instrumentation used to measure the flow of velocity does not obstruct the normal respiratory channels, and thus produce a condition that is not encountered in the normal procedure of breathing. Another example would be in the measurement of the potentials of single nerve fibers. Here, the geometry of the fiber dictates the use of an extremely fine probing electrode. The electrical impedance of the probe itself is of such a high value that, unless special precautions are taken in the design of the processing amplifier, the results obtained are misleading.

A most important factor in the overall performance of the instrument is the relationship between the output and the input to the instrument. This relationship is termed the transfer function. Many instruments have a limited range over which they exhibit a linear relationship between the input and the output signals. In those situations in which a wide dynamic range of input signals is anticipated, the signal processor is designed so that the output is related to the input by a logarithmic function.

Intimately associated with the relationship between the output and the input is the total instrument error, which is the sum of the errors of the individual elements. In some instances, the instrument may be designed such that the individual elements have errors in opposite directions, so that the error in one portion of the overall instrument will be offset to some extent by the error in the other components within the instrument.

Each instrument has a lower limit of sensitivity. In a properly designed instrument, this is set by statistical variations of the parameter being measured. In many situations, the input transducer sensitivity is such that the lower limit of sensitivity is set by the extraneous noise within the signal-processing equipment, as discussed in the section on amplifiers.

The other limit of importance in instrumentation is the saturation point. This is the point at which a further increase in input does not result in an equal increase in the output signal. This saturation point is determined by a variety of conditions existing in the input transducer, signal-processing, or output portions of the instrument.

In summary, the instrument should be designed to be an integrated unit whose overall performance is tailored to the type of measurement desired. Once the application has been determined, the input transducer, the processing amplifier, and the output devices may be selected from a wide range of alternates. Alternate and competing methods of making the same type of measurement will be discussed, in order that the student may have available a variety of possible solutions to the problem at hand.

TRANSDUCERS

Transducers may be classified according to the type of input energy to which they are sensitive. The broad categories under such a classification include: (a) mechanical, (b) temperature, (c) magnetic, (d) electrical, and (e) radiation. In biomedical instrumentation, all five of these classifications are utilized to some degree. The types most widely used, however, would be those classified as mechanical and electrical.

Mechanical-Input Transducers

In the broadest sense, the mechanical-input transducer is designed to measure some type of displacement. This ability to measure displacement is utilized in the transducer proper to permit the measurement of a variety of parameters, e.g., acceleration, flow, or weight.

The electrical methods used for the measurement of a linear displacement may be divided into (a) variation-of-resistance methods, (b) variation-of-inductance methods, and (c) variation-of-capacitance methods. Other means of measuring displacement involve the absorption of radiation, or the measurement of the wavelength and mechanical vibrations encountered in the propagation of ultrasound.

In methods now used, (a) any displacement results in a variation in the electrical resistance of the element proper. In any material, the value of electrical resistance is given by

$$R = P\frac{l}{A} \tag{1}$$

where R = resistance of element in ohms
P = resistivity of element in ohms per unit volume
l = length of element in cm
A = area of element in cm^2

In Eq. (1) it may be seen that three parameters control the measured resistance of the device: the resistivity of the material, which is a strain-related function of the number of free carriers within the material; the length of the material; and the cross-sectional area of the material.

A technique in widespread use is that of measuring variations in effective length by means of a "sliding-wire resistance." The most common form is a three-terminal device in which one of the terminals is moved relative to the two terminals at the ends of the conductor. The principal disadvantage of such a system is the amount of effort required to move the sliding contact across the surface of the resistance element.

Another type of resistance variation transducer that is rather widely used employs a liquid conductor, such as mercury, contained in an extensible structure such as rubber tubing. As the length or degree of stress on the rubber tubing changes, the cross-sectional area of the conductor changes; thus, a variation in resistance is obtained. To a lesser degree, this is the same principle used in what is termed a "strain gauge." In a strain gauge, stress on a conducting metal, such as constantan, causes a change in the resistivity of the material. The behavior of such a strain gauge is commonly expressed by

$$\frac{\Delta R}{R} = S\frac{\Delta L}{L} \tag{2}$$

where L is the length and R the resistance of the unstrained wire, ΔL is the change in length, ΔR the change in resistance resulting from the applied stress, and S is a dimensionless quantity termed the gauge factor. Typical values for S are given in Table 9-1.

One of the difficulties with the strain gauge is that the measured resistance of the material depends not only on the degree of strain to which the material is subjected, but also upon the temperature of the material. The effective resistance as a function of the temperature of the material is given by Eq. (3). Typical values for the resistance temperature coefficient α are given in Table 9-2.

Table 9-1.   Strain Sensitivity S for Different Materials

Material	Composition	S
Manganin	Cu: 84% Mn: 12% N: 4%	+0.4
Nickel	Pure	+12.1
Si	Single crystal	100
Cds FET	Cds-Cw	4000

**Table 9-2.   Resistance Temperature
Coefficients at Room Temperature**

Manganin	0.0001
Nickel	+0.0067
Platinum	+0.00392
Carbon	−0.0007
Semiconductors⎫ Thermistors ⎭	−0.068 to +0.14

$$R_t = R_0(1 + \alpha T) \tag{3}$$

where R_t is the resistance of the element at temperature T, R_0 is the resistance of the element at temperature T_0, α is the temperature co-efficient at T_0, and T is the change in temperature.

In some applications, the change in resistance due to temperature effects will mask the change due to strain. Fortunately, in most bio-logical applications, wide variations in the temperature of the trans-ducer normally are not encountered.

With the advent of the semiconductor technology, a new form of strain-sensitive element has become available. The ones most com-monly used are single fibers of elemental silicon. These behave in essentially the same manner as do the strain-sensitive resistance wires; however, the gauge factor S of Eq. (2) for such devices is in the range of 100 to 300. Within the past year, a strain-sensitive field effect trans-ducer has been reported which promises to have a strain sensitivity factor S approaching the order of 1000 to 5000. The sensitivity of this new device will make possible many transducers which heretofore have not been feasible due to the limited sensitivity of the strain-sensitive devices.

In general, then, the change of resistance associated with the three parameters given in Eqs. (1) and (2) may be used to indicate a physical displacement or change in length.

Variable-Inductance Transducers

A widely used technique for measuring changes in physical length is that associated with a change in the inductance of a coil due to variations either in the magnetic path or in the position of the material constituting the magnetic path. Figure 9-2a illustrates a system wherein the reluctance of the magnetic path is varied by the movement of a member in the magnetic path in such a manner as to change the air

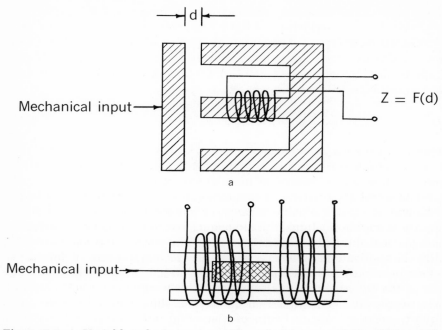

a

b

Figure 9-2. a. Variable reluctance transducer. b. Linear variable differential transformer.

gap existing in the path. An extension of this technique is commonly known as the linear variable differential transformer (LVDT). This device, shown in Figure 9-2b, is able to measure displacements in the order of microinches.

The chief advantage of the inductance distance-measuring technique is that the devices are inherently rugged and exhibit a relatively low impedance. Several methods are available for measuring the change in inductance; the most common is to measure the inductive reactance at a given frequency. This inductive reactance may then be related to the linear distance which the transducer is measuring. The LVDT has the advantage that, if properly designed, the output of the device is zero at dead center, while on either side of center the output is not only a varying potential but, by using a phase reference, the side of the null or zero point on which the measurement is taken may be determined. As nearly as can be ascertained at this time, the majority of the biological transducers utilizing inductance variation today use some form of the LVDT.

Capacitor Transducers

The capacitance displacement transducer is unquestionably the most sensitive device among the various displacement transducer methods. It is based on the capacity of two conducting plates affected by the distance between the plates:

$$C = 8.85 \times 10^{-2} \in \frac{a}{d} \tag{4}$$

where C is the capacitance in picofarads, $\in$ is the dielectric constant of the medium between the plates, a is the area of the plates in cm^2, and d is the spacing between the plates in cm.*

The extreme accuracy of the capacitor-type transducer is due to the accuracy with which frequency may be measured by utilizing readily available equipment. Through the use of the capacitor element as a portion of a frequency-determining element of a resonant circuit, with subsequent measurement of the resultant frequency, displacements in the order of a few microinches may be readily measured. If one of the capacitor plates is formed by means of a thin metallic film evaporated on a plastic membrane, then it may be used to measure extremely small pressure differentials.

The fact that the capacitance for a given area and distance between the plates is dependent upon the dielectric constant may be utilized to extend the utility of the capacitor transducer to the measurement of other systems in which a variation in dielectric constant is encountered. An example of this is the change in the dielectric constant of a gas with the composition.

Velocity or Acceleration Transducers

The output of the mechanical displacement transducer described above can be processed to give electrical signals that are proportional to the velocity, the acceleration, or the higher-order time derivatives of the displacement. Such a system is shown schematically in Figure 9-3b.

An alternate method is to produce a displacement by mechanical means that is proportional to a velocity or an acceleration, and convert this displacement into an electrical signal. This is easily accomplished by use of a mass which, under the influence of acceleration, will

---

*This expression neglects the fringing effect, which is not encountered when the spacing d is small compared to the length or diameter of the conducting plates.

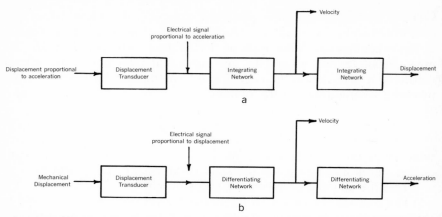

Figure 9-3. Schematic representation of circuit for obtaining output signals proportional to velocity and acceleration from a displacement transducer.

exercise a force and cause a displacement proportional to the acceleration. The force or displacement can then be converted into an electrical signal using one of the displacement transducers described above. Such a signal may be converted into a velocity-proportional signal by the use of an integrating network as shown in Figure 9-3a. When the displacement transducer is coupled with a mass to form a device whose output is proportional to the acceleration, the device is termed an "accelerometer."

In some instances, the motion-sensing element of the accelerometer may be a piezoelectric material. Such a material is usually a ceramic which has the property of producing an electric potential that is proportional to the degree of strain in the material. In certain crystals with a symmetrical charge distribution (quartz, for example), the lattice deformation is, in effect, a relative displacement of the positive and negative charges within the lattice. The displacement of the internal charges will produce equal external charges of opposite polarity on the opposite sides of the crystal. This is termed the "piezoelectric effect." These charges can be measured by applying electrodes to the

Table 9-3. Piezoelectric Constants

Material	Charge sensitivity d $coulomb/m^2/newton/m^2$	Voltage sensitivity g $volt/m/newton/m$
Quartz	2.25×10^{-12}	0.055
Barium Titanate	160×10^{-12}	0.01

surfaces and measuring the potential difference between them. The magnitude and polarity of the induced surface charges are proportional to the magnitude and direction of the applied force as given by

$$Q = dF \tag{5}$$

where Q is the surface charge in coulombs, d the piezoelectric constant given in Table 9-3, and F the applied force in newtons per m^2.

The charge at the electrodes gives rise to a voltage given by Eq. 6. Typical values for the voltage sensitivity of various piezoelectric materials are given in Table 9-3.

$$E_0 = gtP \tag{6}$$

where E_0 is the voltage output of the transducer, g is the voltage sensitivity of the material (Table 9-3), t is the thickness of the material, and P is the applied pressure.

Ceramic materials such as barium titinate are widely used in accelerometers normally encountered in the biomedical field because of ease of fabrication. The material is fabricated as a sintered ceramic which has the internal polarization applied after fabrication. This permits the optimization of the geometric and electrical properties of a given piece of material. Inasmuch as piezoelectric materials are dielectrics, a steady value of strain does not produce an output voltage. This is most easily seen by an examination of the equivalent circuit of a piezoelectric element, as given in Figure 9-4. Since the potential source is coupled to the load through a capacitance, care must be taken that the load resistance is sufficiently high to assure that the capacitive reactance of the element itself at the frequencies involved does not reduce the output voltage. The requirement for high-input impedance of the associated amplifiers is readily handled through the use of field effect transducers as the input stages.

In summary, measurements of velocity and acceleration are normally handled by an appropriate mechanical assembly coupled to a displacement transducer. The latter serves to measure the amount of the displacement resulting from the force on the coupled mass due to acceleration or velocity of that mass.

Pressure Transducers

The pressure transducers most commonly encountered in biomedical instrumentation usually consist of a mechanical element that is subjected to the pressure differential. The displacement of this element as a result of pressure is measured by one of the displacement trans-

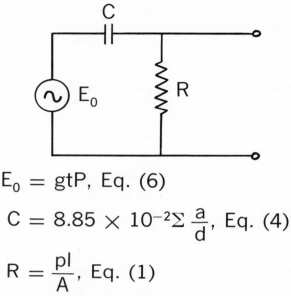

$$E_0 = gtP, \text{ Eq. (6)}$$

$$C = 8.85 \times 10^{-2}\Sigma \, \frac{a}{d}, \text{ Eq. (4)}$$

$$R = \frac{pl}{A}, \text{ Eq. (1)}$$

Figure 9-4. Equivalent circuit of piezoelectric element.

ducers described above. As was stated above, the principal advantage of the inductance variation or LVDT transducer is that it has a low impedance output and good sensitivity to small displacements.

In some instances, the pressure differential encountered may be so small as to require the sensitivity of the capacitor displacement transducer. An example of this is the Golay cell in which the differential pressure, resulting from the absorption of infrared radiation in a given volume of gas, is measured by the change in capacitance of a transducer consisting of a flexible membrane and fixed plate.

Certain piezoelectric crystals, if exposed to pressure from all sides, will develop polarization in a preferred crystal direction giving rise to an output voltage. Generally speaking, the voltages produced at the pressures normally encountered in biological measurements are so low as to make such a device relatively impractical. A further disadvantage of the piezoelectric device is that the output resulting from a steady pressure is zero.

The majority of pressure transducers used for biological measurements consist of a diaphragm which is subjected on one side to the pressure to be measured; the other side of the diaphragm is normally coupled to the atmosphere. The deformation of this diaphragm is measured by a displacement transducer. In all the basic displacement transducers described above, with the exception of the piezoelectric crystal, the steady-state displacement may be readily ascertained; hence, the frequency response of the transducer will be from DC to an upper frequency limit determined by the mechanical properties of the pressure-measuring diaphragm and of the mechanism that drives the displacement transducer. In many cases, the frequency response of the transducer proper is far in excess of that associated with the diaphragm; thus, the mechanical properties of the transducer become the limiting factor on frequency response.

The Measurement of Flow Velocity with Displacement Transducers

Electrical displacement transducers in combination with mechanical systems can be used to measure flow velocity. The flowing medium causes a pressure difference in a tube or restricted orifice which is measured with one of the pressure transducers described above. Another method is to have the flowing medium displace a float in proportion to the flow, or to cause the rotation of a mechanical assembly which is converted into an electrical signal by means of a displacement transducer. In general, however, mechanical displacement transducers are not widely used as a means of measuring flow velocity in biological research, because such devices normally disturb the flow patterns to a greater degree than do the other methods to be described in the following section.

Resistance Measurements of Liquid Flow Velocity

A method of measuring linear flow velocity which has been used in some biological investigations is termed the "electrolytic tracer method." This method, suitable only for discontinuous measurements of linear flow velocity, is shown schematically in Figure 9-5. It is essentially a transit-time measurement and, since the body fluids normally have a fairly high conductivity, can be applied to biological measurements without too much difficulty. Errors in the system are due to the fact that the conductance discontinuity existing in the system depends on a large number of factors that are not easily controlled. The method assumes a uniform concentration throughout the cross section of the vessel through which the liquid is flowing, which is not encountered in actual practice.

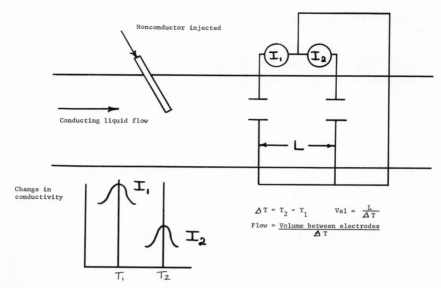

Figure 9-5. Flow velocity measurement system utilizing electrolytic tracer method.

A second method, utilizing a change in conductivity, involves the introduction at a constant rate of a liquid having a conductivity differing from that of the main stream. After mixing, the concentration of the two liquids can be determined at a point downstream by conductivity measurement (Eq. (7)). This method has not been widely used.

$$C_2 = C_1 \frac{F_1}{F_2} \tag{7}$$

where C_1 is the initial concentration of salt introduced into the flowing medium, C_2 is the concentration measured downstream, F_1 is the flow rate of concentration C_1, and F_2 is the unknown flow rate.

Liquid Flow Measurements by Induction Transducers

Probably the most widely used flow-measuring device in biological research is the electromagnetic flowmeter. This technique is particularly applicable to the measurement of the flow of body fluids, especially blood, which is an excellent conductor. A practical configuration for the electromagnetic flowmeter is shown in Figure 9-6. The vessel in question lies in a magnetic field of flux density B. Two electrodes are inserted into the vessel walls, their surfaces arranged to be flush with the inner surface of the vessel. When the liquid flows through the tube with an average velocity V, charges within the liquid are moved to the electrodes, and a potential difference arises between them equal to the value given by Eq. (8):

$$E = \frac{BF}{d} \frac{4}{\pi} \times 10^{-8} \text{ volts} \tag{8}$$

where B is the flux density in gauss, d is the distance between the electrodes in cm, and F is the average flow of liquid in milliliters per second.

The magnetic field is usually in the order of 1000 gauss and may be either a continuous or an alternating field. Since the output voltages are derived from the magnetic field, there will be a DC voltage if a direct current field is used or an AC voltage in an alternating field. In the majority of the instruments used, an alternating field is utilized to prevent electrolytic polarization at the electrodes. An added advantage of AC excitation is that amplifiers may be AC coupled, thus eliminating the drift normally associated with DC amplifiers. One of the principal difficulties with the use of an AC magnetic field is the "shorted-turn" effect resulting from the conductivity of the liquid whose flow is being measured. This effect produces a voltage in the output signal that is independent of the flow and displaced 90° in phase from the flow signal. Some of the instruments eliminate this signal through electrical bridge-balancing methods, others use a timed gating.

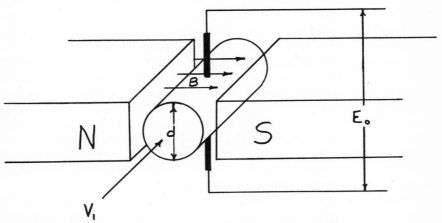

Figure 9-6. Schematic representation of electromagnetic flow meter.

Probably the simplest method of eliminating the signal is to energize the magnetic field in such a manner that a square wave is produced. Thus, the nonflow-dependent signal, which is proportional to the rate of change of the magnetic field, occurs only at those times when the magnetic field is changing, and hence may be readily gated out.

The advantages of the electromagnetic flowmeter are that it is reliable, simple, and extremely rugged. There are no moving parts or constrictions imposed on the vessel being measured. The response is fast and linear, and the output is independent of the physical properties of the liquid, assuming, of course, that the electrical conductivity of the fluid is sufficient to eliminate effects of the internal impedance of the associated amplifiers. The chief disadvantage of the method is that the output voltage generally is small, requires amplification, and, unless care is taken in regard to the magnetic circuit, spurious voltages may be introduced that are dependent upon the geometry between the vessel whose flow is being measured and the magnetic probe.

Flow Velocity Measurements by Thermal Methods

The use of the thermal characteristics of a liquid to determine its flow finds wide application in biological systems. That is, the thermal regulation of many biological systems is such that relatively small temperature differentials may be introduced and readily discerned.

Several methods have been reported for the use of thermal gradients in flow measurement. In the determination of the flow output of the heart a known amount of solution having a given temperature difference from that of the fluid within the heart is introduced into the system. The subsequent mixing on this known amount of liquid at a given temperature with the larger volume of fluid contained in the heart produces a change in the temperature of the fluid at the output of the heart. From a measurement of this change, one may readily calculate the total volume of fluid involved. This technique, known as thermal dilution, has been quite widely investigated and found to be sufficiently accurate for use in diagnostic procedures. Such a thermal dilution technique is useful as it is often sufficient to withdraw a known amount of blood, permit it to cool to room temperature, and then reinject it. Thus, the organism itself is not in any way disturbed by the addition of a foreign substance to the bloodstream.

Another method of measuring flow utilizes the cooling effect of the flowing liquid on an object that is heated to a temperature above that of the liquid. A thermistor, which is an extremely sensitive indicator of temperature described in the following section, is used as the temperature-measuring device. A given amount of heat is applied to the thermistor, causing its temperature to rise. The amount by which the thermistor increases in temperature for a given heat input is dependent upon the amount, and hence the velocity, of the flowing fluid. The thermistors have response times in the order of a fraction of a millisecond; hence, instruments designed along these lines are capable of reproducing the pulse profiles encountered in biological systems.

The same heated-transducer technique has been applied to the measurement of gas flow in which the gas constitutes the cooling medium for the thermistor. Advantage is taken of the fact that the thermal conductivity of the gas is dependent upon its composition; thus, a heated object in the stream of gas having a constant flow will experience various degrees of cooling and, hence, a change in resistance dependent upon the thermal conductivity of the gas. Such a principle is very widely used in the analysis of the effluent streams of gas chromatographic instruments. This is essentially a flow-measuring system if one considers that the measurement of flow is dependent not only upon the amount of material passing the object per unit time, but also upon its thermal conductivity. Thus, for a fixed amount of flow, a variable thermal conductivity can be used to indicate the composition of the fluid, whether gas or liquid, that is flowing past the temperature-sensing device.

Sonic Methods of Flow Measurement

Although the electromagnetic flowmeter is undoubtedly the most widely used in biological systems, the use of sonic methods for measuring flow has increased due to the introduction of the Doppler flowmeter. This instrument utilizes the change in frequency that is associated with the velocity of the liquid through which the sound is propagating (Eq. (9)).

$$F_d = \text{Frequency difference in hertz (Hz)} = \pm \frac{2V\ F_0\ \text{cosine}\ \theta}{C} \quad (9)$$

where V is the flow in cm/sec, F_0 the oscillator frequency in Hz, θ the angle between the ultrasound beam and the direction of flow, and C the velocity of sound in the flowing medium.

A typical instrument is shown as Figure 9-7.

The commonest form of the Doppler meter is the back-scatter device, which has the advantage of requiring only one contact to the organism whose flow measurement is desired. Another use for the Doppler flowmeter is in the determination of the relative movement

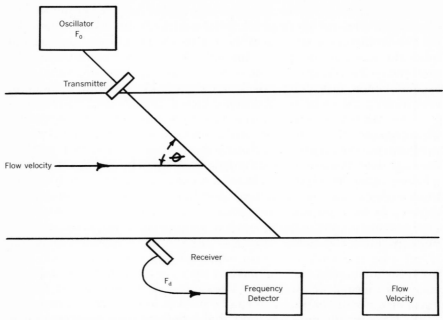

Figure 9-7. Schematic representation of ultrasonic Doppler flow meter.

of various organs within the body, which is directly related to a change in the frequency of the reflected sound. A major limitation to the use of the Doppler flowmeter, however, is the inability to accurately quantitate the measurements obtained.

Temperature transducers may be further classified into those indicating temperature by a change in mechanical properties, and those using electrical properties. For example, a liquid such as mercury, or a combination of metals forming a bimetal strip, may be used. A change in the physical parameters of these materials is utilized to produce a visible or measurable deflection which is dependent upon the change in properties of the material with temperature.

In some experiments, an electrical output directly related to temperature is desired in preference to a mechanical indication such as that obtained by a thermometer. For these situations, an element whose electrical properties are a function of temperature is usually used as the transducer. Within the past five years, the use of semiconducting elements termed "thermistors" has become so widespread that for all practical purposes they are the only ones encountered in biomedical instrumentation.

Characteristics of Thermistors. The term thermistor is applied to elements having a high negative temperature coefficient of resistance. As the temperature increases, the resistance goes down and as the temperature decreases the resistance goes up. This characteristic is opposite to the effect of temperature changes on metals. Thermistors are semiconductors of ceramic material made by sintering mixtures of metallic oxides such as manganese, nickel, cobalt, copper, iron and uranium. Although these materials and their semiconductor character-istics have been known for nearly 150 years, only within the past ten years have techniques of producing thermistors become well enough developed to permit production of reproducable and stable units. Various mixtures of these metallic oxides are formed into useful shapes. The electrical characteristics may be controlled by varying the type of oxide used and the physical size and configuration of the thermistor. The standard forms that are available are beads, probes, discs, washers and rods.

Resistance-Temperature Characteristics of Thermistors. The resist-ance of a thermistor is solely a function of its absolute temperature. Some electrical power is dissipated within the thermistor when meas-urement is made of its resistance. Unless care is taken to limit the power introduced to the thermistor during measurement, the temperature observed will not be that desired, but rather a temperature that is the sum of the ambient environment plus that due to internal heating of the thermistor by the current used to measure its value. When care is taken to assure that the power of the measuring circuit is of no consequence, the measured resistance of the thermistor is termed R_0, which means that the resistance is essentially at zero power.

The mathematical expression relating the resistance and the abso-lute temperature of a thermistor is

$$R_0(T) = R_0(T_0)e^{B(1/T - 1/T_0)} \tag{10}$$

where $R_0(T)$ is the resistance at temperature T, $R_0(T_0)$ is the resistance at absolute temperature T_0, e = 2.718, and B = 4000.

The temperature coefficient of resistance of a thermistor is ex-pressed by

$$A = \frac{1}{R_0} \frac{dR_0}{dT} \text{ ohms/ohm/}^\circ C \tag{11}$$

The value A is an important indicator of the sensitivity of the thermistor to rather minute temperature changes. Typical values for A with readily available materials are in the order of 6 percent per degree centigrade. This compares with the value in the order of 0.4 percent for metals normally used as temperature-indicating devices, e.g., platinum.

Temperature Measurement Using Thermistors

As a result of the extremely high sensitivity of the resistance of the thermistor to temperature changes, relatively simple circuits may be employed to produce an electronic thermometer. Typical circuits are shown in Figure 9-8, where it is evident that, if great accuracy is not needed, temperature measurement utilizing a thermistor requires only a battery and meter. If more accuracy is necessary, however, the bridge circuits shown in Figures 9-8b and c should be used. Those in Figure 9-8c utilize two thermistors and permit accurate temperature differential measurements. If the two thermistors are placed in different locations, the imbalance of the bridge will be dependent upon the temperature difference between the two thermistors. Such a technique is very widely employed in the use of thermistors as flow indicators where, to obtain maximum sensitivity of the technique, as described above, accurate measurement of the temperature of the liquid is necessary prior to the introduction of the cooled liquid.

The circuits shown in Figure 9-8 may also be used as a thermal conductivity instrument. If the two thermistors mounted in the bridge are chosen so that enough current flows to heat them to about 150°C, they may be used for the measurement of various physical phenomena. Two thermistors placed in separate small cavities in a brass block so that gas in the cavities may be circulated become a gas analyzer. If the bridge is balanced with the same gas, and if the air in one cavity is then replaced by carbon dioxide, the bridge will be unbalanced because the carbon dioxide has a lower thermal conductivity than air, and that thermistor will become hotter and lower in resistance and the instrument may therefore be calibrated to read in percentage CO_2 in air. A similar calibration can be devised for any other mixture of two gases. Such an instrument has been made, without using amplifiers, to give a full-scale reading of 0.5 percent carbon dioxide in air. If the same bridge is made with one thermistor sealed in a cavity of a brass block, and the other mounted in a small pipe, the instrument may be used as a flowmeter. When no air is flowing through the pipe, the thermistor is cooled and its resistance increases, which unbalances the bridge. This is similar to the flow device described in the section above. Instruments made along these lines can be made to measure flow rates as low as 0.001 ml per minute. Due to the wide available change in resistance of the thermistors, flow rates over a range of 100,000 to 1 can be measured.

Power Considerations of Thermistors. The accuracy of temperature measurement obtainable by a thermistor depends upon the method used to measure the resistance of the thermistor. In the simple circuitry, the resistance is normally measured by passing a current through the thermistor. This results in a heating of the thermistor element, with subsequent inaccuracy of the measurements. Typical dissipation constants for thermistors are in the order of 1 milliwatt per degree centigrade, i.e., for each milliwatt internal heating induced in the thermistor by the measuring current, the temperature of the thermistor rises 1°C. Thus, if an accuracy of 0.1°C is desired from that transducer, the associated electronic circuitry to measure its resistance value must not result in a power dissipation in the transistor exceeding 0.1 milliwatt.

Although other methods are available for the measurement of temperature in biological systems, the thermistor, due to its extreme sensitivity, ready availability, ruggedness, and simplicity of associated instrumentation, has essentially displaced all other types of temperature transducers in the biological field.

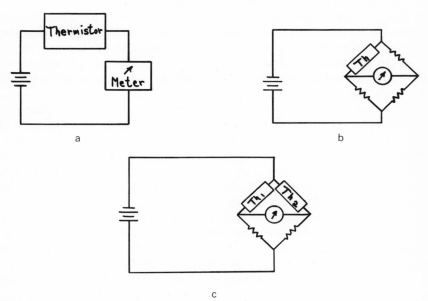

Figure 9-8. Equivalent circuit of active thermistor devices.

Radiation Transducers

The field of radiation transducers may be divided into three categories: (a) transducers responding to visible radiation, (b) transducers responding to infrared and ultraviolet radiation, and (c) transducers responding to ionizing radiation such as α, β, δ, and x-radiation. All these forms of radiation are used in biomedical engineering research instrumentation.

Optical Radiation Transducers. Optical or visible-radiation transducers may be subdivided into (a) those whose operation is dependent upon a photoemissive surface, (b) those whose action is based on a change in the conductivity of a material, (c) those in which the absorption of the radiation produces a voltage, and (d) those in which the absorption of radiation produces a temperature change that is related to the intensity of the incident energy absorbed.

The optical transducers whose basic conversion layer is a photoemissive surface consist of an evacuated chamber in which a semi-metallic surface is deposited. This surface, composed of an alkali metal-silver-oxygen compound, serves to eject an electron into the evacuated space when radiation is absorbed in the layer. In all modern instrumentation, an electron multiplier is combined with the photoemissive surface, forming a "photomultiplier tube."

The sensitivity of the photomultiplier tubes is such that they are capable of detecting a relatively small number of electrons ejected from the photoelectric surface. The use of the intimately coupled electron multiplier structure assures that the original signal-to-noise ratio established by the photoemissive surface is not degraded. Response time of the electron multiplier is in the order of 10^{-9} second. Gains in excess of 10^6 are obtainable.

The wavelength dependence of the response of the tube varies with the composition of the photoemissive surface. Surfaces are available which not only respond to the visible incident energy, but also extend into the ultraviolet and near infrared range. Photomultiplier tubes that are sensitive to the infrared range normally require cooling of the bulb to assure that the ambient temperature of the bulb does not result in the generation of excessive dark current.

The ability of the photomultiplier tube to reliably detect extremely small amounts of incident radiation is widely utilized. The chief disadvantage in the use of the photomultiplier tube is the requirement that a high-voltage power supply in the order of 1000 to 2000 volts be available. This disadvantage is more than offset by the results obtained when low light or radiation levels must be measured.

Photoconductive Cells. Within the past five years, technological advances have resulted in the widespread availability of radiation detectors whose characteristics are such that the resistance changes with the incident radiation. These devices are extremely rugged and, in contrast to the photomultiplier tubes described above, do not require high-voltage power supplies. The principal disadvantage of these detectors is that, due to the inherent nature of the photoconductivity process, the response times are slow compared to the photomultiplier tube. Typical response times for such detectors are in the order of a few milliseconds compared to nanoseconds in the photomultiplier tube.

Depending upon the composition of the photoconductive material, the spectral response varies widely. One of the two most commonly used photoconductive cells is basically composed of cadmium sulfide. Since the response of these cells approaches that of the eye, measurements made using a cadmium sulfide cell are readily correlated with the visible sensations of the eye.

The other widely available cell utilizes cadmium selenide as the basic element. The cadmium selenide cell is characterized by response in the near infrared; it may therefore be used in instruments whose performance is based on the absorption of radiation in that region. An example of such an instrument is one in which the infrared absorption of tissue, which is proportional to the amount of blood contained in the tissue, is measured by the use of a light source whose transmitted radiation is detected by a cadmium selenide cell. Instruments designed using this principle are finding increasing acceptance in many biological measurements.

The photoconductive cells available today exhibit the same sensitivity in amperes/lumen obtainable with the more complicated photomultiplier tubes. The photoconductive cell is a cheap and extremely reliable detector of optical radiation. By tailoring of the composition of the cell, the spectral response may be controlled to solve the problem at hand.

The Photovoltaic Cell. The photovoltaic cell differs from the photomultiplier and photoconductive detector in that it produces a voltage in response to incident radiation. Generally speaking, its sensitivity is considerably less than that of the photomultiplier and photoconductive detectors, yet its response time lies between that of the other two systems. The main advantage of the photovoltaic cell, other than the self-generating characteristics of its output, is that it is sensitive to long wavelength radiation. The photovoltaic cell is used to detect radiation in the 1- to 2-micron region where photoemissive surfaces cannot be made to function reliably, and the photoconductive device has a poor signal-to-noise ratio.

In many instruments concerned with the analysis of the composition in gases, a need arises for the detection of the absorption of radiation in the 5- to 10-micron range. A long established method is the use of the Golay cell. In this cell, a standard volume of gas is subjected to the radiation to be detected. The absorption of this radiation in the confined gas causes the volume of the gas to expand, which in turn displaces one plate of a condenser. This displacement is directly related to the amount of radiation absorbed.

With the advent of the increased sophistication of semiconductor technology, particularly that associated with the manufacture of temperature-sensitive devices (specifically the thermistor), a new type of thermal detector has replaced the Golay cell for most purposes. In the more recent instruments, a semiconductor is used as the radiation detector. Such detectors are termed "bolometers," and are made of a material whose sensitivity to thermal input is such that it permits the detection of the far infrared radiation.

Undoubtedly, the most widely publicized use of bolometer detectors is in the instrument designed for the visualization of the temperature differentials existing on the surface of a biological organism. By proper design of the instrumentation, temperature differentials in the order of 0.1°C can be reliably detected upon the surface of the skin. As a result of the intimate relationship between the skin temperature and the blood flow of the organism, it is possible, through an area scan of the organism as a whole, to obtain an image that is directly related to the spatial temperature profiles of the object. At the present time, the sensitivity of the bolometers available is such that 12 to 14 seconds are needed to produce a complete image. It is anticipated that future advances in bolometer technology will permit the scanning time to be reduced to a few milliseconds. The term "thermography" has been applied to this technique.

Transducers for Ionizing Radiation

The transducers in this classification are based upon the ion-producing effect of the radiation to which they are sensitive. Radiation of sufficient particle energy, such as α, β, δ, and x-ray, when absorbed in a material, will produce both ions and electrons. In one type of ionizing radiation transducers, these charges are collected and processed to constitute the output signal. In another form of ionizing radiation detector, the initially absorbed radiation produces optical radiation, which is subsequently detected and amplified by one of the optical transducers described in the previous section.

Undoubtedly, the most widely used detector of ionizing radiation is the gas-filled chamber which, depending on its mode of operation, may be termed either an "ionization chamber," a "proportional counter," or a "Geiger counter." Ionization chambers and proportional counters measure the energy of the absorbed radiation per cubic centimeter and per second. In contrast, Geiger counters measure the number of particles absorbed, regardless of their energy.

The ionization chamber consists of two cylindrical electrodes, well insulated from each other, enclosing a gas-filled chamber. A voltage source produces an electric field between the electrodes. Ionizing radiation entering the space between the electrodes causes the removal of electrons from some of the gas molecules, which thereby become positive ions. The applied field separates these electrons and ions and moves them toward the appropriate electrode. The current so formed is measured by a meter.

Once the ions and electrons are produced, they will recombine unless they are removed from the site of production. The dependence of the amount of current read by the meter as a function of the applied voltage to the chamber is the result of this effect. As the voltage applied to the chamber, and hence the field strength, is increased, the radiation-produced electrons acquire enough energy from the applied field to ionize additional gas molecules. This behavior amplifies the output current. In such operation, the instrument is a proportional counter. By virtue of the gas amplification mechanism, the total current collected is directly proportional to the number of ion pairs produced by the absorbed radiation. As the applied field is increased still further, a region termed the "Geiger region" is encountered. In this region, the field strengths are sufficiently high so that, once an electron is produced in the gas-filled region, it is able to acquire enough energy to produce additional ionization in the order of a million times. Thus, the characteristic of the proportional counter, i.e., that the height of pulse present in the output is proportional to the energy of the electrons produced by the initial ionization, is lost in the Geiger counter. On the other hand, sufficient amplification is present, due to the gas amplification in the Geiger region, so that the output pulses are in the order of volts.

When the chamber is run in the ionization region with sufficient electric field to assure collection of all the ions and electrons produced by the ionizing event, the currents measured are in the order of 10^{-12} ampere. This low value of current requires that special precautions be taken regarding the insulations surrounding the chamber, as well as assurance that the associated amplifiers have an input impedance approaching 20 megohms.

In the proportional region, by virtue of the increased output current per ionizing event detected, the amplifier requirements regarding input are not as severe. When the chamber is run in the Geiger region, the output pulses are all of uniform height and of such magnitude that levels in the order of a volt or two are achieved.

The price paid for such high amplification in the Geiger region is that, once the discharge has occurred, a period of relative insensitivity ensues. This behavior limits the maximum counting rate of the Geiger tube to the order of 10,000 pulses per second.

In the proportional counter, such a limitation does not occur, and pulse-counting rates in the order of 10^5 to 10^6 per second may be readily handled.

The gas-filled ionizing radiation detectors are unquestionably the simplest of the radiation detectors available. Their principal limitation is the fact that absorption occurs in the gas-filled region or results from photoelectrons ejected from the wall of the chamber by the ionizing radiation. Thus, for the more energetic radiation, the efficiency of absorption, and hence detection, is extremely low. On the other hand, the equipment is simple and, for the most part, reliable enough to have found very widespread utilization in the detection of ionizing radiation.

With the availability of the photomultiplier tubes described above, a new type of ionizing radiation detector has been more commonly used. This consists of a crystal phosphor in which the ionizing radiation is absorbed and subsequently produces a light flash that is coupled to the photomultiplier. Such a system is termed the "scintillation counter." Its chief advantage is that the crystals that convert the incident radiation to light are readily available and in a form such that, when coupled to the photomultipliers, they produce an extremely sensitive and reliable detector of high-energy ionizing radiation.

The photomultiplier tube, by virtue of its extremely high-speed response, does not suffer the serious limitation in counting rate encountered with the gas amplification mechanism of the Geiger tube. Thus, the scintillation counters, though basically more complex and expensive than the simple Geiger counter, nevertheless are more efficient and have a much wider dynamic range of input levels. The specific choice of the Geiger tube, the ionization chamber, the proportional counter, or the combination scintillator and photomultiplier tube depends upon the problem at hand.

Problems Associated with Accuracy of the Data of Counters

With the availability of ionizing radiation detector systems capable of recording the detection of an individual event, it becomes necessary to determine the lower limit of reliability of the measurements obtained. In other words, in all practical applications, unless extreme precautions are taken, all detectors will exhibit a background counting rate. This is particularly important in some of the studies in nuclear medicine where a counter is used to detect the presence of, and to localize a spatial distribution of, extremely weak radioactive sources in the presence of other radiation.

The lowest level of radiation that can be measured with the more sensitive particle counters such as the Geiger tube or the scintillation counter depends upon the background counting rate and upon the error that can be tolerated. If the number of background counts in any time interval t is termed C_b, and the number of counts resulting from the incident radiation is termed C_r, then the total number of counts measured in a given interval of time (C_t) is the sum of these two: $C_t = C_r + C_b$. The probable statistical error of C_t is shown by

$$\text{Statistical error} = \frac{\sqrt{C_t + C_b}}{C_r} = \frac{\sqrt{C_r + 2C_b}}{C_r} \qquad (12)$$

If N is the number of ionizing particles entering the counter in 1 second, M the counter efficiency, and B the background rate in counts per second, then the probable error of the data is given by

$$\text{Probable error of the data} = \frac{\sqrt{MN + 2B}}{MN \sqrt{t}} \qquad (13)$$

where t = observation time.

The error, at a given level of radiation, diminishes in proportion to the square root of the observation time. The minimum radiation intensity that can be measured with a given error, or the time required to obtain a measurement, diminishes with the reduction in the background counting rate. An increase in the counter efficiency does not necessarily produce the same result, since it will lead, for the most part, to an increase in the background counting rate. It may be seen that an increase in the accuracy of measurements may be obtained by an increase in the number of particles being measured from the source as compared to the background. The upper limit is the tolerance of the biological system to the radioactive source.

AMPLIFIER DESIGN

The electrical signals encountered in most biomedical engineering research are of such a nature as to require special considerations in the design of suitable amplifiers. Characteristics such as input impedance, low frequency response, noise performance, and the ability to discriminate against interfering signals are of utmost importance in assuring that the amplifier will perform satisfactorily.

Classification of Amplifiers by Operating Point

Amplifiers are normally classified according to the position of their quiescent point of operation. With few exceptions, most of the amplifiers used with biological systems are of the type termed "class A" amplifiers. In a "class A" amplifier, the operating point and the input signal are such that the current flows at all times of the cycle. Such an amplifier generally operates over a linear portion of its characteristic.

The term "class B" is used to describe an amplifier in which the operating point is at one extreme end of its characteristic; thus, the quiescent power is extremely small. In the "class B" amplifier, the quiescent current or quiescent output voltage is approximately zero. If the signal voltage is sinusoidal, amplification takes place for only half a cycle. "Class B" amplifiers are normally used in a "push-pull" configuration. In most instances, the power amplifier systems used to drive mechanical recorders of the galvanometer or direct-writing type are "class B."

The "class A-B" amplifier is one biased to operate between the two extremes defined for "class A" and "class B." "Class A-B" amplifiers are used to provide a more linear operation than "class B," while at the same time retaining the inherent power conversion efficiency associated with operation of the individual amplifiers for less than the entire cycle of input voltage.

Distortion in Amplifiers

The application of a sinusoidal signal to the grid of an ideal "class A" amplifier will result in a sinusoidal output wave. The output wave form in most cases is not an exact replica of the input signal wave form because of various types of distortion that may arise. These are due to the nonlinear characteristics of the active element of the amplifier, which may be either a vacuum tube or a transistor, or to the influence of the associated circuit.

The types of distortion that may exist either separately or simultaneously are nonlinear distortion, frequency distortion, and delay distortion. In many biological amplifiers, frequency distortion is commonly present at the extremely low frequencies approaching DC, but rarely present in the higher frequency range of the biological signal. Most amplifiers have high-frequency response of kilohertz, and do not constitute a limitation to the processing of biological signals.

A very common type of distortion in biological amplifiers is nonlinear amplitude distortion which produces new frequencies in the output not present in the input signal. These new frequencies are harmonics of the input signal and arise due to nonlinearity in the dynamic characteristics of the active device itself. Generally speaking, the circuitry associated with the amplifier behaves in a linear manner. To the first approximation, all distortion encountered in the amplifier is due to its own nonlinear characteristics.

The term "frequency distortion" describes the situation that exists when the signal components of different frequencies are amplified differently. In most biological amplifiers, the principal frequency distortion is that associated with the low-frequency component of the signal. This is due to the frequency characteristics of the associated circuitry, and is distinct from the nonlinear distortion due to the nonlinear characteristics of the active device itself. "Delay distortion" may be regarded as a form of frequency distortion.

Specification of Gain and Effects
of Interconnecting Amplifier

In many instances it becomes convenient to express the gain of the amplifier on a logarithmic scale. The unit on the logarithmic scale is termed the decibel, dB. The number of decibels N by which the output power exceeds the input power is defined by

$$N = 10 \log_{10} \frac{\text{Power out}}{\text{Power in}} \text{ dB} \tag{14}$$

A negative value of N indicates that the output power is less than the input power. If the input and output impedances are equal, then Eq. (14) may be rewritten as Eq. (15), in which A is the magnitude of the voltage gain of the unit.

$$N = 20 \log_{10} A \text{ dB} \tag{15}$$

In general, the input and output resistances are not equal; however, the expression in Eq. (15) has been adopted as a convenient definition for the decibel voltage gain of an amplifier, regardless of the magnitudes of the input and output resistances. As will be seen later, in the section covering the noise performance of amplifiers, a term "noise figure" is commonly used. The value given for the "noise figure" is based on the power relationship, and hence becomes independent of the impedance of the associated circuitry of the amplifier.

When the amplification of a single amplifier stage is insufficient, then it becomes necessary to cascade amplifiers. The output voltage from the first amplifier serves as the input for the second amplifier and so forth. The resultant voltage gain equals the product of the individual voltage gains at each stage. The overall voltage gain of a multistage amplifier, when expressed in decibels, is the sum of the decibel gains of the individual stages.

Equivalence of Active Devices

All amplifiers have in common some type of active device which serves as a means of converting low-frequency power, usually DC, to power in the frequency spectrum desired. In practice, the active device

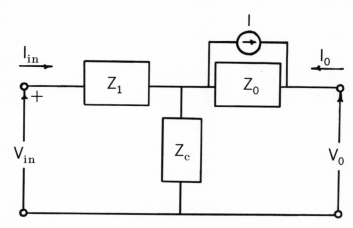

$I = g V_{in}$ for tube or FET
$I = I_{in}$ for transistor in common emitter configuration
See Table 9-4 for typical values of Z_1, Z_c, and Z_0.

Figure 9-9. Typical biomedical amplifier used with recording galvanometer.

Table 9-4.   Values of Parameters for Figure 9-9

Device	Z_1	Z_c	Z_o	I
Vacuum Tube	20 megohms	0	10^4 to 10^6 ohms	$gm\ V_{in}$
Transistor—Common Emitter Configuration	5K ohms	200 ohms	10^5 ohms	$\beta\ I_{in}$
Field Effect Transistor	1000 megohms	0	10^5 ohms	$gm\ V_{in}$

NOTE: These values do not include shunting capacitance which normally need not be considered for biological amplifiers except as noted in text.

may be either a vacuum tube or a solid-state device such as a transistor or a field-effect transistor. For purposes of analysis, the same equivalent circuit may be used, provided suitable consideration is given to the unique characteristics of the element employed.

When viewed in the context of an active element, all three devices may be represented schematically as shown in Figure 9-9. Here it may be seen that the active element can be regarded as a three-terminal device having an input impedance which serves to load the driving circuit, a common impedance through which the input current and the output current pass, and an output impedance which serves to load the output circuit, or determines the type of output circuit that may be used. Table 9-4 gives typical values of these three impedances for the vacuum tube, transistor, and field effect transistor.

As a result of the flow of current from the input and output circuits through the common impedance of the device, the effective input impedance and effective output impedance both depend upon the circuit configuration used. As will be seen in the sections that follow, in the case of the transistor, the effective input impedance of the device may be made many times that of the device itself by proper design of the associated circuitry.

Generalities Relating to Biological Amplifier Design

A typical biological amplifier is shown schematically in Figure 9-9. Viewed as a whole, the amplifier serves either to increase the power available from the signal source to a level sufficient to operate a low-impedance graphic recording device, or to increase the potential level of the signal from the source to that necessary to produce an observable deflection on some type of high-impedance recording instrument. Actually, both of these functions are the same if one considers the power involved to actuate the various recording devices. In general, graphic recorders require a relatively large amount of power to produce a recordable trace, whereas visual-display recorders utilizing cathode ray tubes generally require a large potential difference at low power levels to produce the deflection of the display tube. In any event, the function of the amplifier is first to amplify the signal from the source and then to match this power output to the appropriate recording device.

Generally speaking, the first stage of the amplifier is "untuned," i.e., the overall frequency response is not peaked at a given frequency. A number of factors influence the number and characteristics of the individual stages which must be used to meet certain previously specified requirements. Among the factors that must be considered in the amplifier design are the total overall gain required, the shape of the frequency response characteristic, and the overall bandwidth. Many factors impose limits to the maximum sensitivity that may be achieved, the most important being the inherent noise generated by the active devices themselves. The requirements for stability of operation impose severe practical restrictions on the techniques of construction. Because of a combination of these factors, overall amplifier potential gains in excess of 120 dB are extremely difficult to achieve. Due to bandwidth considerations, amplifiers seldom exceed 6 to 9 stages in cascade and achieve stable operation. In amplifiers utilizing more than 4 stages, extreme precautions must be taken to avoid the possibility of unwanted oscillations, producing unstable operation.

Specific Design Procedure

To calculate the overall gain and frequency response of a multistage amplifier, the equivalent circuit of the amplifier must be drawn. A systematic method of arriving at the equivalent circuit which may be used to determine the alternating-current response evolves from the following procedure.

1. Using the actual circuit of the amplifier, indicate the input, output, and common terminals of the active device. Between the two output terminals of the device, indicate the current source representation as shown in Figure 9-9. Between the input terminal and the common leg of the device, indicate an input impedance the value of which will be determined in subsequent calculations.

2. Transfer all circuit elements from the actual circuit to the equivalent circuit without altering the relative positions of these elements.

In the above procedure, the assumption is made that the impedances of the power supply sources are small enough not to seriously
affect the equivalent circuit. This is easily achieved in practice; if it
is not, it may be a source of possible oscillation in the overall amplifier.

Once the equivalent circuit has been drawn, then conventional AC
network practice may be used to analyze the circuit. An example of
this is as follows: In many biological amplifiers, the "mid-frequency
response" of the amplifier is the only one of concern. The "mid-frequency response" is defined as the response of the amplifier at those
frequencies at which the reactance of the interstage coupling capacitors
is so small as to be neglected. In some instances to be described later,
this "mid-band response" calculation may not be used alone, due to
the need for low-frequency response. In such cases, the effect of the
reactance of the coupling condensers must be considered.

At "mid-band frequencies," the amplifier gain is calculated by first
determining the output voltage of the amplifier. This is obtained by
multiplying the output current of the active device by the total shunting
impedance. This impedance is the parallel combination of all the
resistive elements connected at the output of the amplifier in parallel
with the output impedance of the device. The output load of the
amplifier includes not only the load resistor of the amplifier, but also
the input impedance of the following device which the amplifier feeds.
In the schematics shown, the total impedance of the load of the active
device is made up of the parallel combination of the output impedance
of the device, the resistive load of the device, and the input impedance
of the following circuit.

Once the output voltage is obtained in this manner, it is then
divided by the input voltage to obtain the "mid-band" gain given by

$$A_m = \text{Mid-band gain} = \frac{e_0}{e_{in}} = \frac{I \, Req}{e_{in}} \tag{16}$$

where I is given by Table 9-4, and Req is the parallel combination of
Z_0, Z_L, and Z_{in} of the following stage. At frequencies of interest in
biological amplifiers, only the resistive component of Z_0, Z_L, Z_{in} need
be considered. Typical values are given in Table 9-4.

A complete analysis of a practical amplifier must include not only
the mid-band gain, but the total bandwidth of the amplifier.

It is common practice to define the bandwidth of an amplifier as
the frequency difference between the two frequencies F_h and F_L, which
are those frequencies at which the amplifier response is 0.707 when
measured at mid-band. This dropping off in response is due to (a) the
shunting effect across the output circuit, and (b) the effect of series
capacitance between the amplifier stages.

The gain of the amplifier above and below the mid-band frequencies is given by Eqs. (17) and (18).

$$A_{high} = \frac{A_m}{\sqrt{1 + \left(\frac{f}{f_h}\right)^2}} \tag{17}$$

where $f_h = \dfrac{1}{2\pi C_s\, Req}$,

and Req and A_m are as determined by Eq. (16).

$$A_{low} = \frac{A_m}{\sqrt{1 + \left(\frac{f_L}{f}\right)^2}} \tag{18}$$

where $f_L = \dfrac{1}{2\pi C_c R_{in}}$,

and R_{in} is the value of Z_{in} given in Table 9-4.

The input impedance of the active device is the parameter that constitutes the distinguishing feature of the vacuum tube, the transistor, and the field effect transistor when used in biological amplifiers. Generally speaking, vacuum tubes and field effect transistors are characterized by input impedances so high that they may be neglected. It should be noted, however, that in certain special applications, e.g., those involved with microelectrode recording, this statement is not valid.

The transistor normally has such a low value of input impedance that it must be considered in all calculations involving multistage amplifiers. The output impedances of the various types of active devices may, in most instances, be considered as pure resistances over the range of frequencies encountered in biological amplifiers.

To summarize, the equivalent circuit shown in Figure 9-9 may be used quite accurately in computations associated with any of the amplifiers encountered. The chief distinction among the three types of active devices lies in the input impedance, which in turn is dependent not only on the device but also, as will be seen later, upon the manner in which the device is connected.

In the analysis of the multistage amplifier, each stage is analyzed individually, with the overall amplifier gain obtained by multiplying the gains of the individual stages (Fig. 9-10). This technique will give the mid-band gain of the amplifier if Eq. (16) is used. In the event that low-frequency response is of concern, the reduction in gain at the frequencies in question may become excessive in the multistage am-

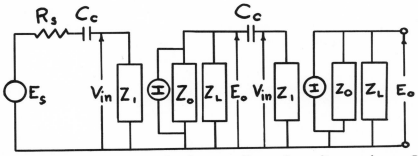

Figure 9-10. Two stage amplifier showing effects of coupling condenser C_c.

plifier. To achieve a given gain at a given frequency in the multistage amplifier, it is necessary that each individual amplifier exhibit low-frequency response in excess of that desired. Amplifiers having good low-frequency response are described in the following material.

Low-Frequency Response of the Amplifier

In many biological instrumentation problems, it is desirable to amplify signals to a low-frequency limit in the order of 0.1 Hz. For a single stage of amplification, the inherent long-term drift of the characteristics of the active element for a given quiescent operating point is generally of such a low magnitude that it may be neglected, provided a vacuum tube or field effect transistor device is used. The transistor amplifier exhibits a shift in characteristics that may be as high as 6 to 10 percent per degree centigrade. Rather elaborate temperature compensation schemes have been evolved for minimizing the change in transistor amplifier characteristics with temperature. In general, this compensation is handled by the use of biasing networks or in a circuit configuration as shown in Figure 9-11. Here it may be seen that a second transistor with characteristics essentially identical to those of the first is used to compensate for operating-point drift. This configuration, commonly known as the "differential amplifier," has an added advantage in that it may be used to discriminate against so-called "common-mode" signals, which are those spurious signals induced in the signal leads from the source. This feature will be discussed in detail under the subject of "differential amplifiers."

Response to frequencies in the order of 0.1 Hz in multistage amplifiers can be achieved through the use of rather large-value coupling condensers and input impedances for the active stage. The equivalent

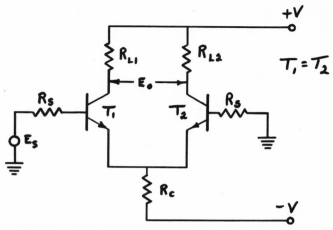

Figure 9-11. Basic differential amplifier circuit used to minimize temperature
effects in active elements and to eliminate stray signal pickup in
input leads.

circuit for such an amplifier must take into account the reactance of
the coupling condensers, since at low frequencies these values are such
that they may no longer be neglected. When this rises to infinity, the
response of the amplifier falls to zero.

One method of obtaining amplifier response to DC is through the
use of "carrier amplifier" techniques. These will be described in the
following section.

In general, response to frequency in the order of 1 Hz may be
readily achieved through the use of the resistance-coupled configuration
shown in Figure 9-11. If response to frequencies lower than 1 Hz is
desired, then a carrier amplifier or a properly compensated differential
amplifier should be used.

Carrier Amplifiers. One system widely used in biological amplifiers
for the amplification of a low-frequency signal is that classified as a
"carrier amplifier." This is shown schematically in Figure 9-12. Here
it may be seen that, through the addition of an external carrier f_0, a
modulated signal is produced whose modulation is directly related to
the input signal. Since the carrier is an alternating one, it may be
amplified using conventional AC amplifiers. In the detection of the
modulated carrier, the DC level which was inherent in the input signal
is restored. The carrier may be generated by an external oscillator, or
by means of a mechanical commutating device commonly known as

a "chopper." In some instances, a field effect transistor may be used in lieu of a mechanical chopper, with subsequent reduction in the minimum detectable input signal. Mechanical choppers introduce spurious voltages, due to contact potential, in the order of a few microvolts. Field effect transistors, owing to signal leakage inherent in the configuration of the device itself, generally will not operate reliably at input signal levels below 100 microvolts.

Differential Amplifiers. In many problems associated with the recording of microvolt potentials from a biological system, the object itself is subjected to stray electrostatic and electromagnetic fields. These stray fields result in the introduction within the recording system of spurious potentials far in excess of those generated by the organism itself. Fortunately, the large potentials due to the stray field are generally induced in the organism and the connecting leads in such a manner as to have equal potential in both leads. The potential desired is that due to the biological system and will usually have a phase compared to that of the spurious signal such that, through suitable amplification techniques, it may be retrieved in the presence of a rather large signal generated by the stray fields. This discrimination between the two signals may be accomplished by the use of either a balanced shielded transformer for the input circuit or a circuit configuration called a "differential amplifier."

These differential amplifiers, shown schematically in Figure 9-13, are widely used in applications concerned with the amplification of very low-level signals in the presence of extremely high electrical and magnetic fields. Operation of the amplifier is as follows. The signal at

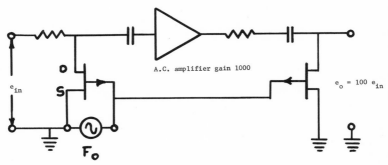

Figure 9-12. Carrier amplifier used to amplify frequency down to direct current. Maximum upper frequency limit is usually $\frac{1}{4} F_0$, the carrier frequency.

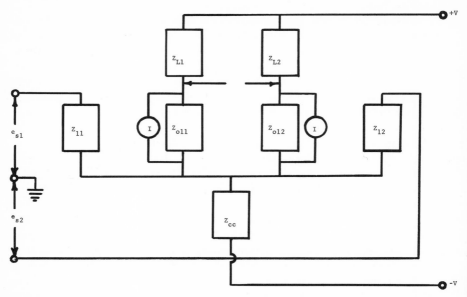

Figure 9-13. Schematic representation of differential amplifiers.

the input of each of the active devices consists of the desired signal as well as the undesired signal due to the interference field. The latter is an "in-phase" signal at both inputs, whereas that due to the desired object is out of phase. The "in-phase" signal, that due to the interference field, is amplified by each of the devices having the large common impedance in series. The effect of this common impedance is to reduce the gain of the amplifier when amplifying the "in-phase" signal by a large amount. The difference or desired signal is amplified by the devices in such a manner that the effect of the common impedance is nullified. Thus, at the output, the difference signal is amplified while the common signal, that due to interference, is reduced in amplitude. The degree of discrimination between the desired and interference signals is dependent on the value of the common impedance. Larger values of common impedance result in higher discrimination ratios. By proper balance, discrimination ratios between the desired and undesired signals as high as 60 dB may be readily obtained.

High-Input Impedance Amplifiers. In amplifiers used to probe individual cells, the impedance of the electrode may rise to the order of tens of megohms. These single-fiber recording applications require the use of amplifiers having input impedance in the order of 100 megohms.

By definition, the nominal gain of the amplifier is given by the output voltage divided by the input voltage. Since a portion of the output voltage is returned through the feedback network and combined in phase or out of phase with the input voltage, the effective input to the amplifier is no longer the value without feedback. The resultant gain of the amplifier with feedback is still defined as the output voltage divided by the input voltage, and therefore the gain with feedback is given by

$$A_f = \frac{A}{1 - BA} \tag{19}$$

where B is the feedback factor, and A is the amplifier gain without feedback.

From Eq. (19) it may be seen that, if the parameter $1 - BA$ is greater than unity, then the amplifier gain with feedback is less than it would have been if feedback had not been applied. If, on the other hand, the phase of the feedback voltage is such as to result in the parameter $1 - BA$ being less than unity, the gain of the amplifier will increase and approach infinity, at which point oscillations occur.

It is of great importance to examine how the input impedance of an amplifier is affected by the presence of feedback. The input impedance of an amplifier increases, depending upon whether or not the signal feedback is proportional to the output voltage of the amplifier or to the output current of the amplifier. The input terminal impedance of the amplifier with feedback is given by

$$Z_{if} = Z_i(1 - BA) \tag{20}$$

From Eq. (20) it may be seen that the input impedance with feedback is greater than the input impedance without feedback to the same degree that the gain and distortion decrease with feedback. As a specific example, assume that Z_1, the input impedance, is due to the capacity between the input terminals of the device and the common and output terminals. Since the impedance increases with feedback, this means that the effective input capacitance is decreased.

One difficulty encountered in feedback amplifiers is that polarity of voltage feedback is greatly dependent upon the characteristics of the network comprising the feedback loop. Although this network may have a given impedance at one frequency, at other frequencies, due to the nature of the components making up the network, the impedance may change in such a manner that the polarity of the signal feedback will completely reverse. Unless properly controlled, this situation will result in oscillations.

A very common method of achieving feedback for the purpose of increasing the input impedance of the devices is by means of a common resistor between the input circuit and the output circuit. The feedback potential is equal to the output current times the impedance of the common leg. Through a proper choice of the feedback resistor very high values of input impedance may be obtained, not, however, without a severe loss in gain. In many instances, the advantage provided by the inclusion of the feedback resistor in the input stage more than offsets the loss in gain, which can be made up in the following stages.

The addition of feedback to an amplifier not only increases the input impedance of the amplifier, but also acts to modify the output impedance of the device.

Noise Consideration

It is customary to specify the performance of the amplifier not only in terms of the maximum available gain and frequency response but also in terms of its signal-to-noise performance, i.e., the minimum detectable signal at the input of the amplifier. In a properly designed amplifier, the minimum detectable signal level will be determined to a large extent by the noise present in the input stage of the amplifier. There are many sources of noise that contribute to the overall noise performance of an amplifier. In most cases, those of importance are due to random fluctuation or motion of the electrons in both the input circuit and the active device itself. The noise performance of an amplifier is generally specified in terms of its "noise figure," which is an indication of the noise generated within the active circuits. The noise figure F is a measure of the ratio of the actual available noise power output of the amplifier to that of an ideal amplifier which is free of noise sources, except for the thermal agitation noise that has been applied to the input terminals of the amplifier. The term "available" describes the condition when the power source is matched to the load for optimum power transfer. By definition then, F is the ratio of the available signal-to-noise power at the input to the available signal-to-noise power at the output as given by

$$F = \frac{S_s/N_s}{S_0/N_0} \tag{21}$$

where F is the noise figure of the network, S_s the input available signal power, N_s the input available noise power, S_0 the output available signal power, and N_0 the output available noise power.

The noise figure is widely used to specify the noise performance of an active device, particularly the solid-state devices. For example, many transistors will be rated with a noise figure of 2 to 5 dB, meaning that the output signal-to-noise ratio of the properly designed amplifier will be 2 to 5 dB less than the input signal-to-noise ratio. It should be recognized that the source signal-to-noise ratio is that of the source itself, and cannot be improved on by amplification. Thus, in the selection of input active devices, the one with a low noise figure as tabulated by the manufacturer should be used. In a multistage amplifier, the overall noise figure is given by Eq. (22), from which it may be seen that, in a properly designed amplifier, the overall noise figure is essentially that of the input device itself.

$$F_{overall} = F_1 + \frac{F_2 - 1}{A_1} + \frac{F_3 - 1}{A_1 A_2} \tag{22}$$

where $F_{overall}$ = noise figure of complete amplifier
F_1 = noise figure of first stage or network
F_2, F_3, etc. = noise figures of second, third, and following networks
A_1 = available gain of first stage
A_2, A_3, etc. = available gain of succeeding stages

Examples of a Typical Biological Amplifier

In terms of numbers, the amplifiers most widely used in biology are those capable of providing a frequency response from a few tenths of 1 Hz to 500 or 600 Hz. These amplifiers normally have a sensitivity such that full output is produced with input signal levels in the order of a few tenths of a millivolt. If the amplifier is to be used for detection of potentials of single cells, then a special preamplifier is normally added.

The general design of biological amplifiers involves first a consideration of the low-frequency requirement and then, through the addition of filter networks, a tailoring of the upper bandpass frequency limit. An excellent example is the amplifier that was developed for use during the Gemini series of flights (Fig. 9-14).

This design is a universal one whose frequency response is tailored not only by the coupling capacitors associated with the interstage circuits, but also by the bandpass or shunting capacities which determine the high-frequency response.

The published characteristics of the amplifier show that it would not be satisfactory for use with electrodes having impedances in the

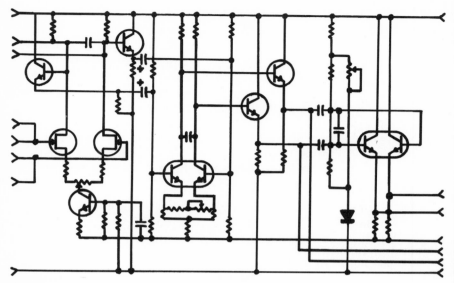

Figure 9-14. High input impedance amplifier utilizing feedback.

order of a few tens of megohms. This requirement, however, can best be handled by the use of a suitably designed head stage.

As mentioned earlier, one of the principal disadvantages associated with the cascading or sequential addition of amplifiers is the possibility that undesired feedback will occur from the output stage to the input. This can be avoided by proper physical layout of the amplifier, and by due consideration to the power supply used with the amplifier.

Since the total current for the amplifier flows through the power supply, extreme care must be taken to provide sufficient isolation between the output and input stages so as to prevent the output current, which is many times greater than the input current, from producing within the internal impedance of the power supply a potential sufficient to cause oscillation when introduced through the power supply leads to the input stage.

Means of preventing this are not only assuring that the power supply impedance is kept as low as possible but also using rather extensive decoupling networks. For the most part, these consist of suitable combinations of R and C designed in such a manner that the effects of the output stage current-produced voltage at the power supply impedance are eliminated.

APPLICATION

A special area of instrumentation, which is increasing rapidly, lies in the application of ultrasound to biological problems. Ultrasound is propagated by mechanical vibrations; thus, the properties of a material measured by ultrasound differ markedly from those measured using other types of probing energy. In a sense, ultrasound technology parallels optical methods, the difference being that the wavelengths associated with the commonly used ultrasound frequencies are greater than those encountered using visible radiation; hence, the laws governing optical propagation as it relates to focusing systems do not apply when ultrasound is used.

Ultrasound applications in medicine and biology embrace fluid flow measurements, therapeutic uses, visualization of soft tissue, and in some instances the modification of the structure and/or the function of the organism. To fully appreciate the potential applications of ultrasound in biology, the fundamental principles governing its generation, propagation, and detection must be examined.

Fundamental Characteristics of Ultrasound

The term "ultrasonics" has been applied to sonic energy having frequencies greater than that to which the human ear can respond. In discussions of the application of ultrasonics to biological problems, it is convenient to divide these frequencies into those concerned with "low-energy" and "high-energy" sound fields. A low-energy field does not result in permanent changes to the medium through which the ultrasound is propagated. High-energy fields, however, are those which use ultrasound energy of such a level that permanent changes in the medium are observed due to the propagation of the ultrasound.

Low-energy ultrasound, by definition, does not in any way affect the physical properties of the material through which it is propagating. Through its use, it is possible to measure the velocity and absorption coefficients of a given material and thus completely characterize the mode of propagation in a given medium. Since the amplitude of low-energy ultrasound waves is such that there is a linear relationship between the applied stress and the resultant strain, the absorption coefficients obtained are essentially based on a series of linear measurements.

The propagation of low-energy ultrasound waves involves the generation of ultrasound by means of a transducer, and subsequent detection. When the sound source vibrates in the direction of the wave motion, the waves produced are termed "longitudinal waves." These waves, in traveling through the medium, give rise to alternate compressions and rarefactions, and as a result are often termed "compression waves." In some instances, the motion of the sound source is at right angles to the direction of wave motion, in which case the waves produced are termed "transverse waves." These waves give rise to alternating shear stresses, and the term "shear waves" is often used to describe them. Generally speaking, shear waves can by propagated only in solids.

The sound transducer applies a periodic force to the medium in which it is coupled. The effect of this force is analyzed in terms of the "mechanical impedance" Z_m of the propagating medium, given by the force F divided by the velocity V:

$$Z_m = \frac{F}{V}\left(\frac{\text{newton-seconds}}{\text{meters}}\right) \tag{23}$$

It is sometimes convenient to relate the equations governing ultrasound propagation to those commonly used in electrical circuit theory. When this is done, the acoustic pressure is equivalent to the potential difference, the particle displacement is equivalent to the electric charge, and the particle velocity is equivalent to the current. Another convenient parameter is the "specific acoustic impedance" Z_s, which is defined as the ratio of the acoustic pressure P to the particle velocity V. This is analogous to the electrical impedance, and is equal to the mechanical impedance per unit area of cross section of the medium as expressed by

$$Z_s = \frac{P}{V} \text{ rayls or } \frac{\text{dyne-seconds}}{(\text{meters})^3} \tag{24}$$

It should be recognized that the acoustic impedance of a material is complex and can be expressed as the sum of resistive and reactive components. The product of the density of the medium P_0 and the velocity of the sound waves C is termed the characteristic impedance Z_c of a particular medium through which the wave is propagating:

$$Z_c = P_0C \text{ rayls} \tag{25}$$

In the case of a plane wave being propagated through a uniform nonabsorbent medium, the principle of conservation of energy demands that the intensity be the same in all points of the wave. This intensity is equal to the energy per unit volume of the wave times the velocity with which the energy passes through a unit cross-sectional area. In an absorbing medium, attenuation of the plane sound wave arises from deviation of the energy from the parallel beam by reflection, refraction, diffraction, scattering, and absorption processes by which the mechanical energy is converted into heat due to internal friction within the medium itself. These absorption losses are a characteristic of the material through which the waves travel, and hence may yield information about the physical properties of the medium, as discussed in the section that follows.

Ultrasound Absorption Mechanisms

The mechanisms underlying the general absorption of plane waves of ultrasound are caused by (a) deviation of energy from the parallel beam, (b) a type of absorption due to the hysteresis effect, and (c) a type of absorption due to the relaxation characteristics of the medium.

The most commonly encountered loss in energy from the parallel beam is that due to scattering of the parallel beam by the structure of the material through which it is propagated. This scattering loss is a characteristic of the structure of the material, and in some instances may be used as a means of identification of that particular type of structure.

In some materials, the application of a periodic adiabatic stress to the medium does not result in strain varying linearly with the applied stress. In this case, the stress-strain curve takes the form of a hysteresis loop. The losses associated with such a deviation from linearity are termed "hysteretic losses," and are characterized by a loss per cycle which is independent of the frequency.

Sound absorption in fluids is determined mainly by the viscosity, heat conduction, and molecular effects within the fluid itself. This type of absorption can be considered as the consequence of a time lag between the variation of sound pressure and a variation of density of the wave in the medium. This lag depends on (a) the time necessary for heat conduction from a high- to a low-pressure region in the sound field, (b) the time necessary for the viscous stress in the liquid to be equalized, or (c) the time necessary for a molecular energy exchange.

The type of absorption most commonly encountered at megahertz frequencies is that termed "relaxational." In order to understand this, a small portion of the medium through which the sound waves are passing should be examined. During the positive half of the stress cycle, energy is absorbed by the medium, and during the negative half of the cycle, it is given up. This energy exchange requires a finite period of time. The time delay is dependent upon the nature of the physical processes involved. When the molecules of this thin layer are subjected to the applied stress and undergo sinusoidal variations of their translated motion, the energy transfer from the sound wave to this particular state takes place in such a manner that coupling occurs with another energy mode, e.g., vibrational motion of the molecules; hence, there is a finite time lag before equilibrium conditions are satisfied. The effect of this time lag is to produce a lag in phase between the motion of the molecules and the resultant stress. The ensuing wave suffers a lowering of its peak amplitude and a raising of its troughs, with a resulting attenuation.

At low frequencies, this phase lag is negligible; hence, the attenuation is very small. As the frequency increases, the phase lag and the absorption will increase. With increasing frequency, the probability decreases that sufficient time exists for all the energy to interact with the molecules; therefore, once a sufficiently high frequency is reached, the attenuation actually decreases.

The net result of this behavior is that, when a given material is subjected to varying frequency ultrasound excitation, as the frequency is increased, the absorption increases to a peak at a frequency termed the "relaxation frequency," and then diminishes to zero at a frequency so high that there is insufficient time for an energy exchange to occur between the sound wave and the molecules of a given cycle. Behavior of a typical sample is shown in Figure 9-15.

In biological tissues, the absorption processes can be identified by measurement of these time constants. Through the observation of changes in velocity and absorption coefficients, which exhibit a linear frequency dependence, materials may be categorized according to the value of the relaxation frequency.

In biological systems, heat conduction loss is relatively unimportant. The most important loss mechanism is that due to the thermal relaxation process which occurs when the temperature of the fluid is changed by the propagation of the sound wave. This process results in a transfer of energy between the external and internal degrees of freedom of the medium, and requires a time interval comparable to the period of the wave. This out-of-phase relationship gives rise to the conversion of acoustic to thermal energy, as was discussed above.

The various types of tissue may be characterized by the type of relaxation processes that are involved. Nerve tissue is characterized by an increase of absorption coefficient with temperature. This is thought to be the result of a low rate of energy exchange between the

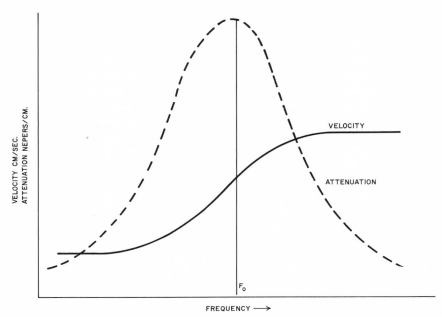

Figure 9-15. Typical variation of velocity and attenuation in biological material as a function of ultrasound energy frequency. F_0 is termed
the relaxation frequency.

external and internal degrees of freedom. Quite in contrast, blood
exhibits a negative temperature coefficient of absorption thought to be
due to the presence of the protein component of the blood.

At least 80 percent of the total tissue absorption is of molecular
origin. The changes in the molecular constituents may be influenced
by a variation in the hydrogen ion concentration. Although the absorption coefficient response to these variations changes very markedly, the
characteristic of the frequency dependence is not affected. By far the
largest part of the ultrasound absorption in tissue occurs on a macro-
molecular level.

It should be apparent from this rather brief description of the
low-level propagation characteristics of biological tissues that by appropriate measurement techniques the structure, and in many cases the
function, of the tissue may be readily ascertained. It has been shown
that low ultrasound levels are nondestructive to biological organisms,
and may therefore be used with safety for elucidation of the characteristics of organisms.

High-Energy Ultrasound Uses in Biology

High-energy ultrasound waves are, by definition, those that result in a deviation from a linear relationship between the applied stress and the resultant strain. These intense waves produce a measurable effect on the medium through which they propagate. One of the chief effects of high-energy ultrasound is "cavitation," i.e., the phenomenon that occurs in regions of a liquid which are subjected to rapidly alternating pressures of high amplitude. The cavitation mechanism depends on the production of minute voids which grow in size with the increasing pressure of the ultrasound field and then collapse on the reduction of that field. The cavities produced in this manner contain only the vapor of the liquid.

The threshold energy necessary to produce cavitation shows a linear characteristic up to a frequency of approximately 10 kilohertz. For air-free water, the intensity necessary to produce cavitation increases rapidly.

The threshold in which cavitation is induced is relatively independent of the applied pressure until a pressure level of 60 atmospheres is exceeded.

Ultrasound radiation produces six distinct effects on living tissue. These include mechanical agitation, cavitation, temperature effects, alteration of the pH, initiation of chemical reaction, and a variation in the permeability of the cell walls.

The physical mechanism by which this high-level ultrasound selectively affects tissue is not clearly understood.

Numerous experiments have been reported which verify that sound levels below 1 watt per cm^2 are nondestructive to biological tissues. In the range from 1 to 20 watts per cm^2, the destructive effects depend primarily on the type of biological tissue being irradiated, as well as on the time over which the radiation is carried out. Above the level of 20 watts per cm^2, the ultrasound produces damaging and destructive effects within the biological tissues, due to a variety of processes that are not well understood at this time.

Muscle and the fluid-filled region of the biological organism have absorption coefficients which are not too different from that of water. However, bone has an absorption coefficient that is considerably larger; also, quite in contrast to the linear variation with frequency exhibited by fluids and other tissue, the absorption coefficient for bone varies with the square of the frequency. It has been demonstrated that, at a frequency of 1 megahertz, one-half of the incident sound energy is absorbed in a 0.5-mm thickness of bone. At a power level in the order of 10 watts per cm^2, absorption of this energy gives rise to a temperature increase of approximately 20°C in bone 2 mm below the surface after an irradiating time of 4 minutes. Thus, extreme care must be taken to avoid seriously damaging the surrounding tissue by the absorption of the incidence sound in the bony structure.

Instrumentation Utilizing Ultrasonic Energy

The majority of instrumentation utilizing ultrasonic radiation involves the measurement of propagation of energy within the object under investigation. This propagation of energy is uniquely dependent on the material composing the medium transversed by the ultrasound energy.

As was detailed above, the propagation characteristics control the absorption and reflection of the incident energy. Thus, the instruments have, as the common basis of their design, features which permit a determination of the relative amplitude between the incident and reflected or transmitted energy. In addition to this feature, means are generally provided to determine the time delay between these energies. In some instances, these parameters will be measured as a function of frequency.

To apply ultrasound in the determination of characteristics of biological tissue, a source of ultrasound energy must be supplied. If the characteristics of the tissue are to be examined in terms of the acoustical impedance, then some means of detection of the ultrasound propagated through the medium must be provided. To measure the propagation characteristics of ultrasound through the tissue, one must determine the amplitude and phase of the transmitted or reflected sound energy. This is accomplished in a variety of ways, as will be discussed below.

In most instruments used in the application of ultrasound to biological tissues for diagnostic purposes, the ultrasound energy is generated by a class of materials termed "piezoelectric." These materials have the property of deforming mechanically when subjected to an electric field. This deformation is in synchronism with the electric field, and may therefore be used to convert an applied electric field into mechanical vibrations, which in turn can be coupled to the biological tissue, thus constituting a source of ultrasound energy. This effect is reversible, that is, the mechanical vibration of the piezoelectric material produces an electrical charge which is directly related to the applied mechanical vibration. In some instruments, single piezoelectric transducers serve both as the source of the ultrasound energy and as the receiver of the reflected ultrasound energy. In other instances, the transducer is used to produce ultrasound energy, while a second or third transducer is used to receive the transmitted or reflected energy.

The power for energizing the piezoelectric transducer may be derived from either a continuous-wave or pulse generator. When the transducer is used as a receiver, it generally is followed by a tuned amplifier, the frequency of the amplifier being matched to the resonant frequency of the transducer.

A continuous-wave generator consists of a tunable oscillator whose frequency is controllable in such a manner as to permit it to be matched to the resonance frequency of the ultrasound transducer. The most efficient conversion of electrical to mechanical energy occurs at the resonant frequency. In systems in which a pulse of ultrasound is used, either a gating device, which controls the output of a continuous-wave generator, or a single pulse generator may be used. In the latter case, one relies on the natural resonance characteristics of the transducer to generate the pulse envelope of ultrasonic energy.

The main difficulty encountered in the design of the driving sources for the transducer is that associated with matching the impedance of the electrical driving source to that of the transducer. The impedance of commonly used piezoelectric materials at frequencies used in biological tissue examination is in the order of a few ohms. Most driving sources are characterized by impedances in the order of 100 ohms or greater. Thus, special impedance-matching networks must be devised in order to assure that maximum power transfer occurs from the driving source to the transducer. In general, the long-term stability of the power output of the driving source is of little consequence; however, in certain applications, particularly those concerned with the application of ultrasound to visualizing tumors of the eye, such long-term stability is of utmost importance.

The transducer most commonly used in biological work is of the piezoelectric type. Commonly used materials in which the piezoelectric effect is prominent are quartz, barium titanite, lead zirconate-titanite, and lead meta-niobate. The principal advantage of the barium titanites and the lead-based ceramics are that they may be fabricated by sintering techniques into unconventional shapes which may be subsequently polarized to produce the desired piezoelectric characteristics. A relatively minor disadvantage associated with the barium titanite material is that the curie point, i.e., the point at which the piezoelectric effect ceases, is in the order of 130°C. In biological applications, this is unimportant because the biological organism itself would be damaged at such elevated temperatures. However, in some instances in which high-energy ultrasound is desired, this temperature limitation may be a disadvantage; in such cases the lead meta-niobate can be used. The factors affecting the maximum electrical-to-acoustical power conversion possible in a ceramic transducer are principally those associated with the dielectric dissipation of the transducer. Some idea of the maximum amount of power produced at resonance for matched transducers may be obtained from an examination of Table 9-5.

For the generation of ultrasound at frequencies above 10 megahertz, quartz is generally used because of its superior mechanical properties, despite its lower conversion efficiency. At a frequency of 20 megahertz, the thickness of a quartz plate operating on the fundamental mode is in the order of 0.15 mm. Such a plate is extremely brittle mechanically, and may shatter under high applied driving voltages. For this reason, it is common practice, when generating frequencies above 10 megahertz, to use a thick quartz plate and to excite it in such a manner that the desired higher harmonic modes are propagated within the plate.

If the system is designed to permit the use of separate energizing and receiving transducers, the quartz receiving transducer is generally preferred since its dielectric constant is approximately 1/200 that of the ceramics. This lower dielectric constant results in a net output voltage (per unit of acoustic energy impinging on the plate) of more than five times that obtained with any of the ceramic materials.

Another very important consideration in the design of any system using ultrasound for the investigation of biological materials is the energy distribution pattern of the transducers used. Since the system is dependent upon the measurement of the intensity of transmitted or reflected energy, it is extremely important to know the spatial pattern of the transducer as it relates to the geometry of the object being investigated. The control of the radiation pattern of a given transducer may be obtained by means of an optical system either separated from the transducer or intimately mounted upon it.

In most applications, the transducer has a thickness equal to an odd multiple of a half-wavelength of the excited energy. The surface area of the transducer is normally many times the wavelength of the sound in the material. The pattern given by such a radiator is most easily understood as being produced by a large number of point sources excited in phase. The resultant intensity pattern from such a transducer may be predicted quite accurately, using methods similar to those developed for the analysis of light beams. Through a graphical construction, the radiation pattern may be obtained.

The near field of the transducer is defined as the distance to the

Table 9-5. Power/Unit Volume at Resonance for Matched Transducer.*

Material and Mode	Temp. °C	Relative Power
PZT-4, Parallel	25	100
PZT-4, Parallel	100	77
PZT-4, Transverse	25	59
PZT-4, Transverse	100	40
PZT-5, Parallel	25	2.3
PZT-5, Parallel	100	3.0
PZT-5, Transverse	25	1.5
PZT-5, Transverse	100	1.8
Ceramic B, Parallel	25	14.3
Ceramic B, Parallel	75	3.8
Ceramic B, Transverse	25	6.3
Ceramic B, Transverse	75	1.6
Mason Comp., Parallel	25	14.5
Mason Comp., Parallel	75	1.9
Mason Comp., Transverse	25	5.0
Mason Comp., Transverse	75	0.7

*Assuming electric field limited to that which gives tan $\delta = .04$.

†Bandwidth $= \dfrac{f_2 - f_1}{\sqrt{f_1 f_2}}$, where $f_2 =$ upper cutoff frequency
$f_1 =$ lower cutoff frequency

‡To obtain power/cm³ at any other frequency f, multiply by f/1000 cps.

NOTE: Dr. D. Berlincourt of Clevite Corp. advises that a new composition PZT-8 produces twice the output given here for PZT-4.

transducer from the first minimum encountered as the transducer is approached from infinity with a point detector. The radiation pattern between the transducer and the first minimum has, as the student may see, been aptly termed "the zone of confusion." Echoes obtained from an object reflecting within this region are extremely difficult to analyze. The region beyond the first minimum is termed the "far field" of the transducer. Provided the object examined is not too large in comparison to the diameter of the transducer, the energy in the far field may be regarded as plane wave energy.

One of the distinct assets in the use of ultrasound is that it may be focused in a manner similar to light. In most instances, rather simple acoustic lenses may be effective. When used with the transducer, the lens assembly is usually an integral part of the transducer itself.

One of the difficulties associated with ultrasound lenses is that the velocity for shear-wave propagation differs from that for longitudinal-wave propagation in most solids. Thus, if a simple lens is constructed using a solid material, it is characterized by having a focal point which is long and cigar-shaped. Solid lenses may be made of materials such

Bandwidth %†	Watts/cm³ at 1 kcps‡	Efficiency, tan δ = .04 Ideal Case %
83	5.6	92.5
72	4.3	91.5
32	3.3	88
29	2.2	87
91	0.13	92.5
80	0.17	91.5
33	0.08	88
31	0.10	88
55	0.80	92
44	0.21	90
20	0.35	82
16	0.09	79
36	0.81	88
34	0.11	87
11	0.28	73
10	0.04	71.5

as lucite, polystyrene, epoxy resins, aluminum, or magnesium. If the immersion medium is water (and recall that biological tissue propagation characteristics are close to those of water), then the important properties for the acoustical lens material are (a) a large index of refraction compared to that of water, (b) acoustic impedance close to that of water, (c) low attenuation of sound, and (d) ease of fabrication.

Acoustic lenses used for work with biological structure must take into account the change in focal length of the lens in water as related to its focal length in the biological material. The velocity differentials between the water coupling medium and the biological material cause the incident sound to bend when it enters the material. This degree of bending may in some cases result in a shortening of the effective focal length below that given for the lens when measured in water. The effect of such shortening is to pull the focal spot very close to the surface of the biological material, and thus distort the apparent-depth measurement being made.

In high-energy ultrasound applications, the absorption loss associated with solid material used as lenses in many cases becomes so large as to prohibit the use of transmission optics. This is due to the fact that many of the materials, specifically plexiglass, which satisfy the characteristics mentioned above are such poor heat conductors that the ultrasound energy absorbed in the lens distorts it and produces mechanical failure. In these instances, it may be advantageous to use a parabolic reflector.

Measurement of Ultrasound Intensity

The measurement of ultrasound intensity is of particular interest in the calibration of transducers, and to some extent in the determination of field patterns associated with the transducer itself. The measurement of ultrasound energy within a given medium is complicated by the fact that the device used to measure the energy may, unless special care is taken, interact with the ultrasound field in such a manner as to give a false reading. The methods used for the absolute measurement of ultrasound energy are mechanical, thermal, and piezoelectric.

In the mechanical method, the radiation pressure and hence the intensity of an interface between a liquid and a solid are measured. This measurement is valid for any type of ultrasound beam, whether pulsed or continuous, provided that the measuring interface is large compared with the wavelength and includes the whole beam from the transducer, and that the field is free of resonances. A very common method is to use a pressure float. The advantages of the float are that it is relatively easy to build and calibrate, and that accuracy is within 5 percent.

Another method of measuring ultrasound field involves the determination of the temperature rise produced by such a field in a known medium. These methods give an accuracy in the order of 2 to 3 percent. A common form of such thermal measurement involves a thermocouple junction, very small compared to the wavelength of sound, embedded in an acoustic-absorbing medium which is retained between two membrane windows. The temperature rise produced at the junction is then calibrated in terms of the acoustic intensity.

Probably the most widely used and most sensitive probe for the calibration of the ultrasound field is the piezoelectric probe. The chief disadvantage is that the electrical output of such a probe is directly related to its size; since the probe is made small so as not to interact seriously with the field being measured, the output suffers. Piezoelectric probes available commercially serve to measure sound intensities to the order of 10^{-4} watts per cm^2 at frequencies below 2 megahertz.

Display Systems

An extremely popular method of visualizing ultrasound fields in biological materials involves the measurement of two parameters simultaneously. These parameters are the amplitude of the signal obtained by a change in acoustic impedance and the time required for the ultrasound energy to travel between these discontinuities.

The basic equipment used in such measurements is shown schematically in Figure 9-16. This approach, termed the "A scan," involves a single-trace method which displays, at the left side of a cathode ray tube, the transmitted pulse and the time at which the pulse was transmitted. By measurement of time to the right of the transmitted pulse, the total elapsed time between the transmitted pulse and the return echo may be obtained. Once the velocity of propagation within the material is known, the exact distance between the sending transducer and the impedance discontinuity causing the reflection may be determined. Conversely, if the material is of a known thickness, the velocity of propagation may be determined by such equipment.

The "A scan" presentation suffers serious limitations in that only those discontinuities on a line normal to the surface of the transducer are displayed. Further difficulty with the system is that spurious lobes of radiant energy from the transducer may give false readings, since the spatial distribution of the sound field from the transducer is not inherent in the system.

These limitations are effectively eliminated by a second type of display system termed the "B scan" which permits both the shape of the discontinuity and its distribution within a sample cross section to be seen. In many instances, the "B scan" is used to complement the

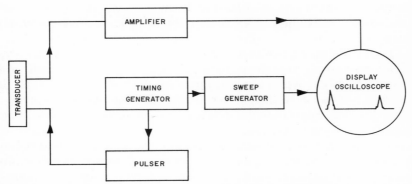

Figure 9-16. Schematic representation of basic "A" scan instrument.

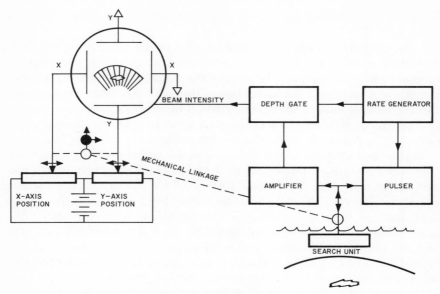

Figure 9-17. Schematic representation of basic "C" scan instrument.

"A scan." In the "B scan" presentation, the directional characteristics of the transducer play an important role. Unless the size and intensity of the sound pattern from the transducer are accurately controlled, false echoes will be obtained which are superimposed on those lying on the axis of the scanning transducer.

These artefacts may be eliminated by the use of the "C scan" display (Fig. 9-17), in which the position of the scanning line on the cathode ray oscilloscope is synchronized with the acoustic transducer scanning motion in two coordinates. In such a display, the plan view of the sample developed is extremely effective in removing spurious echoes.

One of the major problems associated with the A, B, and C type of display systems is that the echo detected by a probe from a reflecting surface depends to a great extent upon the normality with which the ultrasonic waves strike the surface. Thus, unless great care is taken to assure that the sound is striking normal to the surface, confusing and misleading echoes are obtained.

If the probe moves through a path at right angles to the direction of its immergent beam, it is a matter of chance whether a surface within that beam will produce an echo or not. It is only when the beam is caused to rotate at a considerable angle as it moves through its path that it is possible to obtain a satisfactory indication of the surfaces which lie at an angle to the tissue.

A system has been developed whereby the sound beam is moved in such a manner that each interface is effectively scanned from many angles. This method, termed "compound scanning," is accomplished by the superposition of two simple scanning motions. One component moves the average position of the transducer around the periphery of the scanned area on the circular path, while the other causes oscillation along a line tangent to the circle.

In order to display the echoes received, the range sweep on the display oscilloscope must follow the motion of the sound beam in both the starting position and angle. If the display system is perfectly linear mechanically and electrically, only those portions of the echo signals that are not affected by refraction, multiple reflection, and pulse stretching will plot in the same position on the tube face.

The position of the transducer is monitored and used by the deflection system to position the trace on the screen of the main display unit, where the return echoes are represented by an illumination of the trace. The principal design criterion for all echo instruments of this type is the spot size of the display unit, which usually dictates use of a high-quality display tube.

Ultrasound Visualization Systems

In many instances, it is desired that the transmission or reflection pattern of the ultrasound energy be recorded and displayed as a complete image field. This method of display of ultrasound energy parallels quite closely conventional practice of visual and x-ray image-forming systems. To simultaneously detect ultrasound energy across an entire field, many methods have been used, including photographic film. Table 9-6 summarizes the various means employed to detect extended-area ultrasound images, as well as the sensitivity of these methods.

Undoubtedly, the most sensitive method of detection of the ultrasound field intensity in distribution is that utilizing an electronic ultrasound image converter group with closed circuit television display techniques (Fig. 9-18). In this system, the point-to-point voltage generated on the piezoelectric surface, usually a quartz plate, will correspond to the ultrasound intensity pattern striking the piezoelectric material. These ultrasound-induced potential variations may be observed by the use of point probes on the back of the transducer, or by scanning the transducer with an electron beam. The latter approach has been inten-

Table 9-6. Summary of Ultrasonic Imaging Detection Methods

Technique	Approximate Threshold Sensitivity (w/cm^2)
Photographic and Chemical Methods	
Direct action on film	1 to 5
Use of photographic paper in developer	1
Film in iodine solution	1
Starch plate in iodine solution	1
Color changes caused by chemical action	0.5 to 1
Thermal Effects	
Thermosensitive color changes	1
Phosphor persistence changes	0.05 to 0.2
Extinction of luminescence	>1
Stimulation of luminescence	—
Change in photoemission	0.1
Change in electrical conductivity	0.1 to 0.2
Thermocouple and thermistor detectors	0.1
Optical and Mechanical Techniques	
Optical detection of density variations	3×10^{-4}
Optical detection of acoustic birefringence	10^{-1}
Optical detection of liquid surface deformation	10^{-3}
Mechanical alignment of flakes in liquid	2.8×10^{-7}
Electronic Methods	
Mechanical movement of transducer or object to form an image	—
Probe detection of potential on back of piezoelectric receiver	—
Electronic scanning of piezoelectric receiver	10^{-7} to 10^{-9}
Electronic scanning of piezoresistive receiver	10^{-7}

sively investigated and is finding increasing usage in many problems associated with ultrasound visualization.

The heart of the system is the ultrasound image converter tube consisting of an extended piezoelectric plate, one surface of which is scanned by an electron beam. The electron scanning beam is modulated by the ultrasound image-induced potentials and produces a signal which is subsequently displayed on a conventional television monitor. In some instances, the display is in color to increase the sensitivity of the system to minor changes in acoustic impedance.

The sensitivity of the system is such that a usable image may be obtained with input sound power levels to the ultrasound image converter in the order of 10^{-6} watts per cm^2. This sensitivity is such that the object being examined is subjected to ultrasound energy well below the level deemed harmful to the object.

The system is essentially instantaneous in time response; therefore, rapidly moving objects may be viewed. This characteristic is particularly advantageous in the visualization of the cardiovascular system where the motion of the various portions of that system, as well as fluid flow patterns, may be visualized.

The detail resolvable by the ultrasound visualization system is limited by the wavelength of the sound frequencies themselves. These wavelengths in water, a commonly used coupling medium, vary from 1.5 mm at a frequency of 1 megahertz to 75 microns at a frequency of 20 megahertz. The characteristics of the image-forming properties of the ultrasound image converter, as well as the points of maximum sensitivity of the system, are such that these resolution figures are reduced in practice to a value of approximately one-half those set by the wavelengths of the sound in water. In spite of this, however, the maximum resolvable detail is such that the equipments are extremely useful in biological applications.

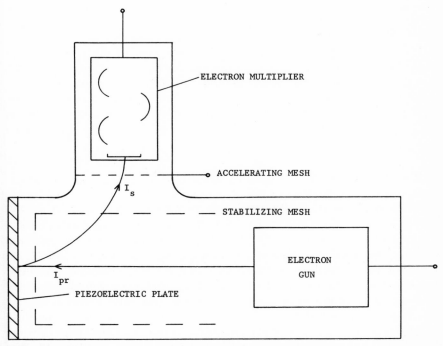

Figure 9-18. Schematic representation of ultrasound image converter systems. The ultrasound energy that is transmitted through the object produces an ultrasound induced potential image on the face plate of the image converter tube. This potential image is detected by the scanning beam of the image converter tube, amplified and displayed on a television monitor utilizing conventional techniques.

Physical Medicine Applications

The use of ultrasound to produce localized heating in the body is increasing rapidly, due to the fact that ultrasound is capable of good penetration in biological tissues with practical beam definition as compared to that obtained using electromagnetic radiation. A further advantage is that the velocity of sound is nearly the same in all soft tissues; therefore, very little of the incident ultrasound energy is reflected at the interface separating two different types of tissue. The distribution of the heat resulting from the application of a plane wave of ultrasound energy traveling through skin, subcutaneous fat, and muscle tissue can readily be predicted from the absorption coefficients of these tissues. Quite in contrast, when one uses microwave energy, it has been determined that the electromagnetic radiation absorbed by the body is determined by a variety of interactions of extremely complex nature which tend to preclude the possibility of prediction of the amount of radiation absorbed in a particular area.

For therapeutic uses of ultrasound, care must be taken to keep the absorbed energy well below values in the order of 1 watt per cm^2. Another parameter that must be considered is that mentioned previously of bone absorbing the incident beam in a very small surface layer. Unless care is taken, the possibility exists of damaging the surrounding tissue by unwanted absorption in the bony structure of the organism.

Flow Measurement

One of the most widespread uses of ultrasound in biology is that concerned with determining the flow of fluids, especially blood, within the organism. Initially, flow measurements were made by determination of the difference in transmission time resulting from the velocity of the fluid as compared to transmission time when the fluid is stationary. Equipments designed along these lines require such a high accuracy of time measurement that they have been essentially discarded.

At the present time, the equipment most widely used for flow measurements is the Doppler flowmeter, which was described in some detail under the subject of transducers. The chief difficulty with this flowmeter is an inability to quantitate the results; however, in many instances, a relative indication of flow is essentially all that is needed. Another use for the Doppler flowmeter is to ascertain the movement of various organs within the body. Here, the impedance discontinuity resulting from the difference in ultrasonic propagation characteristics of the organs is detected by the Doppler flowmeter, with the result that relative movement of organs deep within the body may be determined nondestructively.

Diagnostic Techniques Using Ultrasound

Ultrasound is finding increasing use as a diagnostic tool. In most instances, the instrumentation utilizes the pulse echo technique with either an A, B, or C type of display. In some studies associated with the heart, the Doppler method is used to obtain indication of relative movement of the heart structure in relation to the rest of the organism.

The pulse echo method, in contrast to the Doppler method, is an extension of the sonar technique for underwater detection that was developed during World War II. It is based on the principle that, when pulse ultrasound waves are projected into human tissues, they propagate through the tissues with part of the energy being reflected from the tissue interfaces having varying acoustic impedances along the path of propagation. These reflected waves are detected, amplified, and displayed in relation to a controllable time base by the use of a cathode ray tube. Both the wave form and the time delay associated with the reflected waves are analyzed to diagnose the composition of the tissues and the existence of pathological conditions.

The exact frequency chosen for the ultrasound energy depends to some extent on the depth of penetration desired. Since the attenuation and propagation characteristics of the ultrasound are dependent on frequency, a variation in the frequency used to examine the same biological structure will produce differing results. Most equipment available today utilizes frequencies in the range of 1 to 2 megahertz for diagnosis of intracranial and abdominal regions, and frequencies in the range of 5 to 15 megahertz for diagnosis of breast, body surface, and eye diseases. The principal advantage in using high-frequency ultrasound is that fine structure may be resolved; however, the high absorption of the ultrasound energy at these frequencies limits the diagnostic range to a few centimeters.

Conclusions

Ultrasound is finding increased use in biology, due to its ability to nondestructively ascertain details of organisms that are transparent or invisible to other types of radiation. Ultrasound, by virtue of its ability to visualize soft-tissue structure, offers a technique that is not readily available with any other type of probing radiation.

The exact mechanism by which ultrasound acts destructively on tissue is not completely understood as of this date; however, sufficient experiments have been run to verify the statement that the power levels used with the presently available ultrasound diagnostic equipment are so low as to be completely nondestructive to the biological organism. It is anticipated that, as instruments utilizing ultrasound become more available and their operation is better understood, many new and useful diagnostic techniques based on ultrasonic absorption will be devised.

References

1. Lion, K. S.: *Instrumentation in Scientific Research: Electrical Input Transducers.* McGraw-Hill, New York, 1959.

2. Motorola, Inc.: *Analysis and Design of Integrated Circuits.* McGraw-Hill, New York, 1967.

3. Alt, Fred: *Advances in Bioengineering and Instrumentation,* Vol. 1. Plenum, New York, 1966.

4. Grossman, C. C.: *Diagnostic Ultrasound.* Proc. of 1st Int. Conference. Plenum Press, New York, 1966.

5. Pettit, J. M., and McWhorter, M.: *Electronic Amplifier Circuits: Theory and Design.* McGraw-Hill, New York, 1961.

AUTHOR'S NOTE: These texts give a clear explanation of the fundamentals. Various current applications are reported in:

IEEE Transactions on Engineering in Biology and Medicine. Institute of Electrical and Electronics Engineers, New York, N.Y.

Medical Instrumentation. Williams and Wilkins, Baltimore, Maryland.

Bio-Medical Engineering. United Trade Press, London, England.

10
Microminiaturization and Telemetry

With the advancement of technology, instruments to measure and control human physiological parameters can be made small, reliable, and implantable. The advantages of exploring and controlling the function of a living subject while it assumes "normal" activity need no explanation. The reader can easily suggest many research experiments or clinical applications *if* implant instruments are available. Implant instruments can generally be grouped into the following categories:

1. Telemetering: to measure the physiological signal in vivo.
2. Stimulation: to generate certain desired response of body organs by electrical signal in vivo.
3. Remote Controlled Manipulation: to control implanted machinery by remote signal to interact with the body organs.
4. Closed-Loop Control Systems: to use one part of the body to control or alter the function of other parts of the body, e.g., to control a paralyzed limb by implanted EMG transmitters in shoulder muscles.

The field of implant instrumentation is an "underdeveloped" area, with the exception of telemetering; other groups are mostly unexplored. They are technologically feasible, but before they can be realized many problems in materials, technology, physiology, and psychology must be solved.

The Present Situation

A survey of the work in implant instrumentation indicates that various telemetering devices have been developed and tested. These are implanted or surface mounted to transmit from one to eight channels of signals from a living subject to a remote receiver a few inches to several hundred feet away. The subjects include humans, birds, fish, and other animals. The heart, bladder, muscle, peripheral and sensory nerves, brain, and spinal cord have been stimulated with electrodes from a single pair to about 100 units in parallel. Small motors and other electromechanical devices have been implanted to generate motion or displacement in the body and, in the laboratory, the one-dimensional motion of a limb has been controlled by EMG from the shoulder muscle.[1] Table 10-1 lists some of the desired implant instruments suggested by research groups in the biomedical engineering field; most of these are still to be developed.

With the present microminiaturization technology, these instruments can be studied for feasibility. The key problems common to all the implant instruments at this time are:

1. Microminiaturization techniques, especially the modification of the technology for biomedical applications.
2. Power supply and heat removal of the implants.
3. Transducers that are compatible with the instrument and the body.
4. Packaging materials and techniques.
5. The implant technique and physiological effect to the body over the life of the implant.

Principles and Techniques of Microminiaturization

The idea of implantable biomedical instrumentation is very old, but until the development of microelectronic and miniaturization technology, these ideas appeared only as science fiction or laboratory curiosities. Since the discovery of transistors in 1948, the advances in solid-state electronic theory, devices, and techniques have been phenomenal. Integrated circuit packages are now generally available at low cost. These technologies have made it possible to extend man's control to millions of miles away in space, and will also make it feasible to probe inside the body of the living organism. Therefore, before we discuss telemetry, stimulation, and other implant instruments, let us

Table 10-1. Desired Implant Instruments

Name	Function
1. Heart Pacemaker	Adjust rate on demand, with redundancy for reliability, and telemeter alarm signals when difficulties develop.
2. Bladder Stimulation	Controlled by pressure inside, or can telemeter the critical pressure to outside and be controlled externally.
3. Blood Pressure— Controlling Device	Delivers stimulation automatically to control pressure, but can telemeter the condition to outside.
4. Pain Suppressor	At the spinal cord or peripheral nerve ending.
5. Brain Stimulation	For sleep and tranquilizing effect, controlled externally.
6. Sensory Stimulation	Hearing, vision, touch, etc., for prosthesis.
7. Muscle Stimulation	To have analog control of force or displacement by external command.
8. Stimulation of Glands	To control the secretion of hormones.
9. Telemetered Measurements	Pressure, blood flow, pH, and chemical compositions inside the body or at the organs.
10. Telemetry of Electro- physiological Signals	Such as ECG, EMG, EGG, EOG, ...
11. Implant Manipulators	To release chemicals or mechanical action upon external command, and to adjust the position of implant instruments such as brain electrodes or implant assistive devices.

inquire briefly about the principle and characteristics of microelectronics and miniaturization.[2] The development of present microelectronics may be divided into four stages:

1. Conventional circuits such as those in television sets and radios built ten years ago.
2. High-density packaged circuits as in hearing aids and present pocket radios and television.
3. Film circuits.
4. Integrated circuits.

The last two stages developed in parallel rather than in sequence, and are used in new computers and aerospace equipment. For comparison, we define packaging density as the number of components per cubic centimeter. In the four stages mentioned above, the packaging density increased from 0.1 to 10 to 10,000 to 100,000 per cm^3. The equivalent packaging density of the human brain is about 10 million, which is only 100 times greater than present-day small-scale integrated circuits.

Table 10-2 lists the developmental steps in microelectronics. For example, the development of a single-channel telemetry transmitter is

Table 10-2.   Microelectric Development

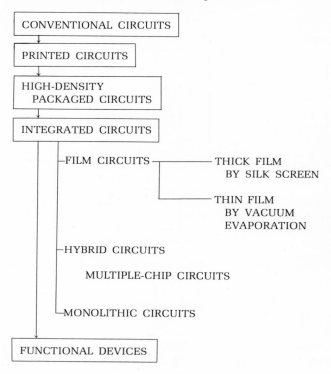

illustrated in Figures 10-2, 10-3, and 10-4. The three circuit diagrams designated K-1, K-5, and K-6 in Figure 10-1 were developed between 1961 and 1967[3] for single-channel biotelemetry with input sensitivity from 1 to 10 μv. Figure 10-2 shows the high-density packaged K-1 and K-5 transmitters. The discrete components are arranged closely together. The popular term "cord-wood structure" is used to name a way of arranging components like the piling of firewood into cords. By using the printed-circuit technique and microdot components, the K-5 circuit is packaged into a volume of less than 0.2 cm³. The high-density packaging is conventional circuit construction scaled down to a smaller size. At present, all the common electronic components (resistors, inductors, capacitors, transistors, relays, and switches) are being made commercially in miniature sizes. High-density package circuits require very little equipment cost.

Figure 10-3 illustrates the thin-film circuit of the passive components used in K-1 circuits. The components and connections are in the

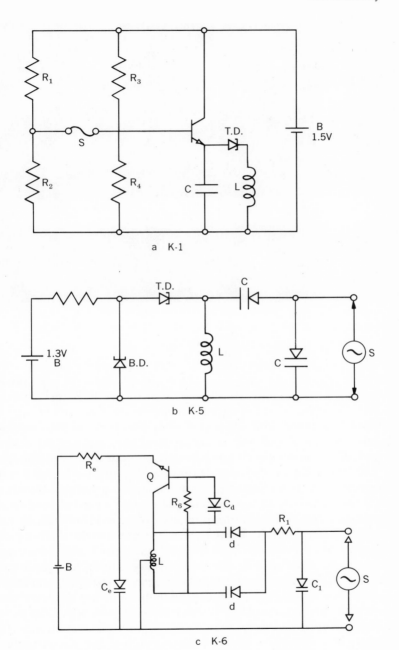

a   K-1

b   K-5

c   K-6

Figure 10-1. Circuit diagrams of K-series single-channel telemetry transmitters.

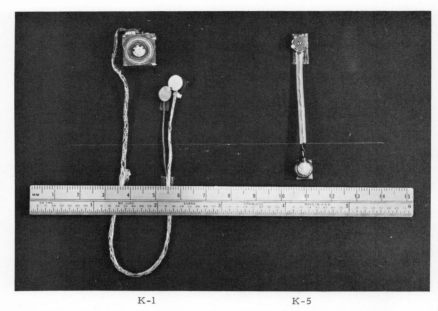

K-1   K-5

Figure 10-2. High-density packaged K-1 and K-5 telemetry circuits.

form of thin films of metals, insulators, and semiconductors with thicknesses of less than 1 micron. These components are deposited on an insulating ceramic substrate in specific geometric patterns so as to give them predetermined electrical properties to form electronic circuits. The first layer is aluminum evaporated to form the base conductors of the capacitor and some connections; layer 2, the four resistors; layer 3, the SiO insulator layer of capacitance and the protection layer for the resistors; layer 4, aluminum again, to form the upper conduct for capacitance and other connections; and layer 5, the inductor coil and antenna. The complete K-1 circuit is illustrated in Figure 10-3b. In this way, many components can be made simultaneously. After the transistor and diode are bound into the circuit, the transmitter will operate exactly as the conventional structured one. The dimensions of the film circuit unit are 1 cm square and 0.2 cm thick.

Figure 10-4 shows a monolithic integrated circuit K-6 transmitter. The whole wafer is about 2.5 cm in diameter and has about 80 transmitters on it. Each transmitter is approximately 0.2 × 0.3 cm in area and 0.025 cm thick. Here, all the circuit components including transistors and diodes were made simultaneously on the silicon wafer by photolithographic, diffusion, evaporation, and oxidation processes.

Figure 10-5 illustrates how transistors, resistors, and capacitors were made by these processes.

A hybrid of film circuit and integrated circuit can be made of components or a group of components in silicon chips of 1-mm^2 size by the integrated circuit method assembled on a film circuit. Figure 10-6 shows a hybrid circuit version of the K-6 transmitter.

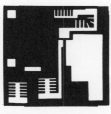

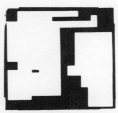

Mask 1 Aluminum   **Mask 2 Nichrome**   **Mask 3 Silicon monoxide**
conductors and first   **resistors**   **protection for resistors**
plate of the capacitor   **and dielectric for capacitor**

Mask 4 Final aluminum   **Mask 5 Copper coil**
conductors and capacitor   **plated on the back of**
plate   **the substrate**

a

b

Figure 10-3. a. Masks used to fabricate the passive components for K-1 transmitter by thin-film technique. b. Completed passive circuits.

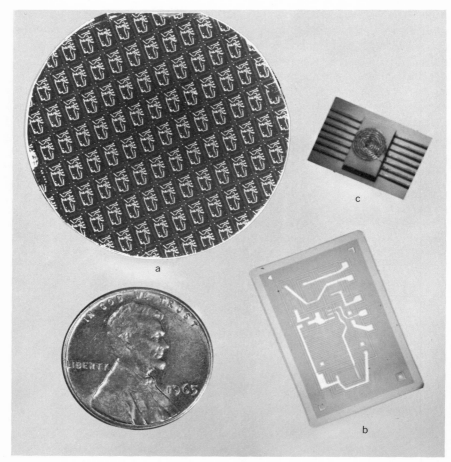

Figure 10-4. Monolithic integrated circuit K-6 transmitter: a. the wafer, b. the micrograph of a transmitter unit, and c. the finished unit.

If the circuit is simple, the reduction in size produced by the use of integrated circuit techniques is not great. The justification for making the circuit integrated must be other than size reduction. However, if the circuit is complex, such as a four- to eight-channel telemetry system that may contain thousands of components, then the integrated circuit technique is necessary to make it implantable.

There are many advantages of integrated circuitry (film, hybrid, and monolithic). Reduction in volume and weight can be accomplished with ease. The mechanical strength-to-mass ratio of integrated circuits is very great. They are tested to withstand 20,000 times the gravitational pull of the earth, and the failure rate of monolithic integrated circuits is shown to be below 10 parts per billion element hours. Better reliability makes complex systems possible. When the speed of electronic circuits is increased from one operation per millisecond (10^{-3} sec) to microsecond (10^{-6} sec) to nanosecond (10^{-9} sec), the time that the signal takes to travel through the circuit can no longer be neglected. There will be a time delay due to the propagation of the signal in the circuit. This propagation time is greatly reduced by making the circuits close together and very small. Although integrated circuits are made by costly, elaborate, and precise processes, many circuits can be made at one time; and once a circuit is developed, it can be mass produced. At present, many integrated circuits are available at fractions of the cost of equivalent conventional circuits.

The integrated circuit also has its disadvantages, however. From the point of view of biomedical application, the component tolerance is large and individual trimming is very expensive. The initial cost is very high, and the yield of integrated circuit decreases rapidly with the circuit complexity. As the number of components increases, the problem of power supply and heat removal of a large system is not solved. Handling and servicing are more difficult because of the small size and complexity. Finally, advanced engineering effort is required for its design. When such a circuit is justified by either the quantity, performance characteristics, or size reduction, the integrated circuit designer must also consider general factors in addition to conventional circuit design.

Tolerance, stability, and limited range of component values may cause restrictions in design. Relative cost of active and passive elements

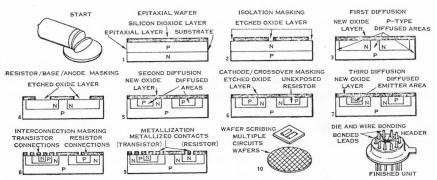

Figure 10-5. Sequential steps of integrated circuit fabrication.

1 cm

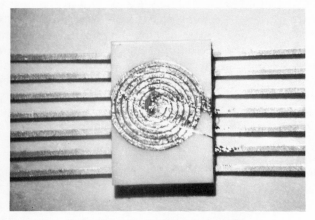

Figure 10-6. Hybrid integrated K-6 transmitters.

is important when the cost of an active element is the same as, or even less than, that of a passive element. Due to the close proximity of elements in an integrated unit, the leakage conductance, parasitic capacitance, and mutual inductance between circuit elements and circuit blocks present serious problems in circuit design. The power supply, heat removal assembly, handling, and servicing of an integrated circuit are severe design problems.

The technology used to make microelectronic circuits can be applied to fabricate other components. Switches, relays, quartz crystals, and mechanical resonant structures have been miniaturized to milli-meter or micron size. The sources of these miniature components can be found in electronic catalogs or buyers' guides.

Telemetry

Let us define biotelemetry as the biomedical instrumentation tech-nique for obtaining information from the body of an unrestrained living organism at a remote location through wireless transmission linkage. The transmitter can be implanted or surface mounted. When implanted, the transmitter is located totally within the body of the organism. The location may be intracavitary, such as within the intestine, mouth, bladder, etc., or subdermal, or deep in the tissues. The following dis-cussion is centered on implant transmission, but will be equally appli-cable to surface telemetry.

Although radio transmission of analog signals has been known since 1884, and FM radio links were used to transmit pneumograms in 1948,[4] extensive development of biomedical telemetry techniques did not proceed until the transistor was discovered in 1948 and made commer-cially available after 1954.

In the late 1950's and early 1960's, Mackay, Noler, Wolff, Zworykin, and others developed active radio transmitting units for use in the gastrointestinal tract and other body cavities. Subcutaneous and deep body implantation of telemetry units was initiated recently to measure physiological information in animals as well as humans. A good histori-cal review of areas of biomedical telemetry has been given by Caceres[5] and Mackay.[6] Surveys as well as bibliographies on biotelemetry are also available.[5-12] Many types of physiological signals have been tele-metered. The characteristics of these signals are summarized in Ta-ble 10-3.

In implant-telemetry systems, the receiving units are generally standard and are commercially available. Therefore, nearly all the research and design efforts are concentrated on the transmitting units, which must be designed to meet the special requirements of each experiment. The common design criteria of an implant transmitter can be stated as follows:

1. Small size and weight: less than a few percent of the subject size and weight.
2. Packaging: packaged with "nontoxic and nonpermeable" mate-rials, and with proper shape to reduce tissue reaction and to protect the implant from body fluid.

Table 10-3.   Transducers and Signal Characteristics of Some Physiological Parameters

Physiological Parameters	Transducer	Amplitude Range	Frequency Range
Electrocardiogram (EKG)	Electrodes	0.75–4 mv pulse	0.1–100 cps
Electroencephalogram (EEG)	Electrodes	10–75 μv	0.5–200 cps
Electrogastrograph	Surface electrodes	10–350 μv	0.05–0.2 cps
Electromyogram (EMG)	Electrodes	0.1–4 mv pulse	2–10⁵ cps (10–500 cps clinical)
Eye Potential EOG or ERG	Electrodes	500 μv	0–250 cps
Nerve Potentials	Electrodes	3 mv peak	up to 1000 pulses/ second rise time 0.3 μs
Bladder Pressure	Strain gauges	0 to 100 cm H_2O	0–10 cps
Blood Flow	Electromagnetic flowmeter; ultrasonic flowmeter	1–300 cm/sec	1–20 cps
Blood Pressure	Strain gauge on artery; hydraulic coupling to transducer; cuff gauges	0–400 mm Hg	0.5–100 cps
Gastrointestinal Pressure	Variable inductance	20–100 cm H_2O	0–10 cps
Intestinal Forces	Strain gauges	1–40 g	0–1 cps
Respiration Rate	Electrode impedance; piezoelectric devices; pneumograph		0.15–6 cps
Stomach pH	Glass electrode; antimony electrodes	3–13	0–1 cp-min
Temperature	Thermistor; thermal expansion	90°–100°F	0–0.1 cps
Tidal Volume	Impedance; pneumograph	50–1000 ml per breath	0.15–6 cps

3. Low power consumption and long lifetime.
4. Reliability and ease of handling.
5. High sensitivity (microvolts) and wide dynamic range, to handle signals ranging from microvolts to millivolts.
6. Good fidelity: signal frequency response from DC to several KHz or higher.

7. Suitable transmission range to permit the free movement of the subject and the use of units with the same frequency at nearby locations without interference.
8. Compliance with F.C.C. and local regulations for radio transmission.

A block diagram of a typical radio telemetering system is shown in Figure 10-7. Nearly all of the present telemetry transmitters are assembled with high-density packaging techniques. Because of the size limita-

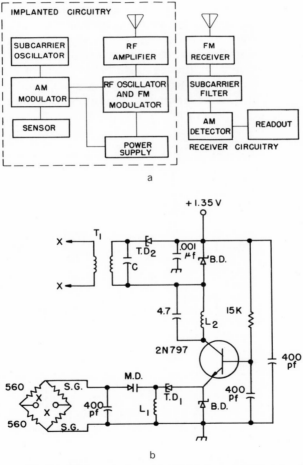

Figure 10-7. a. Block diagram of a biotelemetry system. b. Nonpowered system.

tion, these transmitters are designed to use one transistor and minimum components, with the sacrifice of circuit performance. This limitation will be removed once the integrated circuit technique is adopted.

The passive implant class of circuits does not contain energy sources. One form of implant contains a resonant circuit in which the resonant frequency, made to vary with body signal, can be detected by a grid dip meter.[13,14] Other passive implants contain transmitters powered by pulses of radio energy stored on a capacitor.[15] Powered implants include those in which the carrier may or may not be modulated, and a narrow band receiver is generally used (Fig. 10-8).[16,17] Other units use a pulsed carrier in which the carrier signal of the transmitter is interrupted pulses, and the information may be represented by pulse frequency or duration.[18-20] This has the advantage of low average power and simplification, but the disadvantage of a wide frequency band that causes interference and frequency instability. In increasing order of complexity is the continuous-carrier single-channel unit in which the signal may be amplitude or frequency modulated on the carrier. Frequency modulation is generally preferred.[3,21,22]

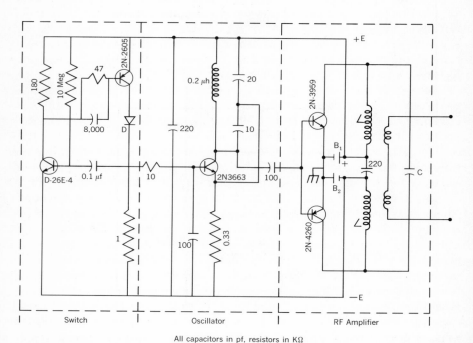

All capacitors in pf, resistors in KΩ

Figure 10-8. Diagram of a subcarrier telemetry system.

Table 10-4. Performance of K-1 and K-5 Transmitters

Item	K-1	K-5	K-6 Hybrid
Size, less battery	1.3 × 1.3 × 0.4 cm	0.8 cm diam × 0.2 cm	0.9 × 0.6 × 0.2 cm
Weight, less battery	1.7 g	0.44 g	0.6 g
Power Consumption	1.3 volts, 1.2 ma	0.2 volt, 1.2 ma	1.3 V, 0.5 ma
RF Frequency	80–100 mhz	100–250 mhz	100–250 mhz
System Noise (in shielded room at 1-kc bandwidth)	0.5 μv	3 μv (at 10-kΩ source impedance)	3 ~ 5 μv
Input Sensitivity (6-db S/N)	1.0 μv	7.0 μv (10-kΩ input)	7–10 μv
Dynamic Range (limited by receiver)	2 × 10³	10³	10³
Frequency Response	0.01–20 khz	0.01–20 khz	0.01–20 khz
Input Impedance	5 to 8 kΩ	300 kΩ to several mΩ	300 kΩ to several mΩ
Transmission Range	5 μv at 3.66 m (with 10-cm lead)	5 μv at 1 to 2 m (without leads)	5 μv at 1 to 2 m (without leads)
Carrier Frequency Temperature Stability	Poor	Better than 0.05%°C	Better than 0.05%°C

Finally, in the multiplex system several channels of physiological signals can be transmitted by a common RF carrier. Frequency division multiplex[1,23] and time division multiplex are both in use.[24,25]

For a detailed explanation of the circuit operation and performance, refer to the original literature and survey papers. The performance of K-1, K-5, and K-6 is given in Table 10-4.

Batteries or means of converting environmental energy to electrical energy serve as the power supply of these transmitters. Both implanted and surface-mounted units require protective coatings. The packaging material is discussed later.

When used properly, these telemetry units can transmit short-term physiological changes with results better than those of wire systems. Because the transmitter is miniaturized and close to the source, the 60-Hz pickup and signal frequency interferences are greatly reduced. The high-frequency response is also generally improved. All the signals listed in Table 10-3 have been telemetered satisfactorily. However, all the simple transmitters have the problem of carrier frequency drift, especially when the subject is moving relative to conductive objects in the environment. This is due to the change of loading reflected to the oscillator. This drift tends to mask any slow variation in the signal. If automatic frequency control is used to stabilize the frequency, the slow varying signal will be lost. To overcome this problem in long-term telemetering of very low frequency (0.5 Hz or less) signals, a subcarrier system (Fig. 10-8) is needed.[3]

An integrated circuit, implantable, time division, multiplex system has been developed to transmit all types of physiological signals.[26] The four-channel test transmitter, without power supply, occupies a volume of less than 0.5 cubic inch and weighs less than 1 oz. It has been functioning for more than eight months in a dog.

The standard IRIG subcarrier frequency bands and optional bands, as well as discussion of various modulation schemes, can be found in telemetry textbooks.[27-30]

In selecting a telemetry system, one should consider the number of signal channels, signal frequency and amplitude, telemetering distance, time duration of telemetry, and funds available. For example, thought must be given to whether several single-channel units or a multiplex system should be used. The common conclusion at present is that the single-channel units may be simpler and cost less for two or three channels, but that beyond four to six channels, the multiplex system is preferable.

In the choice between single-channel units, the pulsed circuits consume much less average power, but may have a limitation on signal frequency and fidelity, and are used for slow signals such as temperature and pressure. For wide-band and high-fidelity signals, the continuous-carrier FM circuits are generally preferred. For short-term AC signals, the simple transmitter is satisfactory; for long-term slow varying signals, either a subcarrier system or a built-in calibration method should be used.[19]

Generally, the selection must be made on an individual basis. More detailed discussion can be found in the telemetry textbooks.[6,7,28,29] However, there are common problems for all telemetry systems that require investigation.

Aside from the power supply and the packaging materials to be discussed later, the most important development is in the area of micropower circuits. It is always desirable to reduce the power consumption if the performance can be maintained. Research in both circuit design and electronic devices will be needed to develop practical micropower systems. As an example, the pulsed-carrier circuit consumes about 10 μw average power by using a low-duty cycle. With careful study, the duty cycle can be reduced to 10^{-5}, and the average power drain will be below 1 μw with the same or better performance.[20]

With the adoption of integrated and micropower circuit techniques, complex circuits can be assembled in smaller space and consume less power than conventional units, thus raising the performance of the unit. The next generation of telemetry devices will be designed for performance, not simplicity.

Implantable transducers that convert the body parameters into electrical signals (potential, current, or impedance) are still to be developed. Many desirable transducers do not exist; others are not compatible with the implant telemetry, either in size or in lifetime. Research and development in biotransducers will be a major field for further engineering activity.

Since the field of application varies, the standardization of final systems is virtually impossible. In order to justify the high setup cost, and to take advantage of the mass production ability of integrated circuits, the reasonable solution will be the modular approach. The major building blocks of all biotelemetry systems will be made in modular blocks which can be assembled to form specific systems with minimal external circuit elements. The circuit blocks in Figure 10-8 illustrate preliminary attempts to develop such modularized systems. Each of the three blocks in Figure 10-8 can be packed in a flat pack, and can be connected externally for[31] (a) short- or long-range applications, (b) continuous- and pulsed-carrier systems, and (c) basic blocks of time division multiplex systems.

Each block performs a separate function. The switch block is a complementary astable multivibrator used so that the circuit consumes no current during the "off" period. The diode D_1 controls the duty cycle of the circuit down to $1:10^{-5}$, and compensates the variation of the battery voltage. The frequency and pulse duration stay constant with the battery voltage change from 2.7 to 2.0 volts. The oscillator block is a standard Copitts oscillator which operates only when the switch is "on." The output amplifier operates in the class c nonlinear region and doubles the oscillator frequency.

The pulse rate or pulse width of the switching circuit can be modulated according to the input information. The frequency of the oscillator tank circuit also can be modulated by signals applied to the silicon capacitor in the resonant circuit of the oscillator.

The power drain of the switch and the oscillator without the amplifier is about 2 μw, and the unit has a range of 15 feet. If the amplifier is added, the drain is 15 to 20 μw, and the range can be extended to 100 feet.

Stimulation, Implant Control, and Manipulators

Electrical stimulation of the nervous system for remote control of animals was tried as early as 1934.[32] However, its widespread clinical application is relatively recent. The heart pacemaker is the best known clinical device. In addition to the heart, stimulation has been applied to the bladder for voiding; the phrenic nerve for diaphragmatic con-traction; baroreceptors for reduction of blood pressure; the spinal cord for pain suppression; and paralyzed muscle for regaining motor functions. With recent developments in brain stimulation, it is possible that it can be used to bypass sensory organs to generate hearing or some degree of vision directly at the central nervous system level,[33] as well as to temporarily alter personality[34] or induce sleep.[32-56]

The stimulating circuit is generally a simple radio-powered passive resonant circuit. Batteries, rechargeable batteries, radio power, and energy converters have been used to power these implant stimulators. The stimulating parameters for heart, bladder, nerves, and muscles are summarized in Table 10-5.

The major problems in stimulators are power supply, packaging, and electrodes. The requirements for electrodes include: (a) minimal corrosion and erosion by body fluid, (b) fatigue resistance to repeated stress, and (c) minimum body reaction to limit the increase of threshold due to peri-electrode fibrosis. Multistranded wires of gold, platinum, stainless steel, and Elgiloy are used; however, coils or helical-wound multistranded wires provide better flexibility and less fatigue. Elgiloy is reported to have the best fatigue resistance, but corrodes seriously when used as an anode. Platinum electrodes have minimum corrosion and erosion when used for biphasic stimulation pulses. Bipolar electrodes are generally more effective than monopolar ones.

Many electrode problems remain to be solved. For brain stimulation, high-density electrodes of the order of 1000 electrodes per cm^2, and the associated electronic circuits and systems, are still to be developed. Although now in its infancy, stimulation is believed to have great potential in the future. Many applications of stimulation to control pain,

Table 10-5.   Stimulating Parameters

	Heart	Bladder	Nerve	Muscle	
Pulse Width (msec)	0.1–10	0.1–10	0.05–1	0.1–5	
Pulse Rate (pulse min)	30–200	15–35	5–90	30–80	
Pulse Voltage (volt)	0–9	0–10	0–7	25 Intramuscular	0–56 Surface
Pulse Current (ma)	4–45	50	—	3–6	0–30 max

hormone secretion, mental disturbance, and to partially compensate for defective organs can be suggested.

The combination of telemetry and stimulation forms a control system that may pick up a signal from one part of the body, such as the brain or a muscle, to control the action of other parts of the body such as a paralyzed limb. This then can serve as a bypass to replace damaged nerve. Some preliminary work in this direction has shown results that would justify further investigation.[53]

Environmental Energy Supply

At present, power supply is the major problem of implant instrumentation. Primary batteries have a limited lifetime, and existing rechargeable batteries have a low energy density; also their lifetime is limited to a few hundred charging cycles.[57] From the literature, other methods of power supply investigated can be grouped as follows[58-60]:

1. Electromagnetic Induction: An electromagnetic field is created by an external source oscillator. The energy is coupled to the implanted resonant circuit and then rectified.
2. Mechanical-to-Electrical Energy Converter: The rhythmic body movement is converted into electrical energy by piezoelectric devices or mechanical springs similar to those used in automatic watches. The stimulated motion of muscles has also been investigated for energy generation.
3. Chemical-to-Electrical Energy Converter: The galvanic reaction between two dissimilar metals has been used to generate electrical energy in an animal. In vitro experiments are being conducted to devise fuel cells that use oxygen and/or other chemical gradients in the body to generate electrical energy.
4. Biological Potential: The body generates many electrical potentials in the heart, muscles, stomach lining, and other parts. If the small voltages generated (10 to 100 mv) can be transformed by electronic means, they can be used as a source for implants.
5. New Types of Long-Life Batteries: Nuclear batteries and solid electrolyte batteries with 5 to 20 years of life are being investigated.

Among all these possible methods, the electromagnetic induction method is the best developed. It has been used in research and clinical applications.[61,62] The mechanical converters have been demonstrated to be feasible and are being evaluated.[63] Other types of power sources are still in the early research stage.

In the electromagnetic induction power supply area, the frequencies used are in the 100 kHz to 20 mHz range. The major considerations in selecting a system are:

1. The maximum field intensity and the power loss in the tissue. The values should be far below the limits that might cause any hazards to the body. The accepted safe level for whole-body continuous radio frequency dissipation is 0.01 watts per cm^2. This is derived from one-half of the field that may cause the temperature to rise 1°C in tissues under radiation.[64]

2. The transmission efficiency of radio energy. It is estimated that below 250 mHz the total loss in the skin and fat layer is smaller than the available power at the muscle-fat interface, and the transmission efficiency through the skin and fat layer is around 20 to 30 percent.[61]

3. The range of transmission and the source power. The required power of the source oscillator increases rapidly with the area or volume in which the power transfer is needed. At 1 mHz frequency, a 30-watt oscillator is able to transmit 7 mw per cm^2 power to an implant resonant circuit in rats over a volume of 14" by 18" by 14", whereas a 200-watt source covers a cage of 24" by 30" by 14".

4. The method to insure power transfer when the subject is changing its orientation. Three or two mutually perpendicular coils may be used either for the implant detector or for the transmitting source loop.[57,58,61]

Figure 10-9 shows the circuit diagram and the structure of a radio frequency induction power detector. The three mutually perpendicular resonant coils form the structure of a 1-cm diameter sphere, within which other circuit components are housed. This unit has been used in rats for six to eight months with good results. Figure 10-10 shows a photograph of such a rat and the ECG telemetered from the instrument. In the mechanical-energy converter area, slabs of sintered piezoelectric crystals that convert deformation to electric charges can be used to convert rhythmic motion into electrical energy. Two such slabs cemented together form a laminated beam referred to as a piezoelectric bimorph. When a bimorph is bent, one side undergoes tension, the other compression. The output of two sides can be connected in series or in parallel externally. Bimorphs have been coupled to aorta, heart, rib cage, and diaphragm to derive electrical energy from motion. The

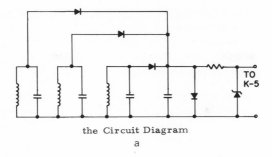

the Circuit Diagram

a

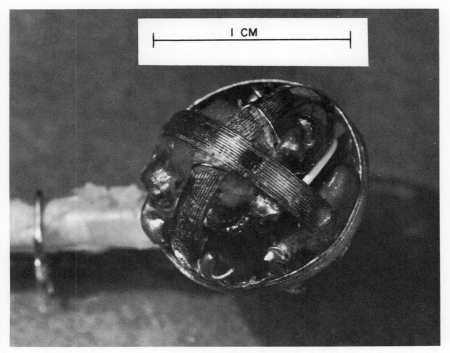

The Structure

b

Figure 10-9. Radio frequency induction power detector. a. Circuit diagram.
b. The structure.

failure to find a flexible insulating package material to protect the
bimorph has limited the life of such devices. Figure 10-11 illustrates
the design of a converter in which the bimorph is an end-loaded
cantilever beam housed in a sealed chamber. The beam is set in reso-

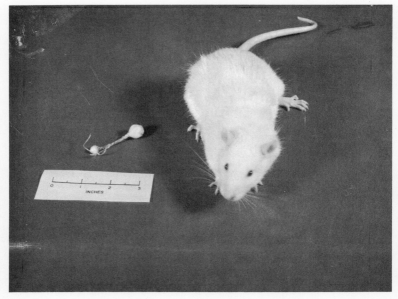

a

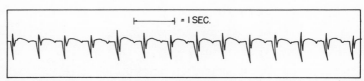

BATTERY POWERED K-5 IN A MOUSE. HEART RATE AFTER OPERATION (9-3-63)

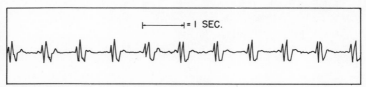

RF POWERED K-5 IN A RAT. HEART RATE AFTER OPERATION (2-2-64)

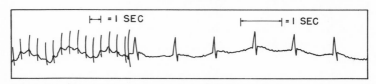

RF POWERED K-5 IN A RAT. HEART RATE AND RESPIRATION RATE (2-8-64)

b

Figure 10-10. a. Rat with a transmitter implanted. b. The ECG transmitted by
the RF-powered K-5 transmitter.

nance by the heart motion coupled to the unit. The AC electric current is rectified by a voltage doubler, using the output capacitance of the bimorphs as one capacitator. The power is then used to power a pace-maker.[63] This design eliminates the need for flexible packaging material, and also raises the converter efficiency by operating in the resonant mode and using the voltage doubler circuit.

Other mechanical converters have been investigated, including the use of air or fluid flow in the body to drive a turbine-generator,[65] and the use of the quadriceps femoris muscle to operate a pump.[66]

Much work has been reported on the implant action of platinum black and steel or zinc electrodes in the body to generate electrical energy.[69] However, a galvanic reaction occurs, and the dissolving metal accumulates in the pancreas and the kidney.[60]

The conversion of chemical energy into electrical energy by fuel cells has not been satisfactory in biological systems. However, many related research projects are being carried on.[67,68]

The use of electrical potential generated in the body for powering implants has been examined. Rehm reported that an energy of 9 μw

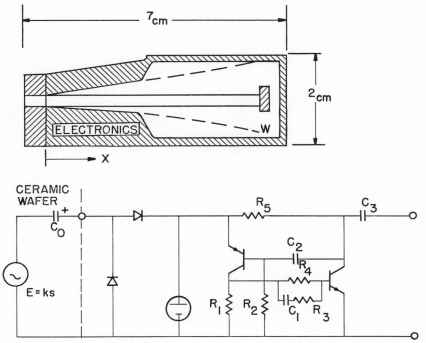

Figure 10-11. The piezoelectric energy converter and the self-powered pace-maker circuit.

per cm^2 can be obtained from stomach tissue.[69] Steel electrodes in the cat's brain were shown to be able to extract power of 40 mv and 0.5 μa.[70] However, the long-term feasibility of these approaches has yet to be established.

The search for long-life batteries is being carried on; the solid electrolytic battery and nuclear battery of small current and long life seem promising.[71]

Packaging Material and Techniques

All implant electronic devices require a suitable package to inter-face the circuit and the body tissue. The packaging coating serves to protect the circuits from the body fluid, which is highly conductive and corrosive, and to present to the tissue a nontoxic and stable surface.

The physical properties of coating materials for the protection of the circuits must include good mechanical strength, low permeability to water and ions in the body fluid, stability under the body environ-ment, easy handling or moldability, ability to be sterilized, and minimal impurity absorption.

Gold, platinum, some stainless steel, titanium, and vitallium have proved acceptable. Silicone rubber, polyethylene, teflon, and wax have been implanted with various degrees of success. There is considerable evidence that the residual monomers are more toxic or antigenic than are the large-molecular-weight polymers. Many of the plasticizers, curing agents, and antioxidants are toxic. Other impurities and addi-tives such as catalysts, tapped free radicals, fillers, inhibitors, and colorants may be toxic, and may vary from batch to batch.

The contact of a large solid surface to body tissue may cause irritation and carcinogenic reaction in animals over a period of years. The hardness of the implant should be matched as closely as possible to that of the tissues. The so-called Oppenheimer effect indicated that, in rats, implanted sheets of many materials caused cancer whereas perforated sheets and small particles did not. Porous metal, plastics, knitted sheets, or velours structure are desirable.

The site of the implant is important in the carcinogenic reaction in animals. Implantation at bony areas generally irritates more than in soft body tissues. The surface condition and cleaning process are reported to affect the blood clotting and immune reactions. A smooth, carefully cleaned surface is desirable. The shape of the implant and the pressure caused by its weight are important in body reaction, due to their role in altering a blood circulation pattern at the implant site.

Finally, the surgical procedure itself may influence the local reaction at the site.

From the considerations of mechanical and chemical protection and body inertness, a three-layer package is used in the author's laboratory. The layer immediately in contact with the electronic circuit is a soft plastic material with high resistivity and low dielectric loss. The intermediate layer is a hard epoxy to provide mechanical protection and a sufficient fluid barrier. The outer layer is medical-grade silicone rubber (Dow Corning No. 782) to insure minimal body reaction. Glass encapsulation has also been attempted.

Implant Instrumentation—Promises and Problems

From the above discussion, the excellent potential of implant instruments can be recognized. If the problems in material, power supply, microminiature technology, and electronic design can be solved in time, the implanted instrument can control organs, assist their functions, and partially rectify abnormality.

The role of implant instruments in biomedical engineering is similar to that of space instrumentation in interplanetary exploration. If one uses a sphere of one unit radius to represent the region, a person can have direct sensory contact. Then going outward, there is the community, the earth, and outer space; going inward, there are the body systems, the organs, and the cells.

The inward world is as vast and complex as the outer one. The implant instrument is a means to reach inward, a tool to probe the phenomena of life. The final result will depend upon how the tool is used, but at this time "tool engineering" is needed to make the tool available for research and clinical investigations.

References

1. Robuck, R. B., II, and Ko, W. H.: IEEE Trans. BME-14, 1967, pp. 40–46.

2. Warner, R. M., Jr., and Fordemwalt, J. N.: *Integrated Circuits.* McGraw-Hill, New York, 1965.

3. Ko, W. H., and Neuman, M. R.: Science 156:351, 1967.

4. Fuller, J. L., and Gordon, T. M.: Science 108:287, 1948.

5. Caceres, C. (ed.): *Biomedical Telemetry.* Academic Press, New York, 1965.

6. Mackay, R. S.: *Bio-Medical Telemetry.* John Wiley, New York, 1968.

7. Slater, L. (ed.): *Bio-Telemetry Symposium.* Pergamon Press, New York, 1963.

8. Murry, W. E., et al.: *Biomedical Sciences Instrumentation.* Plenum Press, 1963–1964, Vols. I and II, 1963.

9. Barwick, R. E., and Fullagar, P. J.: Proc. Ecol. Soc. Aust. 2:27, 1967.

10. Davis, J. F.: *Bibliography List on Biotelemetry.* International Institute for Medical Electronics and Biological Engineering, Paris, France, 1967.

11. Schladweiler, J. L., and Ball, I. J., Jr.: *Telemetry Bibliography Emphasizing Studies of Wild Animals Under Natural Conditions.* Report 15, Bell Museum of Natural History, University of Minnesota, Minneapolis, Minnesota.

12. Namura, S., Ko, W. H.: *A Bibliography of Biomedical Telemetry and Stimulation from 1966 to 1968.* Report 510, Solid State Electronics Laboratory, Case Western Reserve University, Cleveland, 1968.

13. Collin, C. C.: Proc. 19th Ann. Conf. Engrg. Med. Biol. 1966, p. 171.

14. Farrar, J. T., Berkley, C., and Zworykin, V. Y.: Science 131:1814, 1960.

15. Naqumon, J., et al.: IRE Trans. BME-9, 1962, pp. 195–198.

16. Cochran, W. W., et al.: Bioscience 15:98, 1965.

17. Craighead, F. C., Jr., and Craighead, J. J.: Bioscience 15:88, 1965.

18. Mackay, R. S.: IRE Trans. ME-6, 1959, pp. 100–105.

19. Fryer, T. B., et al.: J. Appl. Physiol. 21:295, 1966.

20. Lin, W. C., and Ko, W. H.: Med. Biol. Engrg. 6:309, 1968.

21. Tishler, H., and Fri, E. H.: IRE Trans. BME-10, 1963, p. 29.

22. Young, I. J., and Naylor, W. S.: Amer. J. Med. Electron. 3:28, 1964.

23. Zweizig, J. R., et al.: IEEE Trans. BME-14, 1967, pp. 230–238.

24. Marko, A. R.: Proc. 1965 Int. Telemetry Conf. pp. 253–256, 1965.

25. Thompson, W., Ko, W. H., and Yon, E.: Nat. Telemetry Conf., Houston, Texas, 1968.

26. Ko, W. H., Yon, E., and Ramseth, D.: SSEL Report No. 512, Engineering Design Center, Case Western Reserve University, Cleveland (to be published).

27. *Telemetry standards.* Telemetry Journal 1:11, 1966.

28. Bordon, P. A., and Mayo-Wells, W. J.: *Telemetering Systems.* Reinhold, New York, 1959.

29. Foster, L. E.: *Telemetry Systems.* John Wiley, New York, 1965.

30. Nichols, M. H., and Rauch, L. L.: *Radio Telemetry.* John Wiley, New York, 1957.

31. Lin, W., Yon, E., and Agarwal, A.: SSEL Report No. 511, Engineering Design Center, Case Western Reserve University, Cleveland (to be published).

32. Chaffee, F. E., and Light, R. U.: Yale J. Biol. Med. 17:83; 17:441, 1934.

33. Brindley, G. S., and Lewin, W. S.: J. Physiol. 196:479, 1968.

34. Delgado, J. M. R.: Science 148:1361, 1965.

35. Kantrowitz, A.: Ann. N.Y. Acad. Sci. 3:1049, 1964.

36. Zoll, et al.: Ann. N.Y. Acad. Sci. 3:1068, 1964.

37. Bonnaheau, R. C., et al.: J. Thorac. Cardiov. Surg. 50:857, 1965.

38. Glenn, W. L., et al.: Ann. Surg. 160:338, 1964.

39. Albert, M., et al.: Ann. N.Y. Acad. Sci. 3:889, 1964.

40. Richwien, R., and Millner, R.: Med. Biol. Engrg. 4:193, 1966.

41. Cobbold, R. S. C., and Lopez, J. F.: Med. Electron. Biol. Engrg. 3:273, 1965.

42. Bradley, W. E., et al.: J. Urol. 90:575, 1963.

43. Kantrowitz, A.: Arch. Phys. Med. 46:76, 1965.

44. Kantrowitz, A.: Amer. J. Gastroent. 44:57, 1965.

45. Scott, F., Quesada, M., Cardus, D., and Laskowski, T.: Invest. Urol. 3:231, 1965.

46. Stenberg, C. C., Burnette, H. W., and Bunts, R. C.: J. Urol. :79, 1967.

47. Hald, T., et al.: J. Urol. 97:73, 1967.

48. de Villiers, R., Nose, Y., Neiser, W., and Kantrowitz, A.: Trans. Amer. Soc. Artific. Int. Org. 10:357, 1964.

49. Johnson, E. B.: Proc. Nat. Telemetry Conf. 1966, pp. 178–180.

50. Sheldon, C. H., Pudenz, R. H., and Doyle, J.: Amer. J. Surg. 114:209, 1967.

51. Mortimer, J. T., Reswick, J. B., and Shealy, C. N.: Proc. 20th Conf. Engrg. Med. Biol. 1967, p. 187.

52. Wall, P. D., and Sweet, W. H.: Science 155:108, 1967.

53. Lorig, R. J., et al.: Proc. 20th Ann. Conf. Engrg. Med. Biol. 1967, p. 28.

54. Frei, E. H., et al.: Proc. Conf. Engrg. Med. Biol. 1966, p. 65.

55. Vautrappen, G., et al.: Gut 5:96, 1964.

56. Silgalis, E. M.: M.S. Thesis, Case Western Reserve University, Cleveland, 1967.

57. Kuechle, L. B.: BIAC Module M-10, BIAC-AIBS, Washington, 1967.

58. Konikoff, J. J.: BIAC Module M-9, BIAC-AIBS, Washington, 1967.

59. Long, F. M.: Int. IRE Convention Record 10-9, 1962, p. 68.

60. Noel, B.: M.S. Thesis, Engineering Division, Case Institute of Technology, 1966.

61. Ko, W. H., and Yon, E.: 6th Conf. Med. Electr. Biol. Engrg., Tokyo, 1965, pp. 206–207.

62. Eisenberg, L., Mauro, A., and Glenn, W.: IRE Trans. Biomed. Electron. BME-8, 1961, pp. 253–257.

63. Ko, W. H.: Proc. 19th Conf. Med. Biol. 1966, p. 67.

64. Schwan, H., and Li, L.: Proc. IRE, 1956, pp. 1572–1581.

65. Sarnoff, S. J., and Beglund, E.: Circ. Res. 1:331, 1953.

66. Kusserow, B. K., and Clapp, J. F., III: Trans. Amer. Soc. Artific. Int. Org. 10:74, 1964.

67. Brake, J.: U.S. Dept. of Commerce, AD-619-655, 1965.

68. NASA Research Report, Contract No. NASw-654, Report No. 25144, Marquart Corp., Van Nuys, Calif., 1964.

69. Rehm, W.: Amer. J. Physiol. 154:148, 1948.

70. Pinneo, L. R., and Kesselman, M. L.: ASTIA Doc. AD-609067, 1959.

71. Kestenbach, H. J.: M.S. Thesis, Case Western Reserve University, Cleveland, 1967.

72. Keenjian, E.: *Microelectronics.* McGraw-Hill, New York, 1963.

73. Grove, A.: *Physics and Technology of Semiconductor Devices.* John Wiley, New York, 1967.

74. *Integrated Silicon Device Technology.* Research Triangle Institute, Durham, N.C., ASD-TDR-63-316, Vols. 1–12, 1963–1966.

75. *Medical and Biological Applications of Space Telemetry.* NASA Report, SP-5023, 1965.

76. Haggerty, P. E., et al.: Proc. IEEE 52:1400, 1964.

77. Hittinger, W. C., and Sparks, M.: Sci. Amer. 213:56, 1965.

78. Block, B., and Hasting, G. W.: *Plastics in Surgery.* Charles C Thomas, Springfield, Ill., 1967.

79. Weyer, E. M. (ed.): *Materials in biomedical engineering.* Ann. N.Y. Acad. Sci. Vol. 146, Part I, 1968.

80. *Plastics in Surgical Implants.* Symposium of F-3 Committee, A.S.T.M. Special Technical Publication No. 386, 1965.

81. Bachtol, C. O., Ferguson, B. B., and Laing, P. G.: *Metals and Engineering in Bone and Joint Surgery.* Williams and Wilkins, Baltimore, 1959.

82. Levine, S. N. (ed.): J. Biomed. Mor. Res., 1967.

83. Grau, H. R.: *Bibliography of foreign substance implants in reconstruction surgery.* Transplantation 2:306, 1964.

84. Austian, J.: J. Biomed. Mat. Res. 1:433, 1967.

PART FOUR

Biomaterials

11

Materials for Bioengineering Applications

Historically, the use of engineered materials to replace living tissue from a prosthetic point of view is as old as both medicine and materials technology. Much of the early work pertained to replacement of hard tissue and attempts at reduction of fractures by fixation with metal implants. Ideographs have indicated employment of metals for these applications in the ancient Egyptian and Etruscan civilizations.

Twentieth century materials technology has provided a wide spectrum of materials and processing techniques capable of generating prosthetic materials which can be subcutaneously implanted with varying degrees of success. In addition, engineering technology has created a body of knowledge capable of creating functioning artificial organs and life support systems. These organs and systems, however, must be created from engineered materials which, in turn, must interact either intracorporeally or extracorporeally with the biological environment of the host.

The widespread current usage of artificial engineered materials in the body includes the following:

A. Permanent implants for function
    1. Orthopedics
        a. Fracture reduction, i.e., splints, plates, screws, intermedullary nails, etc.

       b. Orthoplasty, i.e., replacement of femur heads
       c. Complete joint replacement
    2. Cardiovascular
       a. Cardiac valves
       b. Arterial and venal prosthesis
       c. Extracorporeal heart-lung machines
       d. Implantable organ replacements and boosters
    3. Respiratory patches and tubes
    4. Digestive system
    5. Genitourinary system
       a. Extracorporeal artificial kidney
       b. Implantable functioning kidney
       c. Ureter
       d. Urethra
    6. Nervous system
       a. Hydrocephalus valves
    7. Orbital
       a. Corneal and lens prosthesis

B. Permanent implants for cosmesis
    1. Maxillofacial
    2. Breast
    3. Testes

C. Other applications of engineered materials in clinical practice
    1. Electrical probe material for sensors, cardiac pacemakers, etc.
    2. Sutures, dressings, and surgical adhesives
    3. Decompression and drainage

D. Dental materials

The clinical requirements of a successful implant material are simple yet stringent. It must function for the necessary time to effect the required therapy (a period ranging from minutes to the lifetime of the patient) while simultaneously avoiding such reactions with its host as might lead to damage of the host tissue, or to a decrease in the desirable properties of the implant.

    In light of these requirements, therefore, the implant must have adequate strength, good serviceable life resistance to corrosion and biodegradation, adequate tissue compatibility, and, in the case of materials indicated for employment in the cardiovascular system, pronounced antithrombogenicity.

    In view of these requirements, it is not surprising that the number of clinically acceptable materials is small.

Properties of Materials

The choice of implant material is important to successful repair and replacement. No single material, metal or polymer, meets every necessary characteristic for successful implantation in all of the many regions where implants are used. Solids in the pure elemental state do not generally possess the desired combinations of properties necessary for a successful implant. Instead, most implants are made from alloys in the case of metals, and highly polymerized chains in the case of polymers.

Every solid material exists in one of three forms: amorphic, mesomorphic, or crystalline. In amorphous substances, such as glass, atoms or molecules are positioned randomly in space. Mesomorphic substances have some regularity of atomic or molecular arrangement in certain directions but not in others. Crystalline substances are characterized by an orderly three-dimensional array of atoms or molecules making up the crystal. Most polymer substances are either mesomorphic or crystalline, while all metals are crystalline.

When a pure material is in the crystalline state, it consists of crystals having the same three-dimensional array of atoms or molecules. Large temperature changes or additions of other elements cause the three-dimensional array to either distort to accommodate the change or shift to an entirely different three-dimensional array, which is now a second phase. (A phase, in this context, may be thought of as a homogeneous body of matter containing crystals of one three-dimensional atomic or molecular array.) The crystals of any phase may not necessarily be in the same orientation, and where they impinge upon one another a grain boundary is formed.

Alloys are pure base materials to which other elements are added to produce one or more phases. The type and amount of elements added to the base material will determine the number of phases, the characteristics of each phase, and the properties of the resulting alloy. Heating or mechanical deformation of an alloy will cause shifting of the atoms or molecules, which will alter the grain size, the characteristics of the phases, and the alloy.

The ability of the polymer chemist and plastics engineer to produce tailor-made synthetic materials has found a wide range of application in medicine. A polymer is a large molecule built up from repetition of small, simple chemical units called mers. In some cases the repetition is linear, similar to a chain built up from links. In other cases the chains are branched or interconnected to form three-dimensional networks. The repeat unit of the polymer is usually equivalent or nearly equivalent to the monomer, which is the starting material from which the polymer is formed. The length of the polymer chain, called the degree of polymerization (DP), is specified by the number of repeat units in a chain. The molecular weight of the polymer is the product of the molecular weight of the repeat unit times the DP. The control of the DP makes possible the tailoring of properties as the larger chains become materials of increased strength and dimensional stability, thereby allowing a given species of polymer to range from liquid of low DP to very rigid solids of high DP. Most of the useful polymers, such as the man-made plastics, fibers, and elastomers and the few naturally occurring rubbers, wool, and cellulose, have DPs in the range of 500 to 1500.

The mechanisms by which polymerization takes place fall into two general categories, addition and condensation. The prototype for addition polymerization is one in which succeeding mers are added to the molecule to increase its size. In this type of polymerization, the molecule obtains its reaction bonds by breaking double bonds and forming two single bonds. If the addition polymer contains two or more different species of mers, the resulting structure is called a copolymer. In contrast to addition reactions, which are primarily a summation of individual molecules into a polymer, condensation reactions form a second, non-polymerizable molecule as a by-product. Usually the by-product is water or some other simple molecule.

After polymer chains have been formed by either addition or condensation, the chains may be further joined together by cross-linking. In this process an intermediate element joins the unsaturated carbon bonds within several normal chains (e.g., sulfur cross-links natural rubber chains into vulcanized rubber). The result of these various controllable polymerization mechanisms has been a wide range of products, some of which have been found acceptable in the bio-chemical environment.

It is through the control of the alloy phases and the polymer chains that materials can be developed to provide the mechanical and chemical properties necessary for a successful biological implant.

When a solid material is placed under a load (i.e., walking with a leg implant, surge of blood through a prosthetic artery), the material undergoes deformation, as illustrated by the uniaxial tension stress-strain curve in Figure 11-1. The stress, σ, is given by the applied load divided by the original cross-sectional area of the specimen. The strain, e, is given by the change in specimen length divided by the original length. In the first portion of the curve, Hooke's law

$$\sigma = eE \tag{1}$$

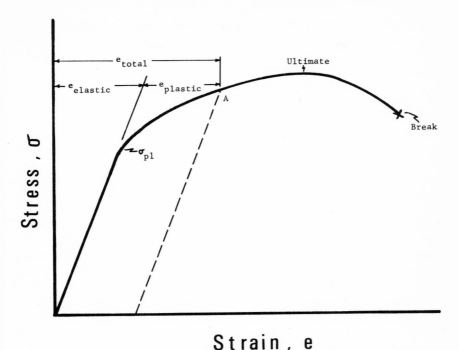

Figure 11-1. A typical stress-strain curve. The stress at point A represents any
stress in the strain-hardened plastic region. The total strain produced at A is shown broken into its elastic and plastic components. Also, the dashed line illustrates the path to be followed if
all loading were removed at A.

where E = elastic modulus, is obeyed up to stress σ_{PL}, the proportional
limit. In this initial region, called the elastic region, any applied stress
will be accompanied by a proportional amount of elastic strain. (The
proportionality constant is simply the Young's modulus E.) Whenever
the stress is completely removed, the strain returns to zero, and there
is no permanent change in the specimen dimensions. Materials with
larger elastic moduli are stiffer or more rigid than those of low E,
because they require much larger stresses to produce the same amount
of strain.

At applied stresses larger than σ_{PL}, plastic deformation occurs.
Upon removal of stress in the plastic region, the strain decreases in
a linearly proportional manner with the same slope or modulus as in
the initial portion of the curve; however, at zero stress, a permanent
strain or permanent deformation results, as illustrated by the dashed

line in Figure 11-1. The yield strength of a material is usually reported in the literature as the stress at the proportional limit, or the slightly higher stress required to produce a permanent deformation of 0.2 percent. In some materials (most notably iron-carbon alloys), there is a sharp drop in stress upon reaching σ_{PL}. This is called the discontinuous yield point, or simply the yield point, and is also referred to as a yield strength. Table 11-1 lists the values of modulus and yield strength for several materials of interest in implant work.

In practice, it is normally desirable to use materials at stresses below σ_{PL}, so that they will not permanently change shape; however, during fabrication it is often necessary to be able to exceed the proportional limit in order to permanently deform the material into the proper shape. Beyond this value, the mechanism by which stress increases with increasing permanent strain in the material is known as "strain hardening." This process occurs up to a maximum or ultimate stress, then the stress decreases as the specimen undergoes necking (i.e., decreases area drastically in a highly localized area), and finally the specimen breaks.

The ability of materials to undergo plastic deformation is known as ductility. A ductile material will exhibit a large amount of plastic deformation before fracture. Brittle materials, on the other hand, show little or no plastic deformation. They usually break in the elastic region or within 1 or 2 percent strain after the proportional limit. Table 11-1 also includes values of plastic elongation.

When a material combines a high proportional limit with good ductility, the material is said to be tough. Toughness is graphically the area under the stress-strain curve up to the point of breaking.

The hardness of a material is a frequently measured property.

Table 11-1.   Typical Mechanical Properties of Materials of Interest to Biological Implantation

Material	Modulus (in 10^5 psi)	Tension Yield Strength (in 1000 psi)	% Elongation	Strength-to-Weight Ratio* in 1000 in
316L				
Annealed	290	35	50	125
Work Hardened	290	115	20	410
Zimmaloy	300	75	10	250
Polymethyl Methacrylate	5	7	1	160
Human Bone	29	13	1	190

*Strength-to-weight ratio is described as $\dfrac{\text{Yield strength}}{\text{Density}}$.

Usually this property is an indication of resistance to plastic deformation measured by penetrating, scratching, or bouncing an object on surfaces of various degrees of polish. Hardness measurements such as Brinell, Tukon, Rockwell, and Vickers can be considered relative indications of hardness. However, the following factors preclude the establishment of a single test to evaluate all materials:

1. Materials strain harden at different rates.
2. They are often not isotropic (same mechanical strength in all directions).
3. Materials encompass such a wide range that no single test arrangement adequately measures all materials.

Therefore, hardness values must be carefully interpreted; however, if confined to a single type of material, increasing hardness will generally indicate increasing strength.

The establishment of the values of proportional limit allows one to calculate the amount of load which can be placed on a specimen of a certain cross-sectional area without causing permanent deformation. If the implant in question is being loaded near its proportional limit, the situation can be alleviated by increasing the cross-sectional area, if space permits, with a resultant reduction in the stress. It is therefore obvious that the employment of stronger and stiffer materials allows smaller implants, which then require less removal of healthy tissue.

On the other hand, size may not be as important a factor as weight in some applications. Values of strength-weight ratios, obtained by dividing yield strengths by density (lb per in^3), are also included in Table 11-1. Materials take on different aspects of usefulness when viewed by this criterion.

Materials in a biological environment usually fail by one or more of the following mechanisms:

1. Loading above the elastic limit.
2. Fatigue.
3. Creep.
4. Wear and abrasion.
5. Corrosion alone or in combination with the above four factors.
6. Rejection by the body via sloughing, or surgical removal due to foreign body reaction.
7. Unfavorable tissue response.

Fatigue is a fracturing caused by repeated cycling of applied loads usually below the proportional limit. Fatigue usually begins with minute crack formations in a small region of high stress concentration. Because the crack is itself a stress concentrator, it can become self-propagating and as a result will grow larger. The two opposite faces of the crack rub against each other with a polishing action. When the crack becomes sufficiently large so that the remaining sound cross-sectional areas can no longer withstand the applied load, there is a cataclysmic brittle failure. The failed portion of the brittle fracture has a characteristic crystalline appearance, while the rubbed section is smoothly polished. This sometimes leads to the incorrect assumption that the material has "crystallized" when, in fact, it always has been crystalline. A typical fatigue data curve is shown in Figure 11-2. The problem of fatigue is usually corrected by using smaller grain sizes of a material, and by avoiding designs which have high stress-concentrating features such as notches, sharp corners, and surface scratches. Fatigue action often accelerates chemical attack, and this combined process is known as corrosion fatigue.

Creep is the slow flow or plastic deformation of materials held for long periods of time at stresses below the conventional yield strength. This mechanism usually occurs only at temperatures near the softening

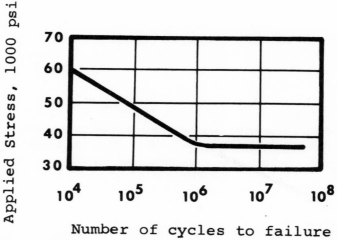

Figure 11-2. The fatigue curve for an annealed medium carbon steel. The number of cycles a material can endure before failure increases with decreasing applied stress. At very low applied stresses, some materials reach a fatigue or endurance limit below which a material can presumably endure an infinite number of cycles without failing.

point of a material. Therefore, this rarely occurs in metals at body temperature; although it can be a problem in some polymers.

Another cause of failure in materials is wear and abrasion. When a part is subjected to abrasive wear, minute particles stress small regions of the surface beyond its endurable limit. That is, portions of the surface are stressed not only beyond σ_{PL} but beyond the breaking point, so that material is broken away from the surface. This process may continue and render the part useless because of changed dimensions, or the cross-section may be reduced so that failure will occur when the load exceeds the elastic limit. In a biological environment, there is the added danger that the freed particle products will damage the surrounding tissue by reacting with it or, in a more dramatic fashion in cardiovascular prostheses, by becoming material emboli. Wear is particularly emphasized in products such as moving prostheses (i.e., ball and cup joints, cardiac valves), and where two segments of different stiffness and hardness rub as load is applied. Two examples encountered in clinical practice are (a) high-modulus metal fastened to low-modulus bone, rubbing under a cycling load, and (b) silastic interacting with a vitallium cage. The solutions to such wear problems include reducing friction by suitable lubricants and by making rubbing surfaces of similar materials (i.e., same hardness and same elastic modulus).

Corrosion, Biodegradation, and Compatibility of Biological Materials

The biological environment into which an engineered implant is placed is hostile to both metals and plastics. Normal physiological fluids can be thought of as highly oxygenated saline solutions (somewhat analogous to sea water) which present an atmosphere conducive to corrosion of conventional materials. The situation is further complicated by the presence of enzymes and amino acids which, in addition to complicating the corrosion phenomena associated with metallic implant materials, may also attack the hydrocarbon bonds associated with conventional polymeric materials. The latter process is usually referred to as biodegradation. Furthermore, the reaction products produced by the corrosion and biodegradation process may interact with normal tissue, affecting many biological processes including thrombogenesis, healing, foreign body reaction with subsequent encapsulation of implant by fibrous tissue, sudden onset of inflammation around an implant in the absence of sepsis, and, perhaps, replacement of normal ions in enzymatic function. In order to understand the electrochemical reactions between the implant and the biological environment, it is necessary to outline some of the basic concepts which govern them.

Usually, the lowest free energy state for a metal in an oxygenated environment is that of its oxide. In addition, the presence of an electrolyte can add to this free energy difference by giving rise to the possibility of lowering the energy for the metal ion (cation) still further by a process of hydration. A standard schematic of an electrochemical cell is given in Figure 11-3.

The cell is characterized by the following important components:

1. Anode: This is the electrode at which positive ions go into solution (i.e., are oxidized) according to the reaction

$$M^0 \longrightarrow M^+ + e^- \tag{2}$$

2. External circuit: The anode must be connected to the second electrode, the cathode, by an external circuit through which electrons may flow in order to equalize the potential. As a result of this potential equalization, the cathode becomes negatively charged due to an excess of electrons and the anode positively charged due to a deficit of electrons.

3. Cathode: Reduction reactions occur at this electrode. These are necessary in order to consume the electrons generated by the anodic reaction. In aqueous oxygenated media, there are three important typical cathodic reactions, all capable of electron consumption:

$$M^+ + e^- \longrightarrow M^0 \tag{3}$$
$$2H_3O^+ + 2e^- \longrightarrow H_2(g) + 2H_2O \tag{4}$$
$$\tfrac{1}{2}O_2 + H_2O + 2e^- \longrightarrow 2OH^- \tag{5}$$

4. Electrolyte: The presence of an electrolyte composed of ionic charge carriers not only produces an appropriate solution potential, but, more important, allows for both completion of the electrical circuit and appropriate mass transfer of ions through the electrolyte under the influence of the electric field produced. It is this mass transfer which allows steady-state operation of the cell as evidenced by a continuously observed corrosion current. Positive ions will of course tend to migrate toward the anode (and are thus called anions).

Mass transfer has important consequences in that the solution around the anode will become alkaline (high pH), the solution around the cathode will become acidic (low pH), and the cation produced by the

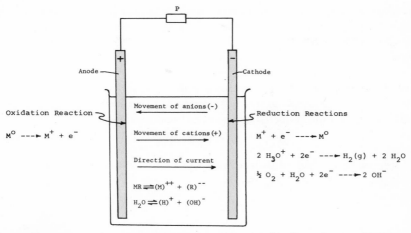

Figure 11-3. Ion and current movements in a simple electrolytic cell showing some typical reactions that may occur. The electrochemical potentials are measured in the potentiometer at P.

anodic reaction (i.e., M^+) may meet the anion produced by the cathodic reaction (i.e., OH^-) to form an insoluble precipitate according to $M^+ + OH^- \longrightarrow M(OH)$. These diffusion-related phenomena in the electrolyte can have far-reaching effects on stifling the reactions associated with the corrosion cell; this must be our goal if we are to preserve the structural integrity of our implanted prostheses.

Consider only the anode reaction postulated in Eq. (2). Since the electrons produced migrate to the cathode, we may express the first law of thermodynamics in terms of heat produced and work done. If the system is at constant temperature, pressure, and volume, and if the anodic reaction is reversible, the change in internal energy ΔE may be expressed as

$$\Delta E = T\,\Delta S + nF\mathcal{E} \qquad (6)$$

where the electrical work term $nF\mathcal{E}$ is the product of the number of equivalents n, Faraday's constant F (96,500 coulombs/equivalent), and the change in electrical potential $\mathcal{E}$; $T\,\Delta S$ is the product of absolute temperature T and change in entropy ΔS. For the same thermodynamic state, the free energy charge ΔG may be expressed as

$$\Delta G = \Delta E - T\,\Delta S \qquad (7)$$

Consequently

$$\Delta G = nF\mathcal{E} \qquad (8)$$

Since energy changes (and consequently electrical potential) are relative, we must refer the equation to an accepted thermodynamic standard state. Thus, Eq. (8) becomes

$$G - G^0 = nF(\mathcal{E} - \mathcal{E}^0) \tag{9}$$

where superscript ^{0} refers to the standard state.

At equilibrium between reactants and products of any reaction, $G_{reax} = G_{prod}$, and therefore $\Delta G = 0$. Consequently

$$\Delta G^0_{reax} = -nF(\Delta\mathcal{E} - \Delta\mathcal{E}^0_{stand}) \tag{10}$$

where the Δ symbol now refers to the change between reactants and products. However, ΔG^0_{reax} can be expressed in terms of the equilibrium constant K

$$\Delta G^0_{reax} = -RT \ln K \tag{11}$$

where R is the gas constant and T is absolute temperature. The equilibrium constant K can be approximated as (Prod)/(Reax) where ( ) signifies concentration in appropriate units. Eq. (10) can then be rewritten as

$$\Delta\mathcal{E}_{reax} = \Delta\mathcal{E}^0_{stand} + \frac{RT}{nF} \ln \frac{(Prod)}{(Reax)} \tag{12}$$

or in the case of the anodic reaction given in Eq. (2)

$$\Delta\mathcal{E}_{anodic} = \Delta\mathcal{E}^0 + \frac{RT}{nF} \ln (M^+) \tag{13}$$

since the standard state of a pure metal is taken at unit concentration. Eqs. (12) and (13) are forms of what is commonly referred to as the Nernst Equation.

The Nernst Equation is derived from equilibrium thermodynamics, and will represent, for a particular metal/electrolyte cell, the driving force associated with the reaction. It will not, however, allow for the prediction of reaction rates, which is an essentially kinetic process.

Given the Nernst Equation, it is possible to draw up tables of relative electrochemical potentials for a particular system to indicate not only base and noble metals, but also some idea of the magnitude of the potential difference between two particular half-cells.

If the concentration of metal ion in the example just discussed is taken as unity, then the Nernst Equation reduces to

$$\Delta\mathcal{E} = \Delta\mathcal{E}^0 \tag{14}$$

and the standard series of electrochemical potentials may be evolved.

Since the Nernst Equation is thermodynamic, one cannot speak in terms of absolute potentials. Instead, an arbitrary zero potential reaction is assumed and all measured potentials are compared with it. Classically, the hydrogen electrode is taken as the zero potential electrode and is so noted in Table 11-2, which presents the standard electrochemical potentials of a wide variety of metals and alloys. The hydrogen electrode is composed of a platinum electrode (covered with platinum black for enhanced surface activity) immersed in an electrolyte saturated with H_2 gas at a pressure of 1 atmosphere.

It must be stressed that these potentials refer only to the half-cell reactions in which the concentrations of reactant ions are unity. Change of electrolyte composition and type will also alter the relative position of elements according to the rationale evolved from the Nernst Equa-

Table 11-2. Standard Electrochemical Potentials

Electrode Reaction	E_0 Volts	Electrode Reaction	E_0 Volts
$Li \rightleftarrows Li^+$	-3.045	$V \rightleftarrows V^{3+}$	-0.876
$Rb \rightleftarrows Rb^+$	-2.925	$Zn \rightleftarrows Zn^{2+}$	-0.762
$K \rightleftarrows K^+$	-2.925	$Cr \rightleftarrows Cr^{2+}$	-0.74
$Cs \rightleftarrows Cs^+$	-2.923	$Ga \rightleftarrows Ga^{2+}$	-0.53
$Ra \rightleftarrows Ra^{2+}$	-2.92	$Fe \rightleftarrows Fe^{2+}$	-0.440
$Ba \rightleftarrows Ba^{2+}$	-2.90	$Cd \rightleftarrows Cd^{2+}$	-0.402
$Sr \rightleftarrows Sr^{2+}$	-2.89	$In \rightleftarrows In^{2+}$	-0.342
$Ca \rightleftarrows Ca^{2+}$	-2.87	$Tl \rightleftarrows Tl^+$	-0.336
$Na \rightleftarrows Na^+$	-2.714	$Mn \rightleftarrows Mn^{3+}$	-0.283
$La \rightleftarrows La^{3+}$	-2.52	$Co \rightleftarrows Co^{2+}$	-0.277
$Mg \rightleftarrows Mg^{2+}$	-2.37	$Ni \rightleftarrows Ni^{2+}$	-0.250
$Am \rightleftarrows Am^{3+}$	-2.32	$Mo \rightleftarrows Mo^{3+}$	-0.2
$Pu \rightleftarrows Pu^{3+}$	-2.07	$Ge \rightleftarrows Ge^{4+}$	-0.15
$Th \rightleftarrows Th^{4+}$	-1.90	$Sn \rightleftarrows Sn^{2+}$	-0.136
$Np \rightleftarrows Np^{3+}$	-1.86	$Pb \rightleftarrows Pb^{2+}$	-0.126
$Bc \rightleftarrows Bc^{2+}$	-1.85	$Fe \rightleftarrows Fe^{3+}$	-0.036
$U \rightleftarrows U^{3+}$	-1.80	$D_2 \rightleftarrows D^+$	-0.0034
$Hf \rightleftarrows Hf^{4+}$	-1.70	$H_2 \rightleftarrows H^+$	0.000
$Al \rightleftarrows Al^{3+}$	-1.66	$Cu \rightleftarrows Cu^{2+}$	$+0.337$
$Ti \rightleftarrows Ti^{2+}$	-1.63	$Cu \rightleftarrows Cu^+$	$+0.521$
$Zr \rightleftarrows Zr^{4+}$	-1.53	$Hg \rightleftarrows Hg_2^{2+}$	$+0.789$
$U \rightleftarrows U^{4+}$	-1.50	$Ag \rightleftarrows Ag^+$	$+0.799$
$Np \rightleftarrows Np^{4+}$	-1.354	$Rh \rightleftarrows Rh^{3+}$	$+0.80$
$Pu \rightleftarrows Pu^{4+}$	-1.28	$Hg \rightleftarrows Hg^{2+}$	$+0.857$
$Ti \rightleftarrows Ti^{3+}$	-1.21	$Pd \rightleftarrows Pd^{2+}$	$+0.987$
$V \rightleftarrows V^{2+}$	-1.18	$Ir \rightleftarrows Ir^{3+}$	$+1.000$
$Mn \rightleftarrows Mn^{2+}$	-1.18	$Pt \rightleftarrows Pt^{2+}$	$+1.19$
$Nb \rightleftarrows Nb^{3+}$	-1.1	$Au \rightleftarrows Au^{3+}$	$+1.50$
$Cr \rightleftarrows Cr^{2+}$	-0.913	$Au \rightleftarrows Au^+$	$+1.68$

tion. However, the more successful implant materials at present generally occur in standard potential tables at greater than $+200$ mv (0.200 volt).

If an attempt is made to compare measured electrochemical potentials, it is obvious that the electrolyte must be specified. One may develop an electrochemical potential series, referred to as a galvanic series, for a specific electrolyte. Table 11-3 shows a galvanic series in sea water, which roughly approximates tissue fluid.

The value of $\Delta\mathcal{E}^0$ itself is subject to a great deal of variation. Anything which affects the internal energy, entropy, or resultant free energy of the material will result in changes in $\Delta\mathcal{E}^0$. High-energy grain boundaries, residual internal energy as a result of cold work, inclusions of foreign matter in processing, and compositional heterogeneities are all factors which may play a drastic role in altering the magnitude of $\Delta\mathcal{E}^0$. In addition, the role of alloying to produce multiphase equilibrium structures will, as we shall see, have profound effects on $\Delta\mathcal{E}^0$ in the biological environment.

Table 11-3. Galvanic Series of Metals and Alloys Arranged in Order of Increasing Static (Steady-state) Electrode Potentials (E_H) in Sea Water

Metal	E_H	Metal	E_H
Magnesium	−1.45	Brass (30 Zn)	−0.11
Magnesium Alloy			
(6 Al, 3 Zn, 0.5 Mn)	−1.20	Bronze (5–10 Al)	−0.10
Zinc	−0.80	Tombac (5–10 Zn)	−0.10
Aluminum Alloy (10 Mg)	−0.74	Copper	−0.08
Aluminum Alloy (10 Zn)	−0.70	Copper-Nickel (30 Ni)	−0.02
		Stainless Steel 403	+0.03
Aluminum	−0.53		
Cadmium	−0.52	Nickel (passive)	+0.05
Duralumin	−0.50	Inconel (11–14 Cr,	
		1 Mn, 1 Fe)	+0.08
Iron	−0.50	410 Stainless Steel (passive)	+0.10
Carbon Steel	−0.40	Titanium (commercial)	+0.10
Gray Cast Iron	−0.36	Silver	+0.12
410 Stainless Steel (active)	−0.32	Titanium (iodide)	+0.15
Nickel-Copper Cast Iron		304 Stainless Steel (passive)	+0.17
(12–15 Ni, 5–7 Cu)	−0.30	Hastelloy C (20 Mo, 18 Cr,	
304 Stainless Steel (active)		6 W, 7 Fe)	+0.17
316 Stainless Steel (active)		Monel	+0.17
Lead	−0.30	316 Stainless Steel (passive)	+0.20
Tin	−0.25	Graphite	
Brass + (40 Zn)		Platinum	+0.40
Manganese Bronze (5 Mn)			
Nickel (active)	−0.12		

Obviously, all metals go to the oxidized (corroded) state if the thermodynamic energetics are favorable. Preserving the structural integrity of an engineered prosthesis in a biological environment will require limiting the kinetics of the corrosion process so that the prosthesis may maintain its function over its projected useful lifetime.

Passivization implies an inhibition of either anodic or cathodic reactions as detailed in Eqs. (2) to (5). This may occur through a variety of mechanisms, including a physical barrier created by products of the corrosion reaction which separates the anodic surface from electrolyte and thus breaks the internal circuit, buildup of concentration of cationic or anionic species, thus lowering the potential in accordance with the Nernst Equation, and interference with the cathodic depolarization reactions as detailed in Eqs. (3) to (5).

1. If the early product of the corrosion process forms a physically dense adherent oxide layer over the entire anodic surface, then the anodic layer will be physically separated from electrolyte by a layer possessing excellent insulating properties. A convenient electrical path across the cell no longer exists, and consequently the kinetics of corrosion and/or oxidation are slowed, if not entirely stopped. The excellent corrosion resistance of aluminum, stainless steel, and Co-Cr alloys to certain types of aqueous media is due to the layers of inert oxide (Al_2O_3) on aluminum, or of CR_2O_3 on the stainless steel or chromium-containing cobalt alloys. In the case of many of these materials, a reducing environment such as a high chlorine concentration may serve to destroy passivity and yield a pitting type of corrosion attack.

2. Other types of passivization may be obtained when concentrations of respective ions build up adjacent to anode and cathode. The increase in ionic concentration, when related to the Nernst Equation, leads to what is called "concentration polarization." Considering Eq. (13), it is obvious that if the concentration reaches a limiting value such that

$$(M^+) = e\,\frac{-nF\,\Delta\mathcal{E}^0}{RT} \tag{15}$$

the reaction will cease.

If the electrolyte forms part of a dynamic system, however, this limiting concentration will never be reached; the buildup of corroding ions will not occur, because the electrolyte will be continually flushed. This type of situation may be expected to occur in biological environments.

3. Finally, there may be effective hindrance of the following depolarization reactions associated with the cathode.

$$H^+ + H^+ + 2e^- \longrightarrow H_2 \tag{16}$$

$$\tfrac{1}{2}O_2 + H_2O + 2e^- \longrightarrow 2OH \tag{17}$$

In order to prevent the buildup of an excess of electrons on the cathode, these cathode depolarization reactions most proceed. However, a discrete amount of energy must be available to associate the hydrogen ions into atoms, and these atoms then into gas; and energy is also needed to dissociate the oxygen gas into oxygen ions, and then to cause these ions to react with water to form hydroxyl ions. The energetics responsible for these processes serve to lessen the available potential for corrosion since, if these cathode depolarization reactions do not occur, the corrosion processes do not proceed.

It is quite possible that different electrochemical potentials may be developed around a single piece of metal immersed in an electrolyte in areas where concentration gradients of appropriate ions exist. This phenomenon is directly related to the Nernst Equation.

As derived earlier in Eq. (13), the Nernst Equation may be expressed at two different cationic concentrations (M_1^+) and (M_2^+) where $(M_2^+) > (M_1^+)$. Since $\Delta\mathcal{E}^0$ is the same value for each concentration, the change in potential $\Delta\mathcal{E}$ may be expressed as

$$\Delta\mathcal{E} = \Delta\mathcal{E}_2 - \Delta\mathcal{E}_1 = \frac{RT}{n} \ln \frac{(M_2^+)}{(M_1^+)} \tag{18}$$

If (M^+) represents standard concentration, the surface adjacent to where $(M_2^+) > (M_1^+)$ will be anodic to the area where $(M_2^+) < (M_1^+)$. Thus, a general rule may be formulated that, in any cell in which concentration differences exist, a potential difference will be created where the areas of high electrolyte concentration will be anodic to the areas of low electrolyte concentration.

In the case of the oxygen depolarization reactions as expressed in Eq. (18), greater concentrations of oxygen will lead to higher concentrations of (OH^-) ions, with higher attendant cathode polarization rates. Therefore, areas of high oxygen concentration will be cathodic to areas of low oxygen concentration where effective cathodic depolarization cannot occur. This manifestation is known as the "differential aeration cell," and may be quite important in the corrosion of orthopedic implants. After the fixation of a fracture by bone and plate, for example, it is usually found that the hard tissue beneath the bone becomes necrotic. Associated with necrosis would be an oxygen partial pressure or oxygen "tension" of essentially zero. At the live tissue interface, however, the oxygen pressure would be the equilibrium one, finite and of the order of 0.2 atm. Thus, a differential aeration cell would be set up across the implant, with the implant-soft tissue interface cathodic to the implant-hard tissue interface.

Pure galvanism is not usually associated with the corrosion of implants indicated for biomedical application. Rare cases of galvanism may occur when the surgeon uses plate and screws of different metals and alloys, or mishandles his materials, using common tool-steel instruments for fixation, or when the dentist places restorations of two different metals in apposition. However, the clinician is now very much aware of these consequences and has designed clinical procedures to mitigate them.

As we have seen in the last section, concentration gradients alone may produce microanode and microcathode behavior on the same material through alteration of the second term in the Nernst Equation. In a similar fashion, any structural compositional or energy gradient can produce microelectrode behavior by changes in the first term ($\Delta \mathcal{E}^0$) of the Nernst Equation. Since these potentials are present in the same piece of metal, electrical contact between microanode and microcathode exists (internal circuit accomplished), and all of the requirements of the corrosion cell are met. The metallic materials of choice for implants (i.e., stainless 316L and Co-Cr alloys) are usually multiphase alloys in which a second phase rich in chromium, for example, corrodes preferentially to produce a passivating layer.

The existence of service stresses, residual stress, sharp corners, or nonequilibrium structures produced by poor manufacturing technology may alter the desired corrosion behavior through changes in $\Delta \mathcal{E}^0$, and thus lead to untoward results.

Following implantation, the host begins laying down a capsule of fibrous tissue of small vascularity to seal off the implant. Optimally, this fibrous capsule should be of minimal thickness in order not to adversely affect movement.

In animal studies, it has been shown that the corrosion of an implant can affect the width of the fibrous tissue capsule and interfere with the complete encapsulation of the implant. In many cases, corrosion products will diffuse into the tissue some distance from the implant and set up foreign body reaction, giant cell formation, etc. Typical tissue reactions to metal implanted intramuscularly in rabbits are shown in Figures 11-4a and 11-4b. Figure 11-4a indicates what might be considered a minimal response to the implant, whereas Figure 11-4b presents large areas of fibrous tissue and inflammatory cells, and typifies foreign body reaction.

a

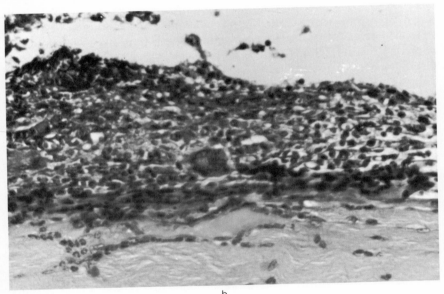

b

Figure 11-4. a. Tissue section of 100X showing mild response to implant. A
fibrous connective tissue capsule surrounds the site of the removed implant. The capsule is in turn surrounded by normal,
healthy tissue. b. Tissue section of 100X showing aggravated response to implant. Large areas of necrotic fibrous connective tissue composed primarily of inflammatory cells.

It may well be that the destruction of fibrous encapsulation by metallic ions produced by corrosion is a function of time of implantation, as well as of service stresses produced by the implant. This problem is now under study in a number of laboratories. The inception of acute inflammation around orthopedic implants in the absence of sepsis after many years of benign performance may be due to this phenomenon.

With the exception of silicone, polymers used as implants have had a C—C, C—N, or C—H chain. These polymers are all biodegradable, since these are the same types of bonds that the body destroys metabolically. Depending on the side groupings of the polymers (phenol, for example), severe foreign body reaction may occur. Although carcinoma has been produced by implantation of polymethyl methacrylate in rats, there has been no firm clinical evidence of the production of carcinoma in humans from any engineered implant.

Silicone, as will be discussed shortly, has been proven to be generally bioinert, because of the existence of the very stable O—Si—O bonds which the body cannot attack metabolically. In addition, it has been widely proven in clinical application that the foreign body reaction associated with silicone has been minimal. Indeed, the claim has been made that the basis for the minimal body reaction to silicone may be that the O—Si—O bond is so completely alien to the host experience that the normal sensor mechanism cannot even detect its presence.

In contrast to orthopedic applications in which metal continues to predominate, polymeric materials are being utilized with increasing frequency in the area of cardiovascular prostheses ranging from arterial grafts to the implantable artificial heart and heart booster pumps. Several heart valves made from combinations of metal alloys and polymers are shown in Figure 11-5. The major drawback with these materials from a functional standpoint is the high frequency of thrombus formation.

The etiology of formation of these thrombi is not understood thoroughly at present, although certain patterns have developed. Effective clinical use of polymers (dacron and teflon) can be made for cardiovascular implants. It has been found that a porous knitted-weave structure (a porosity of greater than 5000 cc/min/cm^2 at 120 mm Hg of pressure) will prove satisfactory. The outer surface of this graft becomes encapsulated with fibrous tissue, while at the same time fibrin is deposited along the inner surface and becomes organized through the interstices in the cloth. The greater the porosity, the easier the fibrous encapsulation of the inner surface of the graft. With little porosity, the tendency is toward loss of patency through occlusion and

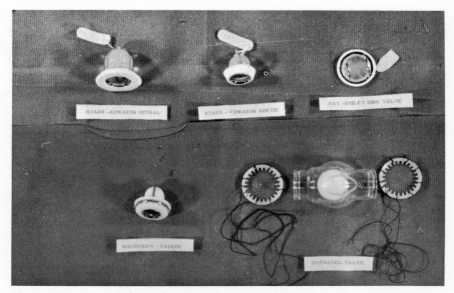

Figure 11-5. Heart valves of various design made from combinations of metal
alloys and polymers.

the sloughing of minor layers of fibrin, or the creation of septa which
shear under cardiovascular flow, leading to embolism formation.

The interaction of materials with the clotting mechanism is probably concerned with both intrinsic and extrinsic clotting factors. A
prosthesis may alter the hemodynamics associated with localized flow,
and create unusual conditions of turbulence and stagnation leading to
nucleation of thrombosis. In addition, the presence of the surface
charge usually associated with polymeric surfaces may play a role in
causing the agglomeration of platelets, which may build to septum
formation.

Joining the prosthesis to normal tissue will usually involve injury
to the normal tissue and exposure of collagen. Fresh collagen appears
to trigger the clotting mechanism, perhaps by activation of Hageman
factor, which leads to the production of fibrin. This is the extrinsic
clotting mechanism.

To obviate the accelerated clotting produced by artificial surfaces,
a number of approaches, including the porous weave already discussed,
are being studied:

1. The use of anticoagulant coatings such as benzalkonium-graphite-heparin. The heparin inhibits the activation of Hageman factor and the production of thrombin.
2. The utilization of surgical adhesives of the cyanoacrylate type to avoid the need for sutures. The product of biodegradation of cyanoacrylate is toxic phenol groups, which limit its use.
3. The use of electrically charged surfaces. It is felt that a negatively charged surface might repel negatively charged platelets and prevent their accumulation.
4. The study of hemodynamics of coagulation and flow to avoid conditions of stagnation and turbulence that promote thrombus and septum formation.

Machinability

Before reviewing some of the specific materials that are actually used for biological implants, an additional factor must be considered. A most desirable characteristic of an implant is that it can be shaped easily, inexpensively, and immediately to the exact dimensions of the section it is replacing or supporting. The common techniques for shaping a part to proper size include machining, grinding, casting, molding, extruding, rolling, forging, bending, weaving, suturing, and welding or joining. Most of these processes are not suited for on-site fabrication, due to the cumbersome nonsterile technology involved. Unfortunately, materials which are easily shaped do not usually possess the proper mechanical and physical properties required for implant in a biological environment. For example, the cobalt-chromium base metals in common use today have good strength and corrosion resistance, but are difficult to machine, bend, or forge. As with any part that is cast, extruded, or molded, the dimensions are limited by those of a die; and because of die fabrication costs, the tendency is to develop a few standard sizes rather than to "tailor-make" parts. Also, inherent in the casting process are such factors as surface finish, internal homogeneity, and internal and external flaws developed during fabrication. With an implant material, the fabrication procedure will determine not only the fit and the speed of obtaining a part, but the useful service life as well. The fabrication procedure will control factors such as surface finish, specimen homogeneity, and internal flaws, which in turn will affect such properties as strength, fatigue life, corrosion resistance, tissue compatibility, and thrombogenicity.

Armamentarium of Biological Materials

Metals. Approximately one-half of the hard metal implants today are fabricated from stainless steels (the other half are most commonly fabricated from alloys of the Co-Cr family). Stainless steels, when handled correctly, are capable of being readily fabricated and show quite good corrosion resistance in biological implantations over the short term.

Stainless steels are basically alloys of Cr, Ni, Fe, and C. The generic term "stainless," however, may cover a wide variety of ratios of the above constituents, as well as considerable amounts of elemental additions such as Mn, Si, Mo, Cb, and Cu to modify both physical properties and microstructure.

In an effort to categorize commercial stainless steels, the American Iron and Steel Institute (AISI) has a classification system which defines the permissible compositions to controllable tolerances. Since the attainable physical properties and corrosion resistance will depend upon structure as well as composition, the designations have further meaning in that they specify the equilibrium structures. The hard stainless steels of the ferritic and martensitic (i.e., 200 and 400) series have not been extensively employed for biological implant applications, since they lack both the ease of fabrication and the saline corrosion resistance of the austenitic (300) series. These harder materials have, on the other hand, been used quite extensively for the manufacture of surgical and dental instruments subject to autoclaving. In addition, it is of importance to note that stainless steels based on the austenitic structural modifications are nonmagnetic, whereas those of the ferritic and martensitic forms are magnetic.

The compositions of the three austenitic stainless steels employed most commonly for hard-metal implantation are given in Table 11-4 along with their appropriate physical properties.

The basis for comparison of composition of all 300 stainless steels is usually taken as #304, and we will discuss the model system in detail. As can be seen, the 304 can be basically considered "18-8" steel: 18 percent Cr, 8 percent Ni by weight, and the balance Fe and C.

The 18 percent Cr is necessary to impart corrosion resistance to the alloy. For a steel to be "stainless," it must contain a minimum of 12 percent Cr. The corrosion resistance imparted by the chromium can best be thought of as the buildup of a stable, dense, adherent Cr_2O_3 film (approximately 400 microns thick) over the surface of the steel, which protects the steel from further attack. The presence of such an inert film on the surface of the metal is known as "passivization." As

Table 11-4. Composition and Physical Properties of Stainless Steels Used in Biomedical Applications

Type	C	Mn	Si	P	S	Cr	Ni	Mo	Fe
				Composition, Weight %					
304	0.08	2	1	0.045	0.03	18–20	8–12	—	Balance
316	0.08	2	1	0.045	0.03	16–18	10–14	2–3	Balance
316L	0.03	2	1	0.045	0.03	16–18	10–14	2–3	Balance

Type	Yield Strength* (psi)	Tensile Strength* (psi)	% Elongation*
304			
Annealed	40,000	83,000	60
Cold Worked	75,000	110,000	60
316			
Annealed	36,000	82,000	55
316L			
Annealed	35,000	80,000	50
Cold Worked	115,000	140,000	20

*Room temperature properties.

long as the coating remains in contact, the corrosion resistance of the steel will be high; however, should the coating break down either mechanically, thermally, or chemically, the passivization will cease and the steel will again become reactive. The passive Cr_2O_3 coating remains quite inert in oxidizing media. However, reducing media such as chlorine ion in aqueous solution tend to locally destroy the layer, and the steel is consequently then subject to pitting corrosion.

The presence of nickel in concentrations of the order of 8 percent stabilizes the face-centered cubic structural modification of Fe known as austenite. The class of steels generically associated with this nickel concentration are termed austenitic stainless steels. Steels of the 200 and 400 series are basically body-centered structures, due to the absence of Ni, which result in high hardness, high strength, and low ductility. The strengths obtained by the austenitic stainless are at present considered satisfactory for biological application, in light of the realized ductility and corrosion resistance presented by this series.

A further advantage of 300 series stainless lies in the enhancement of the mechanical properties as a result of cold working. Since this material has high ductility and may be worked, strain hardening may be expected. In addition, the application of loads may supply sufficient energy to transform some of the face-centered modification to the harder body-centered structure. A combination of strain hardening and structural transformation thus acts to produce the great changes in physical properties accompanying cold working.

As can be seen from Table 11-3, the corrosion behavior of stainless steel in a chlorine environment is markedly enhanced by the incorporation of a few percent of molybdenum. Therefore, AISI type 316 is the indicated stainless steel for application in saline environments. Since Mo tends to favor the stabilization of the body-centered structural modification of Fe, additional nickel is required to assure that the more ductile austenitic form is retained. The best success to date in resisting the familiar pitting attack of chlorine ion has been found with type 316L, which has more and more become a material of choice for design of hard-metal implants.

The major pitfall associated with the service behavior of austenitic stainless steels is the possibility of "sensitization." A stainless steel is said to become "sensitized" when, after a particular heat treatment or temperature change as a result of fabrication into an engineered structure (i.e., welding or brazing), catastrophic intergranular corrosion failure is noted in the affected region. The most critical temperature for the appearance of sensitization is 1200°F.

Based upon current knowledge, the best picture which may be given of this phenomenon is as follows: In the areas immediately adjacent to grain boundaries, chromium and carbon combine to form chromium carbide, which precipitates out into the grain boundary. The result is that the areas around grain boundaries are denuded of chromium by comparison to the average chromium content in the balance of the grain. The region around the grain boundary then becomes more susceptible to corrosion for the following reasons:

1. No chromium exists around the grain boundary to passivate the area through the formation of a protective oxide.
2. An additional galvanic effect aiding corrosion potentials is now possible because of compositional differences between the grain boundary region and the grain itself.

The presence of stress accentuates the catastrophic nature of this type of corrosion, since a corrosion attack produced along grain boundaries soon leads to failure across the structured member.

The reason for the existence of a maximum of sensitization at 1200°F in austenitic stainless is quite simple. At temperatures above 1200°F, the available thermal energy is sufficient to dissolve any chromium carbides which may form at temperatures lower than 1200°F. The diffusion rates of atomic species in the stainless steels are too low to permit sensible sensitization over ordinary time periods.

In an effort to decrease the opportunities for the precipitation of chromium carbides and the resultant sensitization, new grades of stainless steel containing markedly lower carbon contents have been introduced. These new types are designated by the suffix L, standing for low carbon. Thus, 316L contains 0.03 percent C, whereas conventional 316 contains 0.08 percent C. By far the best actor as a biological implant material among the stainless steels has been 316L.

In clinical employment of stainless steels, it is now obvious that, in prostheses containing several metallic parts, care must be taken to use only one type of "stainless." Contact between steels of different types will produce unwanted galvanism, leading to large-scale corrosive attack of the prostheses and enhanced foreign body reaction. For best clinical use, only labeled parts clearly identifying the AISI type, manufacturer, and batch number should be considered. This is necessary because AISI specifications require limits only for Cr, and Ni contents of the order of a few percent. Thus, even though care is taken to match plate and screw steel type, different batch numbers may produce slight compositional differences. Clinical history contains many documented cases of the untoward effects of contact of dissimilar metals in the biological environment.

Finally, the ease with which austenitic stainless steel may be cold worked leads to further possibilities for the construction of electrochemical corrosion cells within the biological environment. The presence of residual cold work will increase the tendency for the metal to react. The same is true for additional hardness produced by structural changes accompanying cold working. Thus, plate and screw of the same AISI type and batch number may be of different degrees of cold work, either prior to or after surgery, as a result of fabricating stresses or stresses induced in the metal by the surgeon during clinical procedures. Little can be done about this situation at present because of the clinical requirement of prosthesis and procedure, other than to be aware of its existence as a possible cause of implant corrosion.

Early in this century, while searching for a spark plug electrode material, Haynes found that additions of approximately 25 percent chromium (by weight) to a cobalt-based material produced alloys of high strength, high resistance to corrosion, oxidation, and wear, and low coefficients of friction with respect to steels and many other materials. He named this alloy "stellite" (a star among metals), and this name has since become a generic term for a wide range of cobalt-based alloys. Much of the development of these alloys has been due to the fact that they also exhibit high strength and corrosion resistance at very high temperatures. While elevated-temperature service is not required for surgical implants, the inherent oxidation and corrosion resistance of these alloys is extremely useful in their remaining benign to the biological environment.

Depending upon their composition, these alloys may be purchased in either wrought or cast forms. Present-day implant alloys have been made from cobalt-based alloys with significant additions of such elements as carbon, chromium, molybdenum, nickel, tungsten, and iron. In general, the high strength of these alloys is derived from the cobalt-carbon base; chromium adds corrosion resistance, nickel adds strength and ductility, molybdenum adds high-temperature stability, and other elements aid in stabilizing various phases. A typical analysis of three different alloys is given in Table 11-5.

The Co-Cr-Mo alloys, which are known commercially by such names as Cast Vitallium, Zimaloy, Stellite 21, and Vinertia, exhibit good strength and fair ductility, as shown in Table 11-5. In addition, they show excellent resistance to stress corrosion and attack in chloride ion environments. However, these alloys do not always withstand high-oxygen environments well, and as with most cobalt alloys, there is a small tendency to displace hydrogen from solution and depolarize at the cathode. Since, in implant applications, galvanic corrosion must be avoided at all costs, other materials such as stainless steels should not be placed in contact with the Co-Cr-Mo alloys.

These alloys are generally cast, since their ductility is only fair and their machinability poor. For high dimensional precision and compositional exactness, they are often vacuum cast by either the

Table 11-5. Typical Compositions and Properties of Cobalt-based Alloys Used in Biomedical Applications

Type	Cr	Mo	Ni	Mn	Si	Fe	C	W	Co
Co-Cr-Mo	27	6	2.5	1	1	0.75	0.25	—	Balance
Co-Cr-Ni-W	20	—	10	1	1	2.5	0.05	15.2	Balance
Co-Cr-Ni-Mo-Fe	20	7	15	2	—	15	0.15	—	Balance

Type	Yield Strength* (psi)	Tensile Strength* (psi)	Elongation* (psi)
Co-Cr-Mo	75,000	150,000	1
Co-Cr-Ni-W	67,000	145,000	50
Co-Cr-Ni-Mo-Fe			
Annealed	129,000	185,000	30
Wrought	260,000	340,000	1

*Room temperature properties.

investment or the lost-wax process. This complicated fabrication procedure adds to the cost of an already relatively expensive alloy. The cast shapes can be arc welded under a protective atmosphere of helium gas. By welding, more complex shapes can be achieved. The final shapes are ground, polished, and electropolished for high-luster surfaces free of pits and scratches. The performance of this material in tissue has been very good. A wide variety of implant shapes and accessories of this type of material is shown in Figure 11-6.

The Co-Cr-Ni-W alloys, which are commercially known by such names as Stellite 25, Neutrilium, and Wrought Vitallium, exhibit good strength and ductility, as illustrated in Table 11-5. They also have excellent resistance to halogen atmospheres and can be cast or wrought, although their high rate of work hardening renders them difficult to machine. They are also subject to crevice corrosion, and therefore carefully polished surfaces must be used in implanting. The degree of tissue reaction accompanying their application is generally small, similar to that with the Co-Cr-Mo alloys. A third group, Co-Cr-Ni-Mo-Fe, known commercially by such names as Elgiloy, was originally developed for use as nonmagnetic watch springs. It has good strength and ductility in the annealed state, and may be work hardened to excellent strength levels. It can also be spot welded or soldered. Although machining is difficult, special techniques are being developed to produce machined parts. Materials of this type are presently being used in drawn-wire form for electrodes in heart pacers and as flat sheet for heart springs, as well as in widespread dental applications in orthodontia and dental prosthetics.

Pure metals have been extensively employed for implant purposes by clinicians, but by and large they have failed, either because of low intrinsic strength or poor resistance to biodegradation, or through a combination of these factors. In addition, many metal ions are toxic to healthy tissue.

The recorded use of metals in surgery begins about 400 years ago when a gold plate was used in repairing a cleft palate. Subsequently, other types of materials were used, and because of availability, the earlier implants were pure metals or simple alloys. The noble metals, such as gold and platinum, were found to be fairly nonreactive to tissue, but they were too weak and soft to meet service requirements. Silver, lead, bronze (copper plus tin), and pure iron also were not sufficiently strong to support fractures of the long bones; moreover, they were not resistant to the corrosive attack of the biological environment. Magnesium was sought as an absorbable implant (i.e., dissolution accompanying healing, so that a second surgical procedure to remove implant

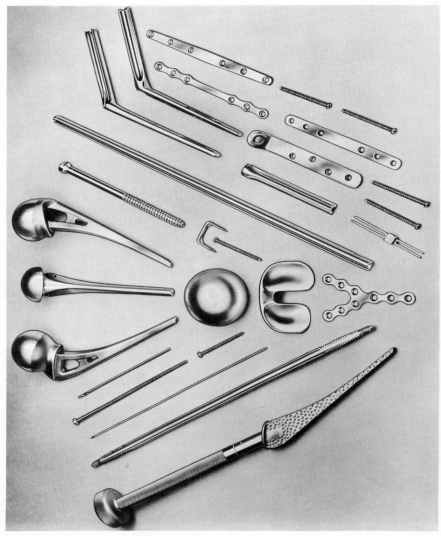

Figure 11-6. Various implants and accessories fabricated from a cobalt-chrome alloy. (Photograph courtesy of the Zimmer Mfg. Co., Warsaw, Indiana.)

after fracture union would not be necessary); however, usually it dissolved before the fracture could heal and the dissolution products were harmful to the surrounding tissue.

Other materials such as Ni-plated steel and vanadium steel provided sufficient strength, but also lacked the required corrosion resistance.

Nickel, copper, zinc, and aluminum fall into the class of pure metals that do not possess sufficient physical properties and do biodegrade to some degree. Copper is normally toxic to healthy tissue, while zinc has been found to interfere with normal bone regeneration. Titanium has shown promise as an orthopedic implant, possessing both desired physical properties and tissue tolerances. Recent advances in manufacturing procedures have allowed for fabrication of titanium into engineered structures without loss of desired physical properties. In addition, the Ti^{3+} ion appears to be quite benign in the biological environment. It has become an accepted material of choice in England, but has not yet had great clinical acceptance in America. A recent alloy modification, Ti—6 percent, Al—4 percent V, has been investigated quite extensively for orthopedic applications, with quite encouraging results in the form of enhanced physical properties and resistance to biodegradation. In addition, the Ti—6 percent Al—4 percent V has a much more favorable strength-to-weight ratio (factor of two) than do conventional stainless steels or cobalt-chrome implant alloys allowing for much more conservative implant design. Zirconium is chemically very similar to Ti and, although heavier, it appears to have similarly good tissue tolerance. The degree of experimental and clinical experience with this material is small, however.

Tantalum has been used for many years as a biological implant material. In its pure state, it is soft enough to be contoured for both maxillofacial prostheses and cranial prostheses. It has also been used as a suture material and as a mesh matrix for various reconstructions. This material enjoys a wide clinical acceptability today, although its biological behavior appears to be ambiguous. The biological behavior and acceptability of this material by host tissue may depend upon such environmental factors as state of stress, local pH, presence of high O_2 tension, and degree of vascularity in the area of the implant. The controversy associated with this material is indicative of the need for carefully controlled systematic in vitro and in vivo research into the problems of acceptability of implant to host.

Polymers. The group of polymers known as acrylics have found uses as bone replacements, corneas, adhesives, hemostatic agents, dentures, contact lenses, and artificial eyeballs. Typical of this group are the polymethyl methacrylates. Their excellent clarity is attractive for optical lenses, while the ability to accept pigmentation makes them desirable in color-matched dentures and prostheses where cosmesis is indicated. As was shown in Table 11-1, their mechanical strength is adequate for only small load-bearing applications. Under large loads, such as for femur heads, they may either fracture or creep. Their adaptability to surface tissue is generally very good, as exemplified by the widespread usage of polymethyl methacrylate contact lenses and dentures, and their subcutaneous reactions are also good. However, they do have strong tendencies to absorb surrounding liquids (with accompanying dimensional change) and to wear away through abrasion.

Another interesting family in the acrylic group is the cyanoacrylates. They polymerize at room temperature without catalyst or hardener, and make excellent adhesives; adhere instantaneously to moist cut tissue; act as intermediate hemostatic agents; accelerate wound healing; can be phagocytosed from the tissues; and are not carcinogenic.

The polymerization of polymethyl methacrylate is representative of an addition polymerization typified by reactions involving the covalent carbon-carbon bond. The methyl methacrylate mer or repeat unit is given by the structure

$$H_2C=\underset{\underset{H_3C}{|}}{C}-COO-CH_3$$

which is of the form

$$\underset{\underset{H}{|} \quad \underset{R''}{|}}{\overset{\overset{H}{|} \quad \overset{R'}{|}}{C=C}}$$

where R' is CH_3 and R'' is $COOCH_3$.

The energy associated with a single carbon-carbon (C—C) bond is 80 kcal per mole; that of the double carbon bond (C=C) is only 145 kcal per mole, however, since the second bond involves electrons not as tightly bound as those in the singly bound state. For this reason, these electrons will react readily in the presence of initiator free radicals. Since two carbon atoms are involved, the double carbon-carbon bond (also called vinyl functional group) is difunctional.

The most common sequence of reactions leading to vinyl group polymerization is:

1. Formation of free radicals which react with the monomer to begin the chain. The most common source of free radicals for initiation is the peroxide bond. Compounds such as benzoyl peroxide can decompose with heat according to

where x is

2. Reaction by addition of monomer molecules to the growing chain, which leads to chain propagation. The initiators may now react with the monomer according to

Chain propagation may then proceed:

until termination of the chain occurs.

3. Termination of the chain may occur through interaction of two growing chains which will mutually destroy their free radicals. In addition, at any time the chain end may interact with a free radical and terminate. However, since there are several orders of magnitude more free monomer than free radical, simple collision probability will favor chain propagation until relative concentrations of monomer and initiator are of the same order.

The presence of impurities may also lead to chain termination through ionization of the impurity. The resultant ionization will also produce additional free radicals which may initiate new chains. This process is known as chain transfer and is used to control the molecular weight.

The principles of addition polymerization may be applied to systems of polyvinyl chloride in which the monomer would have the form and of polytetrafluoroethylene (teflon) in which the monomer would have the form

$$\begin{array}{ccc} F & & F \\ | & & | \\ C & = & C \\ | & & | \\ F & & F \end{array}$$

The advent of a medical-grade silicone has made available to the clinician a material possessing a high degree of biological acceptability in inertness to biological degradation and absence of normal foreign body reaction. It cannot be overemphasized that these commercial silicones as a rule contain highly active side chains which may make them totally unsuitable for clinical application, and indeed even lethal. Because of the need to retain compositional integrity in medical-grade silicones, the available physical properties associated with this material can be changed only through a change in the molecular weight of the polymer. Medical-grade silicone liquid (low-molecular-weight silicone) for implant purposes is currently under controlled study and is not accessible to the clinician. (Injected medical-grade silicone does not appear to remain in situ, possibly because of the absence of foreign body reaction producing encapsulation.) Silicone (high-molecular-weight silicone) is at present too weak and brittle for use in load-bearing applications. Successes in the application of silicones have come through employment of silicone rubber for the desired prosthetic shapes.

Recently, it has been shown that silicone rubber will dissolve lipids and fatty acids, with a resultant decrease in important physical properties such as fatigue strength. Thus, although possessed of the potential for a high degree of biological acceptability, silicone rubber does not represent the universal biological implant material.

Silicone elastomers are pure high polymers based on the unit dimethylsiloxane $((CH_3)_2SiO)_x$.

The reactions depicted below will, of course, be directed toward achieving a chain length (values of x) necessary to produce the desired molecular weights, and hence the desired physical properties.

To do this, pure dimethyl dichlorosilane is hydrolyzed:

$$(CH_3)_2SiCl_2 \xrightarrow{H_2O} ((CH_3)_2SiO)_4$$

The linear polymers can crack with NaOH, so that all of the siloxane product is converted to the cyclic tetramer octamethylcyclotetrasiloxane (b.p. 175°C).

This tetramer can then be distilled for further purification, and finally converted to a linear high polymer by heating with a tiny amount of KOH as catalyst according to

$$((CH_3)_2SiO)_4 \xrightarrow{KOH} ((CH_3)_2SiO)_x \text{ elastomer (m.w. 1 to 4} \times 10^6)$$

In the case of the high-molecular-weight rubbers, the resulting gum is milled with reinforcing agent (finely divided silica gel), and then "cured" or "vulcanized" with small amounts of organic peroxide. As a result of the polymerization, materials result which have the following physical properties:

1. They are more thermally stable than organic polymers.
2. They retain their physical properties over a much wider range of temperatures than do conventional hydrocarbon oils and elastomers, and these properties change more slowly as a function of temperature.
3. Being "unnatural" products of radically different structure, they are not compatible with organic materials and consequently will not, as a rule, dissolve in or adhere to such materials. It is this property which many believe to be paramount in the biological acceptability of silicone rubber. Normally, silicone is not found in naturally occurring organics. Consequently, there may be no mechanism for its detection, and as a result no production of foreign body reaction.
4. Silicone surfaces, even in monolayers, are highly water repellent. This hydrophobic characteristic (no water sorption) results in the use of silicones for many applications in which cosmesis and integrity of surface are important.

The current status of technology allows the fabrication of silicone resins, although, as mentioned previously, their physical properties are not ideal for load-bearing applications. Some resinous material may have use, however, in the encapsulation of implantable materials or devices for which a bioinert outer shell is desired.

Silicones have low solubility in H_2O (are hydrophobic) and vice versa; consequently, silicones may be used to block the passage of water. Additionally, methyl silicone surfaces are not wet by water, that is, any water placed upon them stands with a high contact angle. Since surface tension will keep water drops intact, they will be too large to pass through capillary openings which they cannot wet. The solubility of certain lipid groups in silicone has led to the possibility of controlled-release agents for anesthetic or analgesic treatment through the encapsulation or potting of the appropriate drug in a silicone capsule.

The capsule is then implanted or ingested so that the drug can enter systemically.

Other polymers with biomedical uses include the following:

1. Polyethylene: tubing, syringes, oxygen tents, repair tissue heart valves, contraceptive implants.
2. Urethanes: plastic surgery, vascular and bone adhesive.
3. Nylon: vascular implants, syringes.
4. Polyvinyl pyrrolidine: artificial membranes for filtration of body fluids.
5. Cellulose acetate: nerve regeneration, packaging material.
6. Polypropylene, polystyrene, polycarbonates: syringes.
7. Polyvinyl chloride: tubing, facial repair, surgical drapes.
8. Fluorocarbons: artificial cornea, bone substitution.
9. Polyester fibers: aortic transplants.
10. Hydrogels: corneal implants.

Current Research and New Directions

In this chapter we have attempted to develop a basic understanding of those properties important in the consideration of an engineered material as a biological material of choice, as well as of the evaluation of the potential of a new material to serve in a prosthetic function. Concepts of mechanical properties, corrosion, and compatibility have been developed, and appropriate criteria for their applicability in a biological host have been outlined.

The current armamentarium of materials available to the practicing clinician has been discussed, with particular emphasis on the advantages and disadvantages of particular materials for specific applications. The science of those materials of widely accepted current usage has been considered at some length, because it is felt that an understanding of the properties of materials originates from a knowledge of the microstructural relationships between constituents at the atomic level.

Two classes of materials have emerged for current clinical applications: pure and alloyed metals, and polymers. A fair judgment would have to indicate that the state-of-the-art materials available have performed adequately in their chosen prosthetic roles.

It would be remiss, however, to assume that further progress in biomaterials is unnecessary. The continuing development of new biomaterials is, in fact, a leading concern of scientists and clinicians, and much research energy has been devoted to this problem.

Among the current lines of investigation are:

1. Development of meaningful in vitro and in vivo corrosion tests, both to screen new materials and to adequately simulate clinical conditions.
2. Development of higher strength-to-weight ratio materials compatible in biological environments to replace current hard-metal prostheses.
3. The use of composite technology, i.e., fabrication of an engineered material made up of two dissimilar materials such as a ceramic reinforcing a resin.
4. Membrane development: the use of new bioinert polymers to produce membranes capable of functioning in the renal, cardiovascular, optical, and nervous systems.
5. Study of the interfacial phenomena between polymer and blood in an effort to minimize the thrombogenic process accompanying cardiovascular prostheses. This has been alluded to in the body of the chapter in connection with the discussion of heparin-coated silastic.
6. Research into new production techniques which will allow fabrication of prosthetic appliances with the following enhanced characteristics: new material, lower cost, greater conformity to tissue removed, new-type coatings on conventional materials for increased compatibility and corrosion resistance, new shapes and forms, controlled microstructure for certain applications such as membrane bearing surfaces, etc.

While many of these biomaterials research goals lack the dramatic "breakthrough" character of much of the current work in bioengineering, the success of a large part of the present effort portends a steadily increasing body of knowledge available to the scientist and the clinician for the therapeutic use of engineered materials.

References

1. Van Vlack, L. H.: *Elements of Materials Science,* ed. 2. Addison-Wesley, Reading, Mass., 1964.
2. Bechtol, C. O., Ferguson, A. B., and Laing, P. G.: *Metals and Engineering in Bone and Joint Surgery.* Williams and Wilkins, Baltimore, 1959.
3. Parr, J. G., and Hanson, A.: *An Introduction to Stainless Steel.* American Society for Metals, Metals Park, Ohio, 1956.

4. O'Driscoll, K. F.: *The Nature and Chemistry of High Polymers*. Reinhold, New York, 1964.

5. *Cobalt Monograph:* Prepared by staff of Battelle Memorial Institute. Brussels, Belgium: Centre d'Information du Cobalt, 1960.

6. *Materials Selector, 1965:* Mat. Design Engrg. 62: 1965.

7. Tomashov, N. D.: *Theory of Corrosion and Protection of Metals*. Macmillan, New York, 1966.

8. Reed-Hill, R. E.: *Physical Metallurgy Principles*. D. Van Nostrand, Princeton, N.J., 1964.

9. Wesolowski, S. A., Martinez, A., and McMahon, J. D.: *Use of Artificial Materials in Surgery*. Year Book Medical Publishers, Chicago, 1966.

10. Laing, P. G., Ferguson, A. B., and Hodge, E. S.: *Tissue reaction in rabbit muscle exposed to metallic implants*. J. Biomed. Mat. Res. 1:135, 1967.

PART FIVE
Engineering and Health Care

12

The Artificial Lung

Introduction

The use of respiratory assist devices, partially or totally performing the function of the lungs, is indicated in a variety of medical circumstances. Besides the common usage in heart-lung bypass procedures, auxiliary lungs could play a role in palliative treatment of congenital circulatory defects in newborns, in treatment of wet-lung syndrome following shock, hyaline membrane disease, and organ transplant procedures. No one unit or even one design would be suitable for all cases and the multiplicity of applications undoubtedly would require a variety of units.

A respiratory assist device should perform its function without traumatizing the blood and should be efficient, reliable, safe, inexpensive, and easy to use. In addition, some gross design features might include low priming volume, low head loss, and minimum surface area.

The virtues of imposing a membrane between the gas phase and blood phase have been discussed in numerous places.[1,2,3]. In membrane units the exchange process includes (a) the passage of oxygen from a high concentration gas phase through a membrane and into and through successive layers of the blood and (b) the concomitant passage of carbon dioxide in the reverse direction. Obviously the efficiency of the process depends on the solubility, diffusivity, and geometry of the membrane and the solubility, diffusivity, and kinematics of the blood. The present analysis looks, not at the total process, but rather at two limiting cases: the fluid-limited case and the membrane-limited case. The *fluid-limited case* assumes that the membrane is infinitely permeable (or infinitely thin) and that the process involves only the gas transfer through the blood. This idealization is a valid approximation if the rate of transfer through the membrane is much faster than that in the blood. The *membrane-limited case* assumes that the gas dispersion in the blood phase is infinitely fast (perhaps aided by mixing) and that the process is limited by the gas transfer time through the membrane. Both cases may be experimentally achieved and have been analyzed for several different geometries. The object of analyzing limiting cases is that (a) the analysis is simplified, and (b) the features limiting the gas transfer efficiency are more easily recognized; in design, the selection of the membrane material and thickness is somewhat separated from that of devising the flow kinematics. The analyses provide a basis for excluding much costly testing.

The analysis of the membrane-limited case is a relatively straightforward mathematical problem of diffusion through a solid, with some care required in specifying boundary conditions. The boundary conditions are influenced by the chemical properties of the blood. The fluid-limited case, on the other hand, is a convective diffusion problem with dispersed reaction sites. The problem is mathematically nonlinear and must be solved numerically. The governing equations are derived using a simplified model of the blood: The blood is assumed to be a Newtonian fluid with uniformly distributed sources and sinks for oxygen and carbon dioxide. The nonlinear characteristics of the sources and sinks are determined from the oxygen dissociation curve and the carbon dioxide absorption curve. The reaction kinetics are assumed to be infinitely fast. Details of the model are discussed in previous papers.[4-6] A similar model has been used by others such as Buckles and associates[7] and Spaeth and Friedlander.[8] Experiments[6-8] in some of the geometries have generally substantiated the validity of the model for analysis of gas transfer in respiratory assist devices.

Analytical solutions for oxygen uptake and carbon dioxide removal depend, for any one oxygenator geometry, on a number of parameters. Each design will have its own natural parameters. As in most engineering studies, the results are made concise by choosing parameters optimally to minimize independent variables while retaining a com-

pletely general problem. Advantageously, the parameters chosen usually are dimensionless. The choice of parameters depends on the equations being solved, the boundary conditions imposed, and the information desired. Examples of dimensionless parameters often appearing
in these mass transfer problems include the Reynolds number, Schmidt
number, and Peclet number.

In many of the analyses cited in this paper, a dimensionless length
parameter was found to be especially convenient. Although specific
terms in it may vary from case to case, this parameter is typically a
constant $\times$ diffusivity $\times$ flow length/rate of flow. This dimensionless
length incorporates items easily controlled in the design: length of
flow-path and rate of flow. The parameter for some cases surprisingly
does not include a size parameter, e.g., tube diameter. As shown in some
of the previous studies, the gas transfer rates can be independent of
these size parameters; for example the length of tubing needed to raise
oxygen saturation by some specified amount is independent of tube
diameter in a straight tube oxygenator. Other parameters that may
appear in the results are initial gas concentrations in blood, hemoglobin
concentrations, oxyhemoglobin dissociation characteristics, and carbon
dioxide absorption characteristics. Although earlier papers included
results for various values of these latter parameters, all the following
are for an initial oxygenation saturation of 75%, an initial carbon
dioxide partial pressure of 50 mm Hg, a hemoglobin content of 15 grams
per 100 ml and typical normal oxygen dissociation and carbon dioxide
absorption characteristics. In all cases, the atmosphere outside the
membranes is assumed to be oxygen at about one atmosphere of pressure and free of carbon dioxide.

Fundamentally, the function of any auxiliary lung is to raise the
blood O_2 saturation and reduce the P_{CO_2} from some specified venous
condition to some specified arterial condition for the required rate of
blood flow. Conceivably the operation of any particular unit could be
limited by its O_2 transfer or by its CO_2 transfer. Any unit, however,
may be made either O_2-limited or CO_2-limited by controlling the O_2
transfer rate, as by raising or lowering the O_2 tension in the surrounding
atmosphere. Since CO_2 transfer is limited by the fact that the minimum
CO_2 tension in the atmosphere can be no lower than zero, CO_2 removal
has a maximum rate. By simply using high O_2 pressure in the atmosphere, O_2 addition can be as high as desired without theoretical limit.
Thus any design could be made CO_2-limited by simply adjusting the
O_2 pressure in the atmosphere. However if the atmosphere consists of
715 mm Hg O_2 and no CO_2, it can be generally concluded, as shown

below, that all fluid-limited membrane auxiliary lungs are O_2 transfer-limited and that all membrane-limited lungs, in which membrane CO_2 permeability is five or less times the O_2 permeability, are CO_2-limited.

Parallel Laminar Flow

Perhaps the simplest geometries to analyze and to construct are those designs in which the flow is parallel to the walls of the oxygenator. Three such types are discussed in this section: flow between parallel plates, flow through straight tubes, and axial flow through the annular space between concentric round tubes.

Flow between Parallel Plates. The early Clowes unit[1] and, to some extent, the Bramson unit[9] and Landé unit[10] are examples of the first type in which the blood flows parallel to the parallel flat membranes. The main component of the gas transfer is in the direction perpendicular to the blood flow. An analysis of the membrane-limited and fluid-limited cases using the above described model for the blood is included in other studies,[11-13] and the results are summarized in Figure 12-1.* The increase in O_2 saturation and the decrease in P_{CO_2} are shown as functions of a dimensionless flow-path length, defined as DL/gq, in which D is gas diffusivity in whole blood, L is the flow-path length, q is the flow rate per unit width of channel, and g is the gap distance between membranes. By using this dimensionless length, the fluid-limited process can be expressed as a single curve for oxygenation and a single curve for CO_2 removal; the various combinations of flow rate, gap width, and gas diffusivity are incorporated into this dimensionless length. The flow-path length needed for the membrane-limited process depends on the membrane material used, the membrane thickness, and the gap between parallel membranes.

Curves based on the membrane-limiting assumption are shown in Figure 12-1 for a unit with one-mil thick silicone rubber membranes as with a gap spacing of 3 mm. The curves in Figure 12-1 indicate that (a) if the process is fluid-limited, oxygenating the blood requires more length than that needed to remove the CO_2; (b) if the process is mem-

---

*The membrane-limited curves in Figures 12-1, 12-2, and 12-3 are based on the simplified analyses used by Weissman and Mockros[5] for O_2 and the method for CO_2 described under Analysis of Carbon Dioxide Transfer for Membrane-Limited Case, page 336. More precise calculations (see, for instance, Weissman[11]) reveal that these curves are slightly conservative.

brane-limited, removing the CO_2 requires more length than that needed to oxygenate the blood; and (c) if the unit is made of one-mil thick silicone rubber membranes at a gap spacing of 3 mm, the process is fluid-limited. For this typical unit, the oxygenation process for the fluid-limited case requires about 650 times more flow-path length than the length required if the process were wall-limited. That is, the time needed to diffuse the oxygen through the layers of the blood is much greater than the time needed to diffuse enough O_2 through the membranes. Similarly, the fluid-limited CO_2 process is about 100 times less efficient than the membrane-limited process. An auxiliary lung, then, consisting of laminar parallel blood flow in a 3-mm gap between one-mil thick silicone rubber membranes is, for practical purposes, fluid-limited. The usefulness of this type of analysis is that it indicates steps to be taken to improve gas transfer efficiency. Since the unit illustrated in Figure 12-1 is clearly fluid-limited, it is useless to attempt improvements by using thinner or more permeable membranes. In fact,

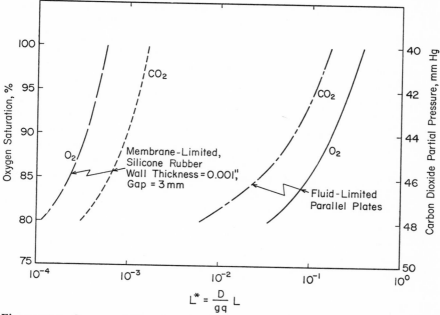

Figure 12-1. Curves of O_2 saturation increase and P_{CO_2} decrease in an auxiliary lung consisting of parallel flat permeable membranes. The abscissa is a dimensionless length (see text). For the fluid-limited case, the flow is assumed to be laminar and parallel to the membranes.

in this case much thicker membranes could be used without affecting the efficiency. On the other hand, the efficiency could be increased considerably by improving the gas transfer in the fluid phase.

The gas transfer in the fluid phase is most easily improved by introducing some form of mixing. If the blood were efficiently mixed as it passed between the membranes, the corresponding fluid-limited curves would shift to the left. If the mixing were very efficient, the corresponding curves may shift three or four log cycles to the left of the membrane-limited process. For practical purposes, such a unit would be membrane-limited. The degree to which the fluid-limited curves shift to the left depends on the mixing efficiency. Various techniques (e.g., placing screens or cones in the gap and pulsating the membranes) have been used to induce such mixing. The effectiveness of mixing with induced laminar secondary circulations is discussed below. If the mixing is only partially efficient, the fluid-limited curves may shift to the left with the fluid-limited curves falling near the membrane-limited curves. In such a case, the unit will be neither membrane-limited nor fluid-limited. The performance will be worse than either limiting case. In general, the actual case will be worse than the worst of these two limiting cases. If the two limiting cases are widely separated, as in Figure 12-1, the actual process should be very close to the worst limiting case.

Flow through Straight Tubes. The results of a similar analysis for flow through straight gas-permeable tubes is shown in Figure 12-2. In Figure 12-2 the dimensionless length is $\pi DL/2Q$, in which Q is the blood flow rate in the tube. The O_2 and CO_2 curves for the membrane-limited process are shown for typical small diameter silicone rubber tubes. The cross-hatched regions encompass the membrane-limited curves for all the standard sized tubes smaller than 0.25 in. i.d. that are readily available from the Dow-Corning Company. The fluid-limited curves are valid for all tube sizes. The details of the analysis and parametric dependence in the results have been previously reported.[4,6] Again, for laminar flow parallel to the tube walls, the process is fluid-limited. For the example tube used for the membrane-limited case in Figure 12-2, the length needed for fluid-limited oxygenation is some thirty-two times greater than the length needed for membrane-limited oxygenation. Indeed, experiments with many of the small silicone rubber tubes indicate the actual process is very close to the fluid-limited process. The membrane is efficient enough to be essentially infinitely permeable as far as the oxygen is concerned. Since, with the given atmospheric conditions, the membrane-limited process is CO_2-limited (i.e., the CO_2 curve is to the right of the O_2 curve) and the fluid-limited process is

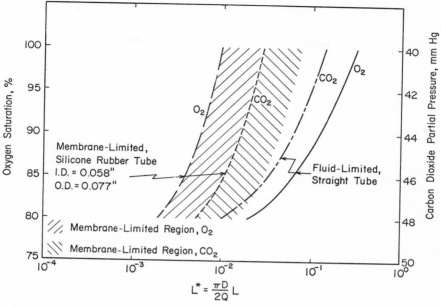

Figure 12-2. Curves of O_2 saturation increase and P_{CO_2} decrease in an auxiliary
lung consisting of gas-permeable tubes of circular section. The
abscissa is a dimensionless length (see text). For the fluid-limited
case, the flow is assumed to be laminar and parallel to the tube
axis.

O_2-limited (i.e., the O_2 curve is to the right of the CO_2 curve), the
corresponding CO_2 curves are relatively close to each other. The fluid-
limited CO_2 process is only five times less efficient than the membrane-limited process. Thus, CO_2 removal experiments in typical tubes
would not be expected to agree with the fluid-limited CO_2 curve because the membrane resistance would slow down the fluid phase
diffusion. The data should fall to the right of the CO_2 fluid-limited
curve. Since the fluid-limited process is O_2-limited, however, some
inefficiency in CO_2 removal can be tolerated without decreasing the
efficiency of the auxiliary lung. As in the flat plate unit, the gas transfer
efficiency could be improved by inducing mixing in the fluid phase.
A variety of mixing techniques were discussed in a previous paper.[6]
One way of gently mixing the blood is to induce laminar secondary
circulations by coiling the tubes and thereby setting up circulation
inducing centrifugal forces. A brief discussion of the effect of coiling
is given below.

Axial Flow through Annular Space between Concentric Round Tubes. Blood flowing axially in the annular space between concentric cylinders will be more easily oxygenated than the same flow through a straight tube. The annular geometry provides more gas exchange surface per unit of blood volume than the corresponding tube and a resulting decrease in necessary diffusion times. The analysis of axial flow in an annular space is a reasonable model for the Kolobow oxygenator.[14] A previous study[15] of blood-gas exchange in an annular geometry used the blood model described above and assumed oxygen atmospheres outside the outer tube and inside the inner tube. The study included a series of parametric variations, including the effects of the gas pressure at the surfaces, the initial O_2 saturation, and the hemoglobin content. For an initial O_2 saturation of 75 percent and an initial P_{CO_2} of 50 mm Hg, the changes of saturation and P_{CO_2} are shown in Figure 12-3 as a function of the same dimensionless length as used for the straight tube. The results parametrically depend on the gap ratio, R_g (the ratio of diameter of the outer surface of the inner membrane to the inner surface of the outer membrane. The O_2 curve for flow through straight tubes is shown as the curve for a gap ratio of 0.0. If the gap ratio is large (greater than 0.9), the curvature effects are small and the results are essentially equivalent to flow between parallel plates. Thus, for a gap ratio of 0.99, the curves are one cycle to the left of the curves for a gap ratio of 0.9; the curves for gap ratio of 0.999 would be one log cycle to the left of those for 0.99, and so on, indicating the linear dependence of the saturation on gap width to be expected for the parallel plate geometry. For smaller gap ratios, such as $R_g = 0.5$, there is a curvature effect. These results are useful in auxiliary lung designs consisting of small concentric tubes, in which case curvature effects are important, or in designs consisting of sheet material wrapped on large porous cylinders, in which case the gap ratios will usually be quite large. The membrane-limited curves in Figure 12-3 are for one-mil thick silicone rubber sheets wrapped on cylinders with a gap ratio of 0.9 and a gap spacing of 3 mm. These membrane-limited curves are three cycles or so to the left of the corresponding fluid-limited curves and, as a consequence, the process for parallel axial flow is strongly fluid-limited.

Improved gas transfer is thus easily achieved by inducing some type of fluid mixing. A variety of mixing techniques are possible. One interesting technique is that proposed by Keller[16] in which the outer surface of the annulus is rotated. The outer surface rotation superimposes a shear field on the axial flow field. If strong enough, the shear field causes the red cells to rotate and thereby generate local transverse

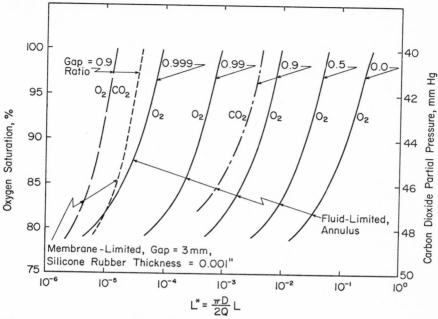

Figure 12-3. Curves of O_2 saturation increase and P_{CO_2} decrease in an auxiliary
lung consisting of concentric gas-permeable tubes of circular cross
section. The blood flows in the annular space between the concentric tubes. The abscissa is a dimensionless length (see text).
For the fluid-limited case, the flow is assumed to be laminar and
parallel to the axis of the tubes.

convection. Keller has induced considerable improvement in efficiency
by this method. He has not, however, been able to approach the membrane-limited curves.

Improved Gas Transfer in the Fluid Phase

As indicated above, all auxiliary lungs in which the flow is parallel
to the membrane surfaces are limited, with presently available silicone
rubber membranes, by the gas transfer in the fluid phase. The obvious
approach to improved efficiency is to induce transverse mixing in the
flow. Although numerous schemes can be used to induce the mixing,
only those that are relatively nontraumatic to the blood are suitable
for consideration. The induced mixing should be efficient in the trans-

verse transfer of the gases and, at the same time, should not be detrimental to the formed elements or proteins of the blood. Turbulent flow and flow induced by roughened walls or screens promote effective transverse gas transfer, but they also promote blood trauma. One especially appealing notion for improving the gas transfer in the fluid phase is to induce a laminar secondary motion transverse to the main flow. The mechanisms available for inducing such a secondary laminar flow depend on the geometry of the unit.

For example, simply coiling the tubes previously discussed into a helix induces a pair of secondary transverse laminar vortices. The main laminar flow traveling axially down the curved tube has a higher velocity at the center of the tube than near the walls. The high center velocity results in a centrifugal force which, in turn, induces the secondary transverse circulation. The effects of these induced laminar circulations were analyzed and reported previously.[5] The effectiveness of the transverse convection depends on the tightness of the coil, the Reynolds number, and the Schmidt number. For typical values of these parameters the fluid-limited curves are slightly to the left of the curves for the typical membrane-limited case. The fluid-limited curves are not generally enough to the left of the membrane-limited curves, however, to expect the process to be completely membrane-limited, and experimental data[17,18] indicates a process less efficient than that predicted by the membrane-limited theory.

An alternative proposal for inducing secondary laminar convection is that suggested by Drinker and associates.[19] They propose that the coiled tube be oscillated about the helix axis with a high frequency and low amplitude. The oscillation also induces two laminar secondary circulations. The oscillation-induced circulations rotate in a direction opposite to that of the axial-flow-induced circulations, but by adjustment of the oscillation frequency and amplitude the former circulations completely dominate the latter. The advantage of these oscillation-induced circulations is that the strength of the mixing can be adjusted independently of the coil tightness and the blood flow rate. The induced circulation can be effectively adjusted by adjusting the frequency and amplitude of the helix oscillation. The fluid mechanics of the oscillation-induced circulations has not been analyzed as yet and, as a result, an analytical analysis of the fluid-limited process for this complex flow is not yet available. Experiments,[19] however, indicated the mixing can be effective enough to achieve a truly membrane-limited process. There are, of course, many other techniques that may be used to induce the required secondary transverse flows. Straight tubes could be made with internal helical fins that induce the circulation.

In the annulus, a secondary laminar circulation can be induced by rotating the inner cylinder about its axis. The superimposed velocity field induces an unstable centrifugal force field, and a secondary flow can be promoted. The resulting secondary circulations, with and without axial flow in the annular space, have been thoroughly studied and are known as Taylor vortices.[20] As yet no one has analyzed the fluid-limited gas transfer effectiveness in an auxiliary lung with such a flow.

A perhaps more appealing approach, in terms of fabrication ease, is to induce an effective and gentle fluid mixing when the blood is flowing between parallel walls. The primary flow is that of flow between parallel plates, as discussed earlier. One simple technique to promote secondary laminar flow consists of curving the walls in the direction of the primary flow. In traveling around the curve, the primary flow generates an unstable centrifugal force field that results in a secondary transverse laminar circulation. The instability of the flow between curved walls has been thoroughly studied, and the conditions necessary for the secondary flows have been analytically delineated[21] and experimentally verified.[22] The strength of the resulting circulations, however, has not as yet been determined by analysis, and some uncertainty remains. The basis of the circulations and the pattern of the circulations are available from analysis, but the strength of the circulations is not as yet known from analysis. If the strength of the circulations in the comparable Taylor vortex case is used as an indication of the strength for this case, an estimate of the circulation strength can be made. Such an estimate was made and used in the analysis of the fluid-limited case for an auxiliary lung consisting of flow along curved parallel walls.[13] The effectiveness of the transverse mixing is estimated to be sufficient to make the process approach the membrane-limited conditions.

Another scheme for introducing convection involves the flow of blood down an open channel cut in a solid block of material. The open channel is "closed" by placing a membrane over its open surface, and the membrane is moved in its own plane with fixed velocity in a direction perpendicular to the channel axis. The lateral motion of the wall causes strong secondary currents to be set up and easily reduces blood diffusional resistance to the point where membrane resistance dominates.[12]

The laterally moving wall, the rotating annulus surface, and the forced oscillations of coiled tubes are examples of *active* convection, i.e., convection requiring the addition of energy other than that derived from the pressure energy of the main flow. The stationary coiled tube, guide fins, and curve walls are examples of *passive* devices, since they need no external energy beyond that pressure necessary to drive the main flow. In general, active methods will have at least one more dimensionless parameter associated with them than passive methods, since only in active devices can the magnitude of convection be controlled independently of that of the main flow.

The actual mechanism used to introduce laminar mixing is a matter of the designer's inventiveness.

Analysis of Carbon Dioxide Transfer for Membrane-Limited Case*

Gas transfer in the membrane auxiliary lung is, in general, retarded by resistance to dispersion in the blood phase and resistance to diffusion in the membrane. If the flow mechanics of a particular design is such that the gas in the fluid phase is very rapidly dispersed, then the gas diffusion through the membrane becomes the limiting process. Calculations of diffusion through the membrane that are based on the assumption that the gas dispersion in the fluid phase is infinitely efficient indicate the bounds put on an exchanger by the membrane. Practically speaking, the gas dispersion in the blood phase is never infinitely fast. If, however, the fluid phase dispersion is, for example, 100 times more efficient than the diffusion in the membrane, such an assumption is quite reasonable.

The assumption that the gas dispersion in the fluid phase is infinitely fast means that any decrease in the blood P_{CO_2} at the inner surface of the membrane, caused by CO_2 diffusion out through the membrane, is instantaneously averaged over all portions of the blood in a cross section perpendicular to the mean flow direction. Thus, the P_{CO_2} in the blood is assumed to be the same for all portions of blood in that cross section. Furthermore, longitudinal gas transport, both in the membrane and in the blood phase, are assumed to be negligible. Since the P_{CO_2} gradients are much smaller in the longitudinal direction than in the direction normal to the mean flow, this assumption is also reasonable.

Parallel Plate Channel. In the membrane-limited case, the lateral mixing is so efficient that the total P_{CO_2} in the blood is independent of the lateral distance from the membrane. The blood P_{CO_2} is assumed

---

*See list of symbols used in this chapter on page 345.

to vary longitudinally only and is some initial venous value, P(O), at the entrance, $Z = 0$; some final arterial value, P(L), at the exchanger exit, $Z = L$; and some intermediate value, P(Z), at each intermediate position, $Z = Z$. The CO_2 is assumed to diffuse laterally through the membrane, and this lateral exiting determines the P_{CO_2} in the blood at any longitudinal position, Z. Thus analyzing the membrane-limited process for flow between parallel plate membranes reduces to analyzing the simple diffusion through a flat slab with the P_{CO_2} maintained at a constant, P_o, on one side, the gas side, and maintained at P(Z) on the other side, the blood side. The P(Z), of course, is a function of the distance, Z, from the entrance to the exchanger, and the object of the analysis is to find the length, $Z = L$, required to reduce the P_{CO_2} from some specified venous value, P(O), to some desired arterial value, P(L).

If the lower membrane of a parallel plate exchanger has a thickness, T_1, and the upper membrane has a thickness, T_2, and if the P_{CO_2} is P(Z) and P_o at the inside and outside surfaces, respectively, of each membrane, the flux of CO_2 out of the blood for a short length, dZ, of exchanger is:

$$CO_2 \text{ flux} = -D_m K_m [P(Z) - P_o]\left[\frac{1}{T_1} + \frac{1}{T_2}\right] b \, dZ, \qquad (1)$$

in which D_m is the CO_2 diffusivity in the membrane material, K_m is the CO_2 solubility in the membrane material and b is the transverse width of the plates. On the other hand, the rate of CO_2 loss from the blood as the blood flows from the section at Z to the section at $Z + dZ$ is:

$$\text{Rate of } CO_2 \text{ loss} = K[P(Z + dZ) - P(Z)] \, bq = Kbq \, dP, \qquad (2)$$

in which q is the blood flow rate per unit width of the plates and K is an assumed linearized proportionality between P_{CO_2} and total CO_2 content of the blood.* Equating the loss of CO_2 from the blood as it flows from Z to $Z + dZ$ to the flux out through the membranes gives:

$$Kq \, dP = -D_m K_m [P(Z) - P_o]\left[\frac{1}{T_1} + \frac{1}{T_2}\right] dZ, \qquad (3)$$

---

* More accurate calculations (see, for instance, Reference 11) can be carried out using a more precise expression for the CO_2 dissociation curve. Such calculations, however, are considerably more cumbersome and linearizing the CO_2 curve for the physiological range is quite reasonable.

and integrating P between P(O) and P(L) and Z between O and L gives:

$$Kq \int_{P(O)}^{P(L)} \frac{dP}{P - P_o} = -D_m K_m \left[\frac{1}{T_1} + \frac{1}{T_2} \right] \int_o^L dZ$$

or

$$Kq \ln \left[\frac{P(L) - P_o}{P(O) - P_o} \right] = -D_m K_m \left[\frac{1}{T_1} + \frac{1}{T_2} \right] L. \tag{4}$$

Thus, if the two membranes have equal thickness, T, and if the atmosphere outside the membrane has zero P_{CO_2}, i.e. $P_o = 0$, the length needed to reduce the P_{CO_2} from P(O) to P(L) is:

$$L = -\frac{KqT}{2D_m K_m} \ln \left[\frac{P(L)}{P(O)} \right]. \tag{5}$$

Tubular Conduit. In a similar manner, as blood flows through a tube the CO_2 exits radially through the gas-permeable wall if the P_{CO_2} in the surrounding atmosphere is less than the P_{CO_2} in the blood. Again the blood is assumed to enter the exchanger section with a P_{CO_2} equal to P(O), leave with a P_{CO_2} of P(L), and have a P_{CO_2} equal to P(Z) at any intermediate position along its length. The blood is assumed to be instantaneously mixed in any cross section and the problem reduces to solving for the rate of diffusion through an annular ring of width dZ. The CO_2 diffuses out radially from the inner surface of the tube wall where the P_{CO_2} is P(Z) to the outer surface where the P_{CO_2} is P_o. If the radius of the inner surface of the tube wall is R_i and the radius of the outer surface of the tube wall is R_o, the flux of CO_2 through such an annular ring (see, for instance, p. 189 of Ref. 23) is:

$$CO_2 \text{ flux} = -\frac{2\pi D_m K_m [P(Z) - P_o]}{\ln (R_o/R_i)} dZ \tag{6}$$

The CO_2 loss from the blood as the blood flows from the section at Z to the section at Z + dZ is:

$$\text{Rate of } CO_2 \text{ loss} = K[P(Z + dZ) - P(Z)]Q = KQ \, dP, \tag{7}$$

in which Q is the blood flow rate in the tube. Equating this rate of loss to the flux through the membrane wall gives:

$$KQ \, dP = -\frac{2\pi D_m K_m [P(Z) - P_o]}{\ln (R_o/R_i)} dZ \tag{8}$$

and integration between P(O) and P(L) and between O and L gives:

$$KQ \ln \left[\frac{P(L) - P_o}{P(O) - P_o} \right] = -\frac{2\pi D_m K_m}{\ln (R_o/R_i)} L. \tag{9}$$

Thus, if the surrounding atmosphere is free of CO_2, i.e. $P_o = 0$, the length of tube needed to reduce the P_{CO_2} from $P(O)$ to $P(L)$ is

$$L = -\frac{KQ \ln (R_o/R_i)}{2\pi D_m K_m} \ln \left[\frac{P(L)}{P(O)} \right] \tag{10}$$

Annular Space Conduit. If the blood flows in the annular space between two concentric gas-permeable tubes, the CO_2 will exit radially outward through the outside tube and radially inward through the inside tube. The analysis is the same as that for the tube flow except that the additional inside surface is available for removing the CO_2. The flux through both the outside tube wall and the inside tube wall is, again, that given by the diffusion-problem solution for the flux through an annular ring with the concentrations maintained at the two surfaces. Thus the flux of CO_2 out of the blood in the annular space is given by:

$$F = -2\pi D_m K_m [P(Z) - P_o] \left[\frac{1}{\ln (R_o/R_i)} + \frac{1}{\ln (r_o/r_i)} \right] dZ$$

$$= -2\pi D_m K_m [P(Z) - P_o] \left\{ \frac{[\ln (r_o/r_i)] + [\ln (R_o/R_i)]}{[\ln (R_o/R_i)] [\ln (r_o/r_i)]} \right\} dZ \tag{11}$$

in which R_o and R_i are the radii to the outer and inner surfaces of the outer tube, respectively, and r_o and r_i are the radii to the outer and inner surfaces of the inner tube, respectively. Equating this flux to the rate of loss of CO_2 from the blood as it flows from Z to Z + dZ and integrating between $P(O)$ and $P(L)$ and between O and L, and setting $P_o = 0$, as in the above two sections, the length required for reducing the P_{CO_2} from $P(O)$ to $P(L)$ is determined to be:

$$L = \frac{-KQ}{2\pi D_m K_m} \left\{ \frac{[\ln (R_o/R_i)] [\ln (r_o/r_i)]}{[\ln (R_o/R_i)] + [\ln (r_o/r_i)]} \right\} \ln \left[\frac{P(L)}{P(O)} \right]. \tag{12}$$

Design Implications

Perhaps the most important aspect of the above mentioned analyses is the indication of a procedure for designing auxiliary lungs. The analyses show that available membranes are efficient enough so that the gas exchange process is limited by the gas transfer in the blood phase. For instance, if the gas transfer process with parallel flow is to be membrane-limited, the diffusional resistance to CO_2 transfer in the blood film must be ten times less than the diffusional resistance through the membranes. Since blood has a permeability (diffusion coefficient $\times$ solubility) of 1.64×10^{-11} moles/(cm²-min-mm Hg/cm) and available one-mil thick silicone rubber sheets have a permeability of 8.33×10^{-11} moles/(cm²-min-mm Hg/cm),[24] the blood film would have to be $(1.64/8.33) \times 1$ mil $\times (1/10)$ or 0.5 μ thick. Inasmuch as this is smaller than the smallest dimension of an erythrocyte, it is not feasible to construct a parallel flow unit that is membrane-limited, i.e., a unit that optimally transfers CO_2 for a given surface area or given priming volume. Increased efficiencies, however, are readily achieved by inducing some mixing. The analyses show that relatively gentle mixing procedures are available that make the gas transfer process membrane-limited for all presently available membrane materials. In the following, sufficient mixing is assumed generated in the flow either by centrifugally induced circulations or by wall perturbations so that the process is truly membrane-limited.

The following are some conclusions as to what can and what cannot be expected from auxiliary lungs if the presently available materials are used. In all such membrane-limited cases, the restricting feature is the CO_2 removal, and the following considerations are based on the requirements for membrane-limited P_{CO_2} reduction. In all cases, the atmospheric oxygen can be adjusted to give a normal respiratory quotient. The analyses discussed above provide the necessary information for optimizing designs of certain types of auxiliary lungs. In particular, the analyses indicate the requirements for lungs consisting of multiple layered plates, parallel tubes or annuli.

Parallel Plates. The considerations of this section apply not only to units in which the blood flows between gas-permeable flat plates but also to units in which the blood flows between parallel curved surfaces (e.g., between concentric cylinders) if the curvature is small such as $Rg > 0.9$.

If an auxiliary lung is to reduce the blood CO_2 partial pressure from $P(O)$ to $P(L)$, the required length of parallel plates is:

$$L = -\frac{Kqt}{2(D_mK_m)} \ln\left[\frac{P(L)}{P(O)}\right], \tag{13}$$

in which K is the assumed linearized proportionality between P_{CO_2} and total CO_2 content of the blood, q is the blood flow rate per unit width of plates, T is the membrane thickness (assumed the same for both top and bottom walls), (D_mK_m) is the product of the diffusivity and the solubility of CO_2 in the membrane material, and $P(O)$ and $P(L)$ are the partial pressures of CO_2 of the venous blood entering the lung and the arterial blood leaving the lung, respectively. Thus, a blood gas exchanger that is designed to carry Q_c cm^3/min and that consists of N identical plate units, each unit processing Q_c/N cm^3/min, requires a blood volume to prime the active exchange section of

$$V = gbLN = -\frac{KgbqNT}{2(D_mK_m)} \ln \left[\frac{P(L)}{P(O)} \right], \tag{14}$$

in which b is the width of the flow channel. For example, if the lung is to reduce the P_{CO_2} from 50 mm Hg to 40 mm Hg, and, as is typical in the physiological range, if $K = 2.51 \times 10^{-7}$ moles/(ml-mm Hg),[4] the required priming volume is

$$V = \frac{-2.51 \times 10^{-7}Q_cgT}{2(D_mK_m)} \ln [0.80]$$

$$= 2.80 \times 10^{-8}\frac{Q_cgT}{(D_mK_m)}, \tag{15}$$

noting that $Q_c = bqN$. Equation (15) indicates that the priming volume required to handle a total flow of Q_c is independent of the flow rate per channel (independent of the number of plate units used). Thus auxiliary lungs can be designed to have as small a head loss as desired by using many channels in parallel. The more channels used, the lower the flow rate per channel and the shorter the length of channel needed to complete the oxygenation and CO_2 removal. With a low flow rate in each channel and each channel of short length, the head loss can be made quite small. The head drop advantage gained by using many short channels as opposed to using a few longer channels, however, must be weighed against the design disadvantage of having to manifold the many-channel unit.

Furthermore, Eq. (15) indicates that the priming volume may be minimized by using thin membranes and small gap spacings. With silicone rubber, i.e., $(D_mK_m) = 8.33 \times 10^{-11}$ moles/(cm²-min-mm Hg/cm), used for the membrane material,[25] the priming volume is

$$V = 336 \, Q_cgT \tag{16}$$

One way to define an efficiency of a lung is the ratio of the flow rate to the priming volume, Q_c/V. The higher the value of Q_c/V the more efficient the process. To put these numbers in perspective, at rest a normal man has a pulmonary flow rate of 5000 ml/min and a capillary lung volume of, say, 100 ml for a value of Q_c/V of 50 min⁻¹. With severe exercise, these values may change to 30,000 ml/min and 200 ml, respectively, for a Q_c/Ψ of 150 min^{-1}. An auxiliary lung of parallel plate construction with one-mil thick silicone rubber membranes and a ratio gap of one millimeter would have optimally a Q_c/Ψ value of 11.7 min^{-1}.

Note that for auxiliary lung designs utilizing flat membranes, the transfer rates depend only on membrane composition, surface area and thickness, and the blood flow rate. Flow-path geometry influences gas transfer only as it affects the discharge Q_c or changes the assumption of a membrane-limited exchange process. Aside from this, the flow-path geometry may be chosen in such a manner as to balance between hydrostatic pressure required for pumping and an acceptable priming volume, because it has no effect on gas transfer efficiency. The membrane area is independent of the priming volume.

Multiple Parallel-Tubes. Similar to the parallel plate analysis, the length of tube needed to reduce the P_{CO_2} from $P(O)$ to $P(L)$ is found to be

$$L = -\frac{KQ \ln (R_o/R_i)}{2\pi D_m K_m} \ln \left[\frac{P(L)}{P(O)}\right] \tag{17}$$

in which Q is the flow rate per tube and R_o and R_i are the radii of the outer and inner surface of the membrane, respectively. Thus a blood gas exchanger that is designed to carry Q_c cm^3/min and consists of N tubes in parallel requires a blood volume to prime the tubes of

$$\Psi = \pi R_i^2 \, LN = -\pi R_i^2 \frac{KQ \ln (R_o/R_i)}{2\pi (D_m K_m)} \frac{Q_c}{Q} \ln \left[\frac{P(L)}{P(O)}\right]$$

$$= -\frac{K}{2D_m K_m} Q_c R_i^2 \ln (R_o/R_i) \ln [P(L)/P(O)]$$

$$= 2.80 \times 10^{-8} \frac{Q_c R_i^2}{(D_m K_m)} \ln (R_o/R_i), \tag{18}$$

in which K is assumed to be 2.51×10^{-7} moles/(liter-mm Hg) and $P(L)$ and $P(O)$ are assumed to be 40 and 50 mm Hg, respectively.

Equation (18) reveals that, as with the parallel plate unit, a tubular auxiliary lung can be designed to have as small a head loss as desired. The priming volume is independent of the flow rate per tube. The small head drop achieved with many short tubes in parallel, however, must be weighed against the resulting cumbersome manifold.

Equation (18) also indicates that the priming volume may be minimized by using small diameter tubes. The required priming volume depends on the square of the tube radius. Equation (15) indicates the required priming volume is reduced by reducing the membrane thickness, i.e., reducing R_o/R_i. The design advantage gained by using thin-

Table 12-1.   Comparison of Optimum Efficiency, Q_c/V

Lung	Q_c/V $(min)^{-1}$
Natural	50 to 150
Tubes, 0.012″ id × 0.025″ od	17.5
Tubes, 0.025″ id × 0.047″ od	4.7
Tubes, 0.058″ id × 0.077″ od	1.93
Tubes, 0.132″ id × 0.183″ od	0.323
Tubes, 0.125″ id × 0.127″ od	7.42
Tubes, 0.250″ id × 0.252″ od	3.70

walled tubes is bounded by the reduction of structural strength as R_o/R_i becomes smaller. Two special cases of design could be considered: (a) a certain minimum structural strength is required and, therefore, R_o/R_i is fixed; and (b) tubes are to be made of sheet material with a fixed thickness, T. If R_o/R_i is fixed, the conclusion is obvious: a tubular oxygenator can be made to have small priming volume by using in parallel many short tubes of as small a diameter as feasible. If the wall thickness, T, is fixed the priming volume varies as $Q_c R_i^2 \ln [1 + (T/R_i)]$. Again, as R_i becomes smaller the priming volume becomes smaller, and the conclusion is the same: use a tube with as small a diameter as possible.

As with the parallel plate design, Q_c/V can be used as a measure of lung efficiency. An auxiliary lung made of silicone rubber tubes would have a priming volume of $336 \, Q_c R_i^2 \ln (R_o/R_i)$, and Table 12-1 gives a comparison of Q_c/V for auxiliary lungs made of typical silicone rubber tubes with that of the adult human lung. The first four tubes listed are shelf items of the Dow Corning Company.[26] The last two tubes listed—$\frac{1}{8}$″ and $\frac{1}{4}$″ in diameter, respectively, with one-mil wall thickness—are not available and are given to show what would be possible with very thin walls. The comparison indicates an advantage of designs with many small-diameter tubes. The unit with a great number of small-diameter tubes, however, presents problems of manifolding and problems in inducing effective enough mixing to ensure a membrane-limited condition. For adult human perfusions, an auxiliary lung should have a Q_c/V of 5 or more to keep the prime to 1000 cc or less. The only practical way to achieve this with tubes seems to be to use very small diameters. In applications to infants or small children these small-diameter tube units offer considerable promise.

Multiple Annuli. In an analysis similar to that for the tube, the length of annulus necessary to remove the excess CO_2 is

$$L = \frac{-KQ}{2\pi(D_m K_m)} \left\{ \frac{[\ln (R_o/R_i)][\ln (r_o/r_i)]}{[\ln (R_o/R_i)] + [\ln (r_o/r_i)]} \right\} \ln \left[\frac{P(L)}{P(O)} \right]$$

in which R_o and R_i are the radii to the outer and inner surfaces of the outer tube, respectively, and r_o and r_i are the radii to the outer and inner surfaces of the inner tube. The priming volume, then, for a unit handling Q_c ml/min is

$$\Psi = \frac{2.80 \times 10^{-8}}{(D_m K_m)} Q_c R_i^2 (1 - R_g^2) \left\{ \frac{[\ln (R_o/R_i)][\ln (r_o/r_i)]}{[\ln (R_o/R_i)] + [\ln (r_o/r_i)]} \right\}$$

in which R_g, the gap ratio, is the ratio of the outer surface of the inner tube to the inner surface of the outer tube. The conclusions are somewhat the same as for the tubular unit: (a) If the wall thickness ratios for the tubes in the annulus are fixed, the smaller the diameter of the unit the better, and the smaller the gap ratio the better; (b) If the wall thickness, T, is fixed and if the wall thickness and the gap distance (i.e., $g = R_i - r_o$) are both much smaller than R_i, the priming volume can be given approximately as

$$\Psi \approx \frac{2.80 \times 10^{-8}}{(D_m K_m)} Q_c g T cm^3.$$

In condition (b) above, unlike the tubular unit and unlike condition (a), the prime is independent of overall size diameter R_i. This case with small gap and thin membranes is the same as the parallel plate case, and the prime depends linearly on the total flow, Q_c; the gap distance, g; and the wall thickness, T. An annular unit with one-mil thick silicone rubber walls and with a gap of 1 mm, for instance, would also have a Q_c/Ψ value of 11.7 min^{-1}. An annular unit with small overall diametric dimensions would be difficult to construct. In annuli with large diametric dimensions and small gaps, the curvature effects are small and the annular space is essentially that of flow between parallel flat plates.

Conclusions

A safe, reliable auxiliary lung device would be useful, if not essential, in the treatment of hyaline membrane disease, alveoli proteinosis, various congenital circulatory defects, traumatic shock, and numerous other cardiopulmonary pathologies. In membrane units, thought to be relatively untraumatic, the gas-transfer process consists of two phases: (a) the gas-transfer through the membrane and (b) the dispersion of the gases in the blood. Recent, detailed analytical and experimental studies of gas transfer provide guidelines for optimal design of membrane auxiliary lungs.

The analysis indicates that with the presently available membranes, gas transfer efficiency is limited by transport in the blood phase. Practical units must incorporate some mixing of the blood to approach optimal design. A number of effective and nontraumatic forms of mixing are discussed. The analysis indicates what optimum design performance can be expected when using most common designs. Optimally, membrane auxiliary lungs, with present materials, will require priming volumes that are numerically about one-tenth the total blood flow per minute.

Symbols

b = width of the flow channel

D = gas diffusivity in whole blood

D_m = diffusivity of CO_2 in the membrane material

g = gap distance between membranes

K = the assumed linearized proportionality between P_{CO_2} and total CO_2 content of the blood

K_m = solubility of CO_2 in the membrane material

L = length from entrance to exit of exchanger

N = number of parallel identical exchange components, e.g., number of tubes in parallel

$P(L)$ = partial pressure of CO_2 of the arterial blood leaving the lung

P_o = partial pressure of CO_2 in the atmosphere outside the membranes

$P(O)$ = partial pressure of CO_2 of the venous blood entering the lung

$P(Z)$ = partial pressure of CO_2 of the blood at some position Z in the exchanger

q = blood flow rate per unit width of channel

Q = blood flow rate in a tube

Q_c = total blood flow rate in auxiliary lung

r_i = radius of the inner surface of the inner tube

r_o = radius of the outer surface of the inner tube

R_g = the ratio of the outer surface of the inner tube to the inner surface of the outer tube

R_i = radius of the inner surface of the membrane

R_o = radius of the outer surface of the membrane

T = membrane thickness

T_1 = thickness of lower membrane in parallel plate exchanger
T_2 = thickness of upper membrane in parallel plate exchanger
V = priming volume of active exchange portion of auxiliary lung
Z = longitudinal distance from entrance to exchanger

References

1. Galletti, P. M., and Brecher, G. A.: *Heart Lung Bypass*. Grune and Stratton, New York, 1962.

2. Dobell, A. R. C., Mitri, M., Galva, R., Sarkozy, E., and Murphy, D. R.: *Biological evaluation of blood after prolonged recirculation through film and membrane oxygenators.* Ann. Surg. 161:617, 1965.

3. de Filippi, R. P., Tompkins, F. C., Proter, J. H., Timmins, R. S., and Buckley, M. J.: *The capillary membrane blood oxygenator: In vitro and in vivo gas exchange measurements.* Trans. Amer. Soc. Artif. Intern. Organs 14:236, 1968.

4. Weissman, M. H., and Mockros, L. F.: *Oxygen transfer to blood flowing in round tubes.* J. Engin. Mech. Div., Amer. Soc. Civil Engin. 93:225, 1967.

5. Weissman, M. H., and Mockros, L. F.: *Gas transfer to blood flowing in coiled circular tubes.* J. Engin. Mech. Div., Amer. Soc. Civil Engin. 94:857, 1968.

6. Weissman, M. H., and Mockros, L. F.: *Oxygen and carbon dioxide transfer in membrane oxygenators.* Med. Biol. Engin. 7:169, 1969.

7. Buckles, R. B., Merrill, E. W., and Gilliland, E. R.: *An analysis of oxygen absorption in a tubular membrane oxygenator.* AIChE J. 14:703, 1968.

8. Spaeth, E. E., and Friedlander, S. K.: *The diffusion of oxygen, carbon dioxide, and inert gas in flowing blood.* Biophys. J. 7:827, 1967.

9. Bramson, M. L., Osborn, J. J., Main, F. B., O'Brien, M. F., Wright, J. S., and Gerbode, F.: *A new disposable membrane oxygenator with integral heat exchanger.* J. Thorac. Cardiovasc. Surg. 50:391, 1965.

10. Landé, A. J., Dos, S. J., Carlson, R. G., Perschav, R. A., Lange, R. P., Sonstegard, L. J., and Lillehei, C. W.: *A new membrane oxygenator-dialyzer.* Surg. Clin. N. Amer. 47:1461, 1967.

11. Weissman, M. H.: *Diffusion in membrane-limited blood oxygenators.* AIChE J. 15:627, 1969.

12. Weissman, M. H., and Hung, T. K.: *Numerical simulation of convective diffusion in blood flowing in a channel with a steady, three-dimensional velocity field.* AIChE J. (in press), 1970.

13. Chang, H. K., and Mockros, L. F.: *Convective dispersion of blood gases in curved channel exchangers.* AIChE J. (in press), 1970.

14. Kolobow, T., and Bowman, R. L.: *Construction and evaluation of an alveolar membrane artificial lung*. Trans. Amer. Soc. Artif. Intern. Organs 9:238, 1963.

15. Chang, H. K., and Mockros, L. F.: *Blood-gas transfer in an axial flow annular exchanger*. (in press), 1970.

16. Keller, K. H.: *Development of a Couette oxygenator*. Proc. Artif. Heart Prog. Conf., National Heart Institute, Washington, D.C., 1969, page 393.

17. Dorson, W. J., Jr., Baker, E., Cohen, M. L., Meyer, B., and Molthan, M.: *A perfusion system for infants*. Trans. Amer. Soc. Artif. Intern. Organs. 15:155, 1969.

18. Eisman, M. M., and Mockros, L. F.: *Tubular Membrane Oxygenators*. In preparation.

19. Drinker, P. A., Bartlett, R. H., Bialer, R. M., and Noyes, B. S.: *Augmentation of membrane gas transfer by induced secondary flow*. Surgery 66:775, 1969.

20. Schlicting, H.: *Boundary Layer Theory*. McGraw-Hill, New York, 1967.

21. Reid, W. H.: *On the stability of viscous flow in a curved channel*. Proc. Roy. Soc. A244:186, 1958.

22. Brewster, D. B., Grosberg, P., and Nissari, A. H.: *The stability of viscous flow between horizontal concentric cylinders*. Proc. Roy. Soc. A251:76, 1959.

23. Carslaw, H. S., and Jaeger, J. C.: *Conduction of Heat in Solids*, 2nd ed. Oxford University Press, 1959.

24. Robb, W. L.: *Thin silicone membranes—their permeation properties and some applications*. General Electric Research and Development Center Report No. 65-C-031, 1965.

25. Dow Corning Corporation: *Gas Transmission Rates of Plastic Films*, 1959.

26. Dow Corning Corporation: *Silastic Medical-Grade Tubing*. Bulletin 14-119, August 1966.

13
Instrument Development

Biomedical engineering is a new science just emerging from a chaotic beginning with hazy definitions of purpose and goals. It is now time to determine the problems which should be solved and to establish a system for solving the problems. Clear-cut, straightforward directions toward goals must be identified most carefully. Such goals lie in the development of biomedical engineering as a system of which instrument development is a part.

Without suitable instruments, it is impossible to collect quantitative data and to assess the variables in a biological system. The design of an instrument to perform such functions presupposes a knowledge of both the engineering and the biological problem. Too often such is not the case. The engineer requires a list of specifications which the biologist is unable to give him, and the biologist is unable to express his needs in mathematical terms which the engineer must have in order to design the needed instrument. The result is often sophisticated gadgetry. It is relatively easy for the engineer to take a well-known principle and utilize it in the construction of a slightly different or slightly improved instrument. Good examples are the many electrocardiographs on the market, all of which are based on the same principles.

There are four distinct levels of sophistication in instrumentation. The least sophisticated is the construction from specifications of a "black box" by a technician without a clear knowledge of what the instrument is to accomplish. At a more sophisticated level is the design of an instrument from information furnished by someone else. The physiologist may indicate to the engineer that he needs a particular transducer and catheter to measure blood pressure. With this information it is relatively simple to design an instrument with the proper characteristics. A step higher is the case of a biologist and an engineer working closely together to develop a new transducer for measuring a new physiological parameter. The fourth and most sophisticated level is the engineer who can develop true conceptual thinking in the application of engineering principles to biological systems.

This type of engineer can take the mathematics of control system theory and apply it to a new theory of muscle interaction, which may in turn lead to a new means of making prosthetic limbs servo-controlled by muscle action potentials; he can use information derived from studies of membrane transport and biological potential to provide power to drive an internal cardiac pacemaker without an external power supply; or he can build mathematical models of the neuron connections in the brain in such a way as to lead to design of new types of computers. There is an acute shortage of such individuals.

It must be stressed, however, that the fields of science, whether engineering or medicine, advance because of the development of concepts rather than instruments. Instruments will not develop a basic physical theory, but once the theory is developed instrumentation is required in order to document it. The application of the principles of systems analysis, feedback controls, and operations research to the many problems of the biologist is much more important than the construction of yet another instrument to measure blood pressure more accurately.

The goals we desire are not achieved through an easy solution by the establishment of a "think tank." Rather, they lie in the need for a complete integration of the biological sciences (from cell biology to medicine) and the engineering sciences (civil, mechanical, and electrical) into a new science of dealing with biological material in a new way. This new science comprises three major areas which have been alluded to above.

The first of these areas is the development of a strong theoretical base. Here the applications of models, systems analysis, and control theory to the problems of the biologist have produced a whole new era of quantitation in biology. We are just beginning to understand how the endocrine, respiratory, and cardiovascular systems behave, and how they are controlled largely through the use of engineering techniques.

The development of new types of instrumentation to measure new parameters is also a part of this first area. Many engineers have built better mousetraps in the form of solid-state circuits and automatic readouts of EKG, for example, but few have asked the really vital question: "Is this the right or the best way to measure cardiovascular response?" It is this kind of approach we seek.

Following the formulation of a theoretical base, the generalization must be put to practical use in the second stage of biomedical engineering, that of development of a prototype of a useful working model, which must be followed by the third stage, that of production of an instrument that is useful in the delivery of a health service.

In the development of useful applications, we must not neglect the theoretical underpinnings of biomedical engineering while we attempt to bring the cutting edge to greater usefulness. We know little about the system of information, communication, and control of the human body, and still less about the uses of engineering science to measure, evaluate, and control bodily processes. There must be an orderly progression in the science from theory to development of practical use to the production of commercial instrumentation. It would appear most logical at the moment for the academic community to develop the theoretical base, for NIH to develop the applications, and for the industry to develop the final products of general usefulness.

For years we have considered the problem of powering internal pacemakers with batteries with the resultant need to perform an operation with attendant risks to replace the power source every year or so; only recently has the question of using an internal power source been raised and studied in depth. Development and design of a practical instrument are yet to come.

One of the major areas in which the engineer can make a contribution to basic knowledge is computer technology. Some doubts exist as to whether advances in computers are as imaginative as they should be. Improvements in computers are needed to aid in simulating biological systems, to help in the analysis of bioelectric signals, and to aid in pattern recognition in the areas of optical images and speech. In medicine and biology, computers are used which were designed primarily to solve problems arising in the physical sciences. Computers need to be designed specifically for biomedical research, since the requirements differ markedly from those in the physical sciences. Much work has been done on the design of large computers, but little advance has been made in the design and use of small computers. The application of minute computers built with microminiaturization of components can significantly advance medical knowledge. For example, a miniature computer could be used in the interior of the body to monitor various functions.

A relatively neglected area in which the talents of the engineer are needed is the field of information storage and retrieval. Handling of the ever-increasing flood of biomedical literature is a problem desperately in need of solution. The computer must also play a large part in the development of cybernetics. Analog computers have already been successfully used for the analysis of control mechanisms in the body. Continued development of transducers and research on control systems of the body are needed to improve our understanding in this area. More engineers with biomedical competence are necessary to bring the computer to its maximum potential as a tool in biomedical research.

It is possible to reproduce the total system. High-frequency sound (ultrasonics) was developed, and the focusing and theory were worked out, on a theoretical basis. The Northwestern University Laboratory at Evanston, Illinois, took this theory and applied it to the point of visualizing the sound on a fluoroscope screen in the same fashion as x-rays. This was then applied to clinical medicine in a direct visualization of blood flow in organs without surgical intervention—and a new tool for the investigation of valvular lesions of the heart had been discovered.

In January 1963, a contract was awarded to the Massachusetts Institute of Technology to support the development and evaluation of the LINC. This program evolved as a natural outcome of the attempt to produce a type of computing equipment of optimum usefulness in the biomedical laboratory. The machine developed was a small, portable, relatively inexpensive, and extremely flexible laboratory-oriented computer, which can be used as a laboratory tool by the individual biomedical investigator within the environment of the research laboratory, particularly for on-line operation. The latter is especially important, since it makes possible the use of the man-machine-biomedical-research experiment complex in efficient, imaginative, varied, and highly useful situations. Working in close collaboration with one another, the computer technology and biomedical research groups refined an earlier prototype model of the LINC, developing it to the stage at which large-scale testing by the national scientific community became feasible. Twelve research scientists were then provided with a machine for intensive use and evaluation. The information was used to modify and refine the current design, which was then turned over to commercial companies for production and sale. The PDP-8 is the outcome.

These few examples illustrate the combination of a strong biological need closely coupled with fine engineering to proceed in a logical
sequence: Problem $\longrightarrow$ Theory $\longrightarrow$ Prototype Instrumentation $\longrightarrow$
Development of Model $\longrightarrow$ Production.

Regardless of the worth of any basic idea or theory, it is of no value
to the public or even to the inventor unless it can be applied to the
solution of specific health problems. This requires development of the
idea from the prototype, or even the theory, into an operating instrument, with the usual problems of design engineering and manufacture.
This is obviously not—nor should it be—the province of the academic
scientist. Industry must assume its rightful place in the scheme of
things. However, costs are high, patent problems are many, and most
small and large concerns do not want to provide this support. The
problem is further complicated by the vagaries of the scientist, who
usually has a particular idea or design in mind and is not willing to
purchase a mass-produced instrument unless it can be modified to his
particular ends. This is particularly true in an era of an increasing
shortage of engineers when it is advisable to apply available engineering talent to the development of items for a more defined and certain
market area. It may be necessary for the government to share in the
development costs of needed instrumentation in order to surmount
some of these problems.

There are often no algorithms in the very areas in which major
development is needed. The development to the stage of commercial
production of apparatus in biomedical engineering creates a very
special problem. The development cost is very high. It is difficult to
find a company that wishes to develop apparatus when it is not sure
of sales or of a return on its investment. The government may need
to initiate development to the point of formulation and construction
of the first prototype, after which the plans can be turned over to
commercial interests. This worked well for the LINC. The basic problem is the determination of which pieces of apparatus are worthwhile
for the government to develop, i.e., which have a general value to the
country as a whole.

It may be that a company must be satisfied to have a head start
in development without patents. In the DOD where there is a one-buyer,
one-seller relationship, it is possible to contract on a cost plus basis
for equipment. This is not reasonable in the medical field where many
types of the same basic equipment exist, where the possible usage may
be nebulous, or where the market is totally unknown. The flame pho-
tometer was once cited as useless on a market survey, but thousands
have been sold. The reverse may also be true. Many individuals may

indicate interest in a given instrument, only to find it less than satisfactory when made, and not purchase it. The automatic suturing machine may be an example.

This brief survey of problems points up the fact that for several years the field of biomedical engineering has been developing in a hit-or-miss fashion. It is now time to consider the best way to form the accumulated knowledge and information into a coordinated thrust which will permit its rapid utilization.

Except in a restricted and specialized sense in some of its national laboratories, the government cannot become a producer of medical devices for the general medical public, since the number and variety are so large that the cost would be prohibitive. Any program developed by the federal government must be selective in two directions. In the first place, the extent to which development will proceed must be carefully ascertained. The development of given instrumentation to the stage of a working prototype model is reasonable; however, beyond this point, private industry must take the responsibility for ultimate refinement and production. Secondly, the choice of which instrument to develop must be made carefully. Evaluation and/or production of every "gadget" wanted by every investigator is unrealistic and impossible. Careful discussions must be held as to the degree of development to be undertaken for each instrument, and as to the type of instrument which would be of the greatest value to the most users and of the greatest benefit to the population as a whole. Such a program, even of limited scope, would require a very complex organization.

The third major area in biomedical engineering includes the scientists who can tie the first two areas together into a system that can deliver health services to the patient in the hospital with a minimum of risk and a maximum of efficiency. This vital area may finally develop as large-scale patient-monitoring systems, drug-reporting systems, automation of the hospital clinical laboratory, etc., because of close collaboration among the three areas of basic investigation, instrument development, and delivery of health services. As an example, a simple spectrometer-computer-gas-chromatograph combination designed in the laboratory has been developed by industrial concerns to a high degree of efficiency, and hospital laboratories are now using this instrument on-line for detecting drug levels in patients in direct clinical care.

Research and development of new technologies which will provide maximum reliability and speed in the clinical laboratory require development and evaluation of instruments suitable for identification and quantification of presently unmeasured metabolites in trace amounts.

We must exploit automation to provide for maximum utilization of professional and technical manpower. Research into uses of computers as control systems in the production of reliable laboratory data, and research into materials and methods suitable for laboratory reference standards for both automated and manual techniques, are badly needed.

As the number of clinical diagnostic tests performed annually by laboratories in the United States climbs towards the billion mark, we are confronted by the dual problem of shortage of trained manpower and poor laboratory technique. Laboratories must be automated since the costs for performing such large numbers of tests manually will be prohibitive. Such automated laboratories are particularly well suited for use in large-scale health care projects.

This is a rapidly advancing field of interest to industrial concerns that manufacture or assemble laboratory and electronic equipment. Many of these concerns have introduced, or have in an advanced state of development, instruments which will permit automation of a large number of laboratory tests.

Equally important, a number of other industrial organizations have been developing equipment for the military services or for NASA, some of which could be readily adapted in civilian laboratories and hospitals.

Programs to develop an artificial kidney or an artificial heart may be cited as other examples of instrument-oriented development research with direct health implications. There are also programs in the computerization of hospital services which are oriented heavily towards systems analysis. These programs with those in the field of automation of clinical laboratories are designed to take recent advances in theory and instrumentation and wield them into a program to provide better, cheaper, and faster laboratory diagnosis.

In any event, industrial development of laboratory instrumentation is undergoing a period of rapid expansion. The end result of this expansion will be better laboratory services and better patient care. Gratifying as all these advances and activities are, the fact remains that they lack coordination. There has been no attempt by any responsible group to outline in detail a rational pattern for the development of biomedical instrumentation. A student of research entering this highly diverse field must decide in advance where he wishes to place his efforts. It is only with difficulty that an individual can move from the academic to the industrial area, or vice versa. The area of development has not yet been defined, but probably lies in a grey region between the two extremes of theoretical approach on one hand and technical production on the other. A well-trained biomedical engineer can be placed anywhere on the scale; however, once his position is determined, it may prove hard to change. Nevertheless, the spectrum is so broad that the field can be "all things to all men." The area of delivery of health services is expanding rapidly with great needs for many kinds of biomedical engineering; and as the federal programs for medical care expand, the demand to automatize and systematize all forms of measurement must increase. Research in this area will not be the broadly based theoretical study of the laboratory, but will deal directly with the doctor/patient relationship and the problems of medical care, ranging from record keeping to new forms of visualization of tissues or disease processes.

Biomedical engineering must be solidified into a single science by the close collaboration of scientists interested in all three areas; this can occur only when the training process produces a group of people who can communicate with each other and with their counterparts in engineering and in the life sciences. It is for this reason that this book attempts to define the whole area of biomedical engineering, in the hope that the entire field will be presented to the student so that he may find what interests him, and at the same time see where he fits into the total universe of a diverse population of skills, interests, and directions of endeavor.

14

Engineering and the Delivery of Health Services

The time has arrived for a more direct assault on the health problems of the country. Our health facilities and services must be strengthened, the adequacy and quality of health manpower must be assured, and the states and communities must be assisted in meeting their local health responsibilities. Currently, the public is spending more than fifty billion dollars a year for health care; in spite of this, it is apparent that many people in the country still suffer, although the technology exists to help alleviate that misery. It has become evident that the knowledge of how to prevent and cure disease has outstripped its application, and that many health needs remain unmet. Contributing to this is the fact that progress has been slow in devising new and more efficient methods of organization, and in applying the technologies called for by the changing pattern of medical care. Accordingly, it is important now, and in the coming decade, to assure that an adequate proportion of the capabilities of scientists and engineers is turned to meet the common everyday needs of men and women in the health area.

Health Care

Over the past twenty years, medicine has gradually emerged as one of the largest industries in the country, and it is projected that by the mid-1970's it will probably become the largest.[1] An interesting aspect of the growth of this health industry, however, is that it has differed in one key feature from that of most commercial enterprises. In the usual enterprise, benefits evolve in reasonable proportion both to the consumer who uses the product and to those who provide the product. However, the characteristics of the health industry are such that this has not been quite the case. In spite of the explosion in medical research that has taken place in the past twenty years, and despite the knowledge that we have accrued relative to disease processes, there has been a failure to bring some of the tangible results of medical research to an ever-increasing and demanding body of public consumers, particularly to those in the ghettos of our cities.

A major question has arisen as to how the universities and their medical schools can accelerate the application of medical science through more efficient and extensive systems for the delivery of health services.[2,3,4] At the present time, a failure to exploit the potential of medical centers and hospitals can partially be laid to inadequate management practices.

It is to be hoped that the thoughtful application of technology and engineering in the areas of the development of instruments, devices, and systems, and of the utilization of engineering techniques for the delivery of health services, will contribute much toward the resolution of these problems. It would appear that our expanding technology will engage in these areas at the same time that we move into a period leading to comprehensive prepaid health insurance for all our citizens.

The matter of medical manpower is difficult, and no one really knows what the ideal physician-population ratio should be in the future. What is needed, rather, is a new approach to the pattern of medical care. The contributions of engineering in the areas of the delivery of health services that are tractable to such technology will be most important. Systems engineers, computer technologists, political scientists, economists, sociologists, and psychologists can all be expected to make key contributions. It should be noted that not all pilot studies in this area need be done in university medical centers. On the contrary, such current projects as those being conducted by the Kaiser Foundation on multiphasic screening techniques will undoubtedly provide important steps forward.

It may reasonably be concluded that, in the final analysis, the demand for more medical services can be met only when the pattern by which medical care is presently delivered is considerably changed. Part of the solution lies in the design of new approaches to medical care. Modern technology can contribute considerably to their success.

Engineering and the Delivery of Health Services

As one considers the complexities involved in the development of more effective and efficient systems for the delivery of health services, it is apparent that an imaginative application of modern technology (instrumentation, automation, computation, communications, systems engineering, etc.) can go a long way toward the solution of the problems. Particular emphasis should be placed on the development and application of instruments, devices, techniques, and systems for the prevention, diagnosis, and treatment of disease, and for the storage and analysis of health data. More specifically, we are concerned here with the following topics:

1. Instruments, devices, and systems for diagnostic testing: Techniques for pattern recognition analysis for electrocardiograms and electroencephalograms, ultrasonics, simulation studies, radiological scanning, graphic record analysis, psychological testing, biotelemetry, etc.
2. The automation of clinical laboratories: The instrumentation of new analytical techniques, the semi-automation and automation of specific methodologies, the computerization of automated information outputs, and the assembling of modular segments of automated clinical laboratories into a totally automated clinical laboratory.
3. The use of computers and systems or sub-systems for machine-aided diagnosis: The processing of configurations of multiple data inputs to develop diagnostic hypotheses or conclusions, initially in such special areas as cardiovascular disease and eventually in general—as with congenital heart disease, blood diseases, and general medical diagnosis.
4. Treatment and patient care: Automated anesthesia, patient monitoring, intensive care units, hospital information systems, etc.
5. Facilities for screening for disease detection: Multiphasic screening, etc.
6. The gathering, storage, updating, and retrieval of those aspects of lifetime medical records for which centralized processing is relevant.
7. The collection and evaluation of significant medical statistics varying from diagnosis and treatment of large populations to research studies on these topics: Demographic studies, epidemiologic studies of a general or categorical nature, large popula-

tion analyses of the effects of treatments, infectious disease analysis, etc.

8. Local, regional, or network facilities for incorporating some of the above techniques and systems.
9. Systems engineering techniques for developing more adequate local and regional systems for the management and implementation of health care.
10. Training related to the above areas.

Clearly, the computer provides a thread of continuity, and the capability of big computers to make generally available the benefits of accumulated knowledge and experience for the delivery of health services has scarcely been tapped. However, applications in this area are underway in the automated and computerized health examination techniques developed by the Permanente Medical Group of the Kaiser Foundation Research Institute in Oakland; in Boston's computerized Columbia Point Health Center, which is being evaluated for its effectiveness in providing medical care for more than 6000 residents in a low-income housing development; in the effort of the Massachusetts General Hospital and other hospitals toward the development of hospital information systems; and finally, in the far-reaching implications for automated disease detection that are reflected in a bill currently before Congress called the Adult Health Protection Act. This measure proposes federal grants for the establishment and operation of regional health protection centers. These centers, using computerized automated equipment, would offer a series of basic screening tests for the detection of disease to people of 50 years or over on a voluntary basis.

Again, and in the same general context, the President's Commission on Technology, Automation, and Economic Progress[5] has indicated the desirability of evaluating the feasibility of regional health computer facilities that might provide medical record storage for perhaps twelve to twenty million people in a geographic region, and that could give hospitals and doctors in the area access to the computer screening, disease detection, and other capabilities via telephone connections. Such regional facilities theoretically might provide regional data processing for some of the activities in automated clinical pathological laboratories, for the automation of certain aspects of disease detection and identification, for the storage and rapid recall of individual health records, and for the collection and evaluation of important and vitally needed medical statistics.[6] Such centers might also possibly be integrated with individual local hospital information systems, with the system for information retrieval and dissemination at the National

Library of Medicine,[7] and with the currently expanding informational needs of the Food and Drug Administration.

The degree to which such regional health computer facilities may be realizable in the future is suggested by the studies of Caceres,[9] who has explored the possible development within five years of a nation-wide computer system capable of providing routine analysis of electrocardiograms and other medical signals. Systems with a processing capacity of 60,000 or more medical signals annually could be sold for $100,000 or rented for about $36,000 per year. These centers would facilitate the creation of storage and retrieval centers at strategic locations, thus providing much needed medical data pools. Although it is possible to construct large, regional, centralized, multipurpose, multi-signal computer facilities from products now on the market, these systems at the moment are impractical for routine clinical medical signal analysis because of staffing difficulties, the continuing high expense of nonlocal analog telemetry, high acquisition and maintenance costs, the absence of a workable time-sharing system for medical care purposes, and the changing character of medical systems due to computer use itself.

Looking forward to the day when the nation's hospitals and other medical facilities will communicate with each other over a computer network providing general and emergency health information at the touch of a button, studies are under way to establish guidelines and develop standards criteria to ensure intercommunication among health facilities, regardless of the type of internal computer system each may have.[8] The present problems must be delineated, the types of information to be communicated must be identified and classified, the varieties of equipment and development efforts being applied must be studied, and an extensive review of current research activity in this area must be undertaken. Particular attention must be given to the privacy of data, validation of patient data, message formats, and communications protocol. If such systems are to develop in time, initial planning steps are most appropriate to determine the type of information to be communicated among the health facilities. Also, the actual and changing requirements of the medical community must be made known to the information-processing industry for appropriate inclusion in the planning phases of large intercommunication networks.

Specific Areas

Computers in Biomedicine. As is widely stressed, there continues to be a severe language gap that impedes the introduction of computers into clinical medicine and the delivery of health services. Mathematicians and engineers have discovered that operating with a computer in the area of basic, fundamental, undifferentiated biomedical research is one thing, whereas the elements of decision making and problem solving in clinical medicine and the delivery of health services are quite another matter, and there are serious problems concerning the need for the computer to be adaptable to the clinician's everyday needs.[9] To a degree, the disappointments to date have been due to an attempt to oversimplify medical problems in the interest of making use of the present capabilities of the computer. This, of course, has resulted in extensive failures, often due to a poor understanding of the complexity of biological situations, especially at the behavioral level. Also, there has been a lack of sophistication evidenced in the direct application of off-the-shelf mathematical techniques and systems to biomedical situations. Mathematical manipulations of various types have been used on biomedical problems in the clinical area at times when the situation was simply not tractable to such manipulation. We have also seen this with problems relating to the delivery of health services, in that assertions about the way to proceed to resolve health care problems have at times shown no cognizance of the real problems that the patient or the physician faces on a day-to-day basis.

At the same time, physicians have not adequately come to grips with the problems and potentials presented by the availability of computers. Not only must they become more intimately acquainted with the specific capabilities of the computer, but they are beginning to realize that their data (as in medical records) must be assembled in a completely different fashion from heretofore, if the most is to be made of the potential inherent in present-day computer systems. Unquestionably, if anyone seriously undertook to automate the information available today in medical records in their present forms, the outcome would not be usable. What is needed, before any such major undertaking is considered, is a new approach to recording medical data. Medical records must basically be made more structured and standardized before excessive funds are inadvertently put into machines that can only be made to attend to tasks that are not resolvable. An important problem for physicians interacting with computers is the need for a more natural language input to these machines. It is striking that at the moment very few important recognizable inroads have been made in this area. The difficulty is of course that natural language depends so much on context, and as such the computer, in its present stage of development, is simply not able to handle it.

There have been continuing discussions as to whether the computer will ever be a better diagnostician than the physician. In summary of the conclusions to date, it is likely that the computer will be able, in the immediate years ahead, to assist the physician diagnostically in the clinical management of his patients. In some situations the computer will be able to manipulate a considerable number of variables in parallel fashion, a technique that the physician cannot hope to master, to the extent that in certain specific instances the computer will be able to render better diagnosis than the physician.[10] This will probably become more obvious and more generally accepted as, with the passage of time, medicine comes to understand more fully some of the subtleties of a variety of disease processes that are now poorly understood.

Automation of Clinical Laboratories

In the near future, due to the increased participation of a variety of federal, state, and local programs in providing essential health services, a sharp increase in demands for clinical pathological laboratory services can be anticipated. It has been shown that the work load in these laboratories has doubled approximately every five years during the past two decades, and all available evidence indicates that this rate of growth will be exceeded in the foreseeable future. (In 1968 the 14,000 clinical laboratories in the country performed some 500 million tests.) The major portion of these increases is due to the development of new and better tests. Although this development shortens the average hospital stay, saves lives, and enables more rapid diagnosis and definitive treatment, there are major problems hindering the provision of better laboratory services.[11]

It is particularly worth noting that the reporting of erroneous results in these laboratories has become a problem throughout the country. It has been estimated that 25 percent of all clinical laboratory tests are erroneous. There are even those who consider these figures conservative; they report erroneous results in as many as 40 percent of the tests performed in a survey of hospitals in specific states. Some of the leaders in the field of clinical biochemistry have stated in testimony before Senate committees that good analytical chemistry is very rarely practiced in the clinical chemistry laboratories of this country. Accordingly, it is apparent that clinical laboratory procedures must be markedly improved. Even in our better laboratories, when tests are done correctly under proper supervision by laboratory scientists, the tests to date have been extremely time consuming and costly to perform. Thus, a major factor in the attempt to provide adequately for these services in the future will be the elaboration of more efficient laboratory methods.

There is general agreement that, if our medical laboratories are to cope effectively with the coming needs for laboratory services, they must be automated insofar as possible. This is based on a comparison of the cost of performing large volumes of tests using manual methods with the cost of doing these same tests by automated methods. Not only would the total cost of the number of tests envisioned for the future be astronomical if done by manual methods, but it would be impossible to find and train the necessary technical and professional manpower to perform these tests. A program of directed research and development for automated laboratory technology and electronic data handling can raise the capabilities of clinical laboratories to high levels of efficiency so that (a) the increasing demands can be met, (b) it will be possible and feasible to use these laboratories for the diagnosis and treatment of disease in ways unheard of only a few years ago, and (c) exciting possibilities will be opened for bringing the full benefits of modern medical laboratory science to the entire population, e.g., by the eventual screening of large groups of patients or segments of the population for the early detection of disease.

The problem is to determine how computerized automation can best be used to provide prompt and reliable laboratory services to the largest number of patients and physicians throughout the country. At the present time, there are very few laboratories in which new concepts of automation or new methodologies are being developed. This is due in part to the high cost of research and development in the field, and in part to a serious shortage of qualified scientists with analytical chemistry and engineering backgrounds who are interested in working in the medical laboratory sciences.

The specific overall objectives in a program to attain the automation of the country's clinical laboratories are as follows:

1. The development of semiautomated and automated instrumentation pertinent to modern medical laboratory science.
2. The development of appropriate computer control and communication for these automated laboratories.
3. The development of techniques for the rapid recall of laboratory data in conjunction with individual health records.
4. The provision of adequately trained scientific and technical personnel who are capable of staffing such laboratories.
5. The establishment of automated clinical laboratories at medical centers and elsewhere throughout the country.

These objectives can be achieved by (a) integrated efforts primarily concerned with the selection of sites for the development of automated laboratories, and (b) marshaling by grants and contracts those industrial, medical center, and university resources that are capable of producing the required instrumentation, techniques, and manpower.

In recent years the patient in a hospital has usually undergone a limited number of tests referable to his particular diagnostic criteria, and the normal individual outside of the hospital has been subjected to various types of gross screening procedures to detect incipient illness. However, the automation of clinical laboratories will result in the performance of a full spectrum of tests and screening procedures on patients in the hospital, in the outpatient department, in clinics, or under the care of a local doctor. The significance of this more extensive application of laboratory science will be considerable in the prevention, diagnosis, and treatment of disease, and in the general maintenance of public health. Costs of laboratory tests are commonly defined in terms of personnel time to perform them, since this represents an average of 80 percent of the cost of most tests. Studies show that, depending on the degree, sophistication, and capability of the automation of the tests and data handling, a cost reduction of up to a hundredfold may be expected. It is also of importance to note that (a) appropriate improved instrumentation and automation of clinical laboratories will produce the sorely needed high degree of accuracy and reproducibility of laboratory tests that is not possible by manual methods, and (b) automation will point up the need for, and force the quality control of, test procedures and the development and standardization of methods, both of which are essential for a major improvement of medical laboratory science itself.

Patient Monitoring and Intensive Care Units

The intensive care and patient-monitoring concept has evolved slowly over the past decade. Its purpose is to provide extensively concentrated care in order to improve the survival of patients affected with life-threatening conditions. These patients are often in shock. The particular disease states and conditions requiring such highly specialized surveillance and treatment include a variety of medical and surgical catastrophes such as pneumonia, drug overdosage, bleeding from a stomach or intestinal ulcer, diabetic coma, body burns, strokes, pulmonary and peripheral emboli, pulmonary edema, serious disorders of cardiac rhythm (particularly when they occur after a heart attack), and the postoperative stresses incident to major surgical procedures.

The past several years have seen the appearance of a variety of specialized intensive care facilities in the larger medical centers. These include trauma units, coronary care units, kidney dialysis units, post-operative intensive care wards, and resuscitation units for the treatment of drug overdosage, asphyxia, and poisoning. Commonly, patients in these units are maintained under intensive care for three to four days. To date, it has been difficult to provide any quantitative documentation concerning increase in survival as a result of the establishment of these units. However, there is little doubt that they have been lifesaving. Probably the most significant statistical information has come from the coronary care units, which have shown that the early recognition and treatment of arrhythmias in patients with heart attacks has definitely produced an increased rate of survival.

In the overview, there are essentially two objectives in the monitoring of acutely ill patients. The first is to provide a more frequent or, when possible, continuous presentation of the patient's condition by monitoring a few basic types of physiological data. It has been proven reasonable in some instances to have these monitoring devices trigger alarms whenever certain variables exceed particular limits. Secondly, other more sophisticated types of monitoring are beginning to emerge which are designed to extend the sensing capability of the physician and nursing staff by providing information that would not otherwise be available. The instruments, devices, and systems of importance here are arranged so that trends and changes that appear can be promptly treated before some irreversible catastrophe, such as stoppage of the heart, occurs. To date, some electronic packages have been provided by manufacturers that are able to continuously monitor effectively the basic data required. However, relatively few of the second type of monitoring devices are available in the form of semiautomatic electronic devices. Clearly, the engineer has much to contribute to this environment in the future in basic and applied research, as well as in the development of the instruments, devices, and systems needed.

More specifically, patient monitoring can be considered generally divisible into the following four main categories.

Staff Surveillance. The strategy arranged for staff surveillance is an important factor in the design of patient-monitoring and intensive care units. In certain types of intensive care, closed circuit television is currently used to supplement staff surveillance when the latter has proved to be a problem.

Conventional Signs. The tracing of the electrocardiogram is easily retrievable, and to date it has been essentially the core of almost all electronic monitoring packages prepared. Numerous automatic devices have been provided for monitoring blood pressure, but they have been found to be of questionable reliability under stressing conditions. The pulse rate is commonly followed from the electrocardiogram; however, there are other mechanisms to recover this. Rectal temperature is usually recorded, although it is probably the least useful of the conventional vital signs. Respiratory rate has been followed by a variety of means, including temperature sensors in the upper airway and arrangements for monitoring the movement of the chest wall.

Additional Signs. Perfusion of various parts of the body is highly significant, and the output of urine is usually considered an indirect measure of body perfusion, as in the renal area and in the cerebral and coronary circulations. Unquestionably, the measurement of central venous pressure is most important. It relates to the adequate replacement of circulating blood volume, and reflects on appropriate balances between peripheral arterial pressure and the central resistance to venous return. The measurement of blood volume has been helpful, particularly in regard to the failure to replace blood that has been lost.

Physiological Systems. There are considerable limits to the information available from the recording of blood pressure, pulse, respiration, and temperature every four hours. Undoubtedly, the attempts to increase the frequency of measuring these variables has been most important, and developments of electronic systems and devices are heading more and more toward finally rendering these measurements continuously available. However, there is dissatisfaction even with the continuous recording of these primary variables, because there is a limit to the important information actually provided, e.g., the knowledge of pressure per se. Often patients get into trouble because of inadequate flow, either in major circulation channels or in the capillaries. It seems likely that, in the immediate future, much of the exciting research in monitoring will be directed toward expanding the knowledge concerning the vital signs for a finer understanding of the overall flow of oxygen and carbon dioxide in the body. We will be mainly concerned here with three areas:

1. Ventilation: Minute ventilation can be combined with airway pressures to give lung compliance, and arterial pCO_2 can indicate the adequacy of alveolar ventilation.
2. Cardiovascular circulation: Cardiac output can provide an indication of total peripheral resistance. Central venous pressure, when plotted against cardiac output, can give a measure of pumping efficiency for a particular filling pressure.

3. Peripheral circulation: It is possible for the total delivery of oxygen to be satisfactory, and yet for the patient to be in shock because of the poor distribution of the capillary blood flow. In this instance, then, selected monitoring of the heart, kidneys, and brain will obviously be of increasing importance. At the present time such techniques are incomplete.

In summary, an interesting technical feature of the patient-monitoring problem is that the sensing capabilities of various instruments that will be increasingly necessary for adequate patient management have not been as highly developed as has our ability to process signals. Actually, this should not be surprising, since the same "effect" has been seen in other industries, such as the computer industry. The high-speed digital computer has proved to be a source of considerable practical and intellectual interest. At the same time, the elaboration of new ideas requires basic invention and then engagement in the extensively difficult process of reducing the ideas to practice. To a degree, what has happened is that the attention of skilled people has been diverted to what appears to be the more sophisticated activity of applying statistical techniques to complex signal analysis, pattern recognition, etc. In the meantime, the considerable needs in sensor instrumentalities have been neglected; accordingly, this is a ripe area for engineering intervention at this time.

Understandably, there is considerable current discussion relative to the selection of appropriate patient-monitoring systems for particular sets of circumstances. In part, the difficulties stem from the opinion that a single system can be a universal one that will satisfy all needs. This is clearly not so. Rather, certain types of intensive care units should be modified for particular types of patients, and in these the monitoring capability will differ depending on the type of patient to be cared for. The use of the electrocardiogram is of central importance in the care of the heart attack patient; however, in situations in which patient monitoring is concerned mainly with general surgical problems, most of the patients will not have life-threatening problems primarily involving the electrical conduction system of the heart. On the other hand, they may have cardiac problems concerned with swings and variations in cardiac performance. Accordingly, in these latter situations, the monitoring of the electrocardiogram becomes of less immediate importance than the frequent measurement of other parameters associated with ventilation, circulation, and acid-base status.

The monitoring equipment involved in patient care units can be considered to include three levels of complexity: (a) some equipment (usually self-contained) is portable and can be brought to the bedside of certain patients when required; (b) some equipment can be built in at each bedside for general use with all patients; (c) additional permanent monitoring equipment can be arranged beside beds with connections to a central nursing station. In general, built-in monitoring equipment can be justified when most patients in a given area are being cared for because of some "central" identifiable problem. On the other hand, if different types of patients are cared for in a certain location, monitoring equipment that can be moved from bedside to bedside may well prove more economical and even more useful in the long run than equipment built in to each bed site. The advantages of transmitting information on monitoring to a central nursing facility can be seen in two instances: (a) a variety of types of signals and alarms that are not suitable to be set off at the bedside of the patient can be geared to alert a central nursing station; and (b) certain equipment for the recording of specific signals can be operated from a central location much more effectively and economically than if it were part of the bedside equipment.

It is worthwhile to note that, in the presence of sophisticated instruments and devices of various types, it is possible for the users of this equipment to be lulled into a false sense of security and to fall into errors in regard to the usefulness of monitoring. There is often a failure to make the best use of the features of both man and machine in a man-machine systems concept when approaching these patients and their problems. It is possible even to eliminate man from the detector-effector loop of a monitoring system by replacing him with three or four relatively unimportant physiological transducers. Too much emphasis can be placed on temperature, pulse, respiration, and blood pressure as measures of the physiological state of the patient. There must be a continuous attempt to bring men and machines together in a more functionally satisfactory way if the most is to be made of the opportunities presented in patient monitoring.

Another problem is that most instruments, devices, and systems have not been designed to respond well to events that are important but that occur infrequently and are often unanticipated. The detection of these so-called low-probability events still requires capabilities that are unique to man. Unquestionably, more basic studies on the continuous analysis of physiological data in postoperative patients are most important. This is especially true if we are to understand and sharpen the so-called pattern recognition capabilities of the computer, as these can result in the development of more adequate patient-monitoring systems in the future. Despite such anticipated sophistications, however, if the situation is such that the variety of instruments and devices reduces the opportunities for direct contact between the staff and the patient, this can sometimes result in an adverse effect on the general postoperative care and management of patients.

An old problem, but one that still remains very much with us, is the matter of electrical hazards. Patients who are extremely ill often show increased cardiac irritability resulting from electrolyte imbalance and from drugs. These patients are most commonly seen in the operating room or in intensive care units, and they are more vulnerable to electric shock than are healthy persons. The frequent use of electronic equipment, and a variety of electronic equipment simultaneously on the patient, coupled with the use of electrodes of various sorts, are increasingly of concern in regard to shock hazards in patient monitoring. This hazard is made worse by the fact that, by and large, the physicians with responsibility for these patients do not understand the extent of the hazards involved. In many instances, neither do the engineers.

If the intensive care unit problem is to be effectively surmounted in the near future, it will be most important for engineers and physicians to isolate and define the following elements of the task, and to jointly assure that these elements are supported individually and in their mutual relationships. This is truly an area requiring an interdisciplinary team effort. Specifically, the segments of this activity requiring definition and support include:

1. The development and validation of primary sensors.
2. The automation of measuring and controlling devices.
3. The development and testing of effectors such as pumps, pacemakers, etc.
4. The development of data acquisition systems that include analog data-filtering techniques.
5. The elaboration of special purpose data management techniques.
6. The provision for informational output procedures concerned with (a) the condensation of data to form that can be presented to physicians and nurses in a functional fashion for decision making, (b) modeling of the decision-making processes of physicians and nurses, and (c) techniques for the exchange of information between intensive care units and other hospital units.

Filtering Techniques. If 10 patients require monitoring, the information from these patients that is delivered to the digital computer as raw unprocessed data can be overwhelming, depending on the continuity of the signals. Even with the most modern machines, memory capacity can be taxed if several patients are to be monitored, particularly if some require an advanced level of monitoring. In these instances, the problem may be managed only by data filtering and digesting in the early stages of data acquisition, before the data is presented to the digital computer. For this purpose, special purpose analog computers are needed which can take in the large amount of broad data presented to the system as analog signals, process this data, and then present it to the central computer, perhaps as brief intermittent digital signals. This is an important area for engineers to focus on.

Functional Output Presentations. There is simply not enough time to print out much detailed information concerning a number of patients. Communication with physicians must be more suited to their interests and demands. A typewriter is probably unsatisfactory both as an input and as an output vehicle. The technique for entry, in the end, should be no more complex than telephone dialing; and the output, from moment to moment, should be on a cathode ray tube with a potential for printed summary data and for the storage of data on a magnetic tape for off-line statistical analysis.

Modeling. Mathematical modeling is potentially capable of making predictions; this is an area of considerable need. Effective models could provide the most logical way to condense the outputs of these systems. The brief predictions and suggestions forthcoming from a system based on a reliable model are far preferable to the presentation of a large amount of numerical data which the physician and nurse can easily learn to ignore.

Though the goals of biomedical engineering in patient monitoring and intensive care units can be reasonably well defined, at least in a general sense, many of the specific problems in attaining these goals are formidable and present important tasks for engineers to engage in. From the design of reliable, easily applied sensing devices to the formulation of suitable mathematical models of the patients in jeopardy, there are many unmet needs. Yet the basic engineering and analytical techniques required are at hand, and the impressive initial efforts illustrate that a generally useful patient monitoring system can be built through a concerted effort in applied biomedical engineering.

Hospital Information Systems

The goal of a successful hospital information system is multi-faceted.[12] While it must collect, record, store, retrieve, summarize, and transmit information relative to the patient population of the hospital, it must also attend to rather extensive accounting procedures involving payroll, patient billing, inventory, etc. In our large medical centers, the amount of paper work in such a broad information system is understandably massive and difficult to manage.

Operational engineering responsibility is an area that must be identified. Here the operator must interpret for the user; he must be sensitive to alterations and changes in the intentions of the user, and he must have a clear idea as to how to produce the respective requirements of the system. Another area of identification is that of the systems design engineer. It is his responsibility to prepare the structure of the system, to arrange the information to be stored in the computer, and to arrange the elements of design of the system in the most efficient manner for the operator and the user. A final area of responsibility is that of the computer manufacturer; ideally, the computer should fit the structure of the design system. Unfortunately, to date the computer has been built apart from the system it is to serve. To a degree, this has been due to the fact that the manufacturer is responsible for providing economical equipment.

As is commonly found in specialized technologies, a principal problem here is that of achieving adequate communication among the user, the operator, the design engineer, and the manufacturer. None of these will be fully knowledgeable of the technologies and problems of the other members of the team. An initial requirement is that at least the languages must be compatible to the extent that symbolization in the computer will be precise and mutually understandable. To a degree, there seems to be a major conceptual problem lying mainly in the area of systems design responsibilities.

Current research specifically directed at moving the field ahead is largely concerned with (a) achieving systematic data acquisition and effective information flow with the hospital, (b) providing all participants caring for the patients with the information needed for rational decision making in diagnosis, therapy, and patient care, and (c) developing a central file containing all the vital clinical data that must be accessible to retrieval. Within the next five years, substantial contributions may be expected in (a) communication activities concerned with the implementation of a "doctor's order," particularly in the area of drug orders; (b) the entry and retrieval of numerical and nonnumerical reports of tests, procedures, and observations; (c) direct entry of data by the patient and by auxiliary medical personnel; (d) storage and rapid retrieval of summary medical records; (e) direct assistance in the decision-making and problem-solving activities of medical care; and (f) a variety of administrative tasks concerned with scheduling, inventory, and record handling.

For some time now, engineers, computer specialists, and physicians have been trying to bring the computer more effectively into the hospital environment. The results have thus far been less than satisfactory, and the prevailing mood is essentially one of impatience on the part of all concerned. In an overview of the progress to date, it is evident that there has been a gross underestimation of the size of the task at hand. The problems of effective implementation of computing capabilities in the hospital include (a) a lack of communication between the engineers and computer specialists on the one hand and the physician and hospital administrators on the other, (b) a failure to develop a mathematical and theoretical base adequate for the conceptual advances needed to come to grips with the problems, and (c) a failure to provide for a sequential series of studies demonstrating to those who will fund the operational phases of computer usage that hospital activities will actually be assisted and improved.

Multiphasic Screening

The purpose of screening for disease processes is to discover those persons who appear to be well but who are in fact suffering from disease. These individuals can then be given treatment and, if their disease is communicable, steps can be taken to prevent them from spreading the disease. Thus, theoretically, screening is an outstanding way of combating disease, because it helps in detection in the early stages, and enables the disease to be treated adequately before it secures a firm hold on the community.

A few centers completely equipped to perform comprehensive examinations are already in operation. For example, the Permanente Medical Group in Oakland and San Francisco, California, provides a multiphasic health checkup for planned subscribers. The Kaiser Foundation hospitals and Permanente Medical Group, through a system of 9 hospitals and 19 clinics, provide a comprehensive medical care program for 760,000 health plan members in the San Francisco Bay area. The Department of Medical Methods Research conducts investigations directed toward developing better methods for the delivery of health services.

In regard to screening for specific purposes, the following areas are important:

Cervical Cancer Screening Projects. More than one million cytological examinations have been made at various hospitals throughout the country. Cervical cancer has been found in more than 6500 women, and over 4000 cases of carcinoma in situ have been detected.

An estimated 26 percent of women over 20 years of age in the United States were cytologically examined in 1966, up from 15 percent in 1963 and 10 percent in 1961. In comparison, the rate in British Columbia is about 65 percent.

Electrocardiography Screening. Studies have been carried out to determine the value of routine screening by electrocardiography in hospital out-patient populations. In one instance, screened electrocardiograms were abnormal in 1,197 of 2,924 cases (41 percent). Screening electrocardiography has produced a yield of abnormalities sufficient for its recommendation as a regular hospital procedure. Interpretation of ECG's and compilation of statistical data were performed by automated methods with the use of computer analysis.

Screening for Heart Murmurs. This technique, using cardiac auscultation by a machine (Phonocardioscan), has shown some promise as a screening procedure in school children. To determine its effectiveness as a screening procedure in adults, results obtained by this means were compared with those obtained by internists in 456 subjects. The overall sensitivity of the new procedure was 75 percent (25 percent false-negative findings), ranging from a 47 percent sensitivity in innocent murmurs to 88 percent in murmurs of congenital or rheumatic heart disease. Overall specificity was 66 percent (34 percent falsely positive). These results were less accurate than those reported in children. At its present stage of development, the machine is considered insufficiently sensitive and specific for incorporation into a multiphasic health appraisal examination of adults.

Cardiovascular Screening to Assess the Risk of Coronary or Heart Disease. Studies have confirmed or identified about a dozen characteristics that are associated, to varying degrees, with an increased incidence of cardiovascular disease. These studies, however, are enormously expensive in terms of both service and research.

Properly managed multiphasic screening operations can be relatively inexpensive, accurate, and effective. A unit equipped for 50,000 to 100,000 examinations a year can be operational in approximately two years, at a cost of about $350,000 including rental of computer and plant. The annual cost to the patient will range from $15 to $30 (depending on volume), which includes operating overhead, physician interpretation of tests, research, and amortization of capital investment in about five years. There are reports of an unusual system being developed overseas to serve an urban population of three million at a projected cost of $5 per patient (including screening, automated patient history, and information storage, retrieval, and transmission at a rather rapid rate).

Engineers have an important role to play in the advance of multiphasic screening techniques. This is particularly evident as one considers the anticipated future needs with regard to (a) improved instrumentation and technology, (b) the need for automated equipment to conserve physicians' time, (c) improved preventive medical methods (as with the instruments, devices, and systems required for modern automated clinical laboratories), (d) improved data-processing capabilities (particularly input and output devices), (e) improved mathematical and engineering methods of analysis for medicine, and (f) improved systems engineering in medicine. As one contemplates and reviews the process by which an engineer visualizes and conceives a dynamic system, it is apparent that this type of thinking and its techniques are applicable to the specific functional systems of the body. Actually, the health evaluation and appraisal process can also be viewed as a similar system. The input of the system is the people, and the processes are represented by the monitoring and testing steps of an evaluation protocol such as a multiphasic screening program. The primary output is represented by the signals or results obtained from the various testing procedures; they serve as feedback signals through an interpreter (the physician) to effect the necessary intervention. This analogy is a true one; thus, it is expected that engineers will have considerable to offer in this area in the immediate future. A further need is the full use of engineering principles as a key to selecting the batteries of physiological and biochemical laboratory tests that can best be used to screen populations for disease. Efforts are being made by leaders in engineering, industry, and medicine to devise economical and reliable health screening systems to separate normal and abnormal populations. A special goal is the selection of a group of physical parameters that can easily be measured with today's technology and that will require a minimum of skilled personnel for operation. Increasing efforts will be directed toward stimulating industrial interest in mass-producing the mechanical and electronic components of such screening systems.

Concluding Comments

Taken in the aggregate, the actual and proposed instruments, devices, and systems mentioned in this chapter constitute a structure to which a more concentrated application of engineering and biomedical engineering technology may be made in the 1970's. Biomedical engineering is a rapidly growing field—no one knows exactly how large it is at present. Studies have predicted that biomedical engineering will be one of seven new industries surpassing the billion-dollar mark in the 1970's. It will in fact come to pervade all aspects of the health sciences. Given the amount of money the country spends for health care and the increasing role that technology plays in this care, it is not hard to see that there is a potential breeding ground here for a great American industry.

It is likely that in time extensions of the active or proposed automated and computer-oriented health activities and facilities mentioned here will prove to be a most important step toward ensuring the health of a nation whose population is growing rapidly. In the future, while helping to afford a high level of health protection and care to all our population, these extensions will particularly allow time for young physicians and engineers to turn their talents to things that machines cannot do. Biomedical engineering is already exerting an increasing influence on a variety of medical and surgical programs throughout the country; it is probable that experience gained in this area relative to the delivery of health services will pave the way for a new breed of physicians as well as engineers.

References

1. Rutstein, D. D.: *The Coming Revolution in Medicine.* Massachusetts Institute of Technology Press, Cambridge, 1967.

2. Glaser, R. J.: *The university medical center and its responsibility to the community.* J. Med. Educ. 43:790, 1968.

3. White, K. L.: *The medical school and the community.* Yale J. Biol. Med. 39:383, 1967.

4. Craft, E. M.: *Health care prices, 1950–1967.* J.A.M.A. 205:91, 1968.

5. *Technology and the American economy.* Rept. Nat. Comm. Tech. Autom. Econ. Progr., 1966.

6. Rappaport, A. E.: *A comprehensive medical data profile system.* Symposium on Recent Developments in Research Methods and Instrumentation. 1965, p. 7.

7. Rogers, F. B.: *The national library of medicine's role in improving medical communications.* Amer. J. Med. Electronics 1:230, 1962.

8. U.S. Department of Health, Education and Welfare: *Report on Regional Medical Programs to the President and the Congress.* U.S. Government Printing Office, Washington, 1967.

9. Caceres, C. A.: *Computer aids in electrocardiography.* Ann. N.Y. Acad. Sci. 118:85, 1964.

10. Dickson, J. F., and Start, L.: *Remote real-time computing system for medical research and diagnosis.* J.A.M.A. 196:149, 1966.

11. Kinney, T. D., and Melville, R. S.: *Automation in clinical laboratories.* Lab. Invest. 16:803, 1967.

12. Barnett, G. O.: *Hospital computer project status report.* Massachusetts General Hospital, Boston, Massachusetts. Memo Nine, University Microfilms, 1966.

APPENDICES

Appendix 1
Control Theory

Differential Equations: Time Domain Description

Linear systems with constant coefficients can be represented equivalently by means of differential equations, weighting functions, or transfer functions. Consider the simple linear network shown in Figure App. 1-1a. This figure represents a resistor-capacitor network with an applied or input voltage $v_i(t)$ and an output voltage $v_o(t)$. It is shown in any elementary circuit analysis book that the charge across the capacitor $q(t)$ and the input voltage $v_i(t)$ are related by the equation

$$R\frac{dq(t)}{dt} + \frac{1}{C}\, q(t) = v_i(t) \tag{1}$$

where R represents the resistance and C the capacitance of the network. If this equation is solved for the charge q, the output voltage can be computed from the expression

$$v_o(t) = \frac{1}{C}\, q(t) \tag{2}$$

The differential equation and the specification of the initial condition comprise a complete description of this network. For any particular values of the parameters and form of the applied voltage, the solution

379

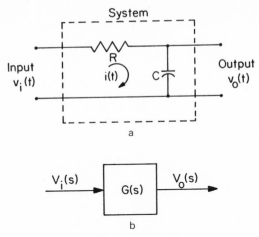

Figure App. 1-1. Basic diagram of a simple system.

of the differential equation yields the specific output voltage. For example, if v_i is simply a constant voltage, say from a battery, applied at $t = 0$, the output voltage becomes

$$v_o(t) = v_o(0)e^{-t/RC} + (1 - e^{-t/RC})v_B \qquad (3)$$

where v_B represents the battery voltage and $v_o(0)$ the initial voltage across the capacitor. Evidently, the solution consists of two parts: a portion due to the initial conditions (the transient solution) and a portion due to the applied or input voltage (the steady-state solution). It is evident that the steady-state solution gradually approaches the applied battery voltage v_B. The rate at which the output voltage rises is determined by the RC product usually known as the "time constant."

Transfer Functions: Frequency Domain Description

In many cases we are not interested in specific transient responses, but rather in the steady-state behavior of the network. In such cases, it is more convenient to employ frequency domain analysis, using Laplace transform techniques. If we apply the Laplace transformation technique to each term in Eq. (1), we obtain the relationship

$$sRQ(s) - Rq(0) + \frac{1}{C} Q(s) = V_i(s) \qquad (4)$$

where capital letters are used to represent the respective Laplace transforms, i.e.

$$\mathcal{L}[v_i(t)] = V_i(s)$$
$$\mathcal{L}[q(t)] = Q(s) \tag{5}$$

In terms of the desired output voltage v_o, Eq. (8) can be rewritten as

$$sRCV_o(t) - RCv_o(0) + V_o(s) = V_i(s) \tag{6}$$

If this equation is solved for the Laplace transform of the output voltage, we obtain the relationship

$$V_o(s) = \frac{RC}{1 + sRC} v_o(0) + \frac{1}{1 + sRC} V_i(s) \tag{7}$$

If appropriate tables are used to obtain the inverse Laplace transformation of each term on the right-hand side of this equation, one obtains a time domain expression of the form of Eq. (7). Let us concentrate for the moment on the steady-state expression, and assume that initial conditions are equal to zero. We can then define the ratio of the Laplace transform of the output voltage to the Laplace transform of the input voltage as

$$\frac{V_o(s)}{V_i(s)} = \frac{1}{1 + sRC} \tag{8}$$

Note that the right-hand side of this expression depends only on the system parameters R and C and the complex frequency variable s. Hence, this form of expression is known as the *system function* or *system transfer function,* which is often designated by G(s). Employing the designation $RC = \tau$ for the time constant, the system transfer function for our example can be written as

$$G(s) = \frac{1}{1 + \tau s} \tag{9}$$

Hence, we can represent the circuit of Figure App. 1-1a by the equivalent block diagram of Figure App. 1-1b. Given any particular applied voltage, we may compute the Laplace transform on the output voltage by means of the expression

$$V_o(s) = G(s)V_i(s) \tag{10}$$

and the time domain response is obtained by evaluating the inverse Laplace transform

$$v_o(t) = \mathcal{L}^{-1}[G(s)V_i(s)] \tag{11}$$

by numerical techniques on computers or by using tables.

Impulse Response as System Description

A third equivalent way of describing the system is by means of its weighting function g(t), which is defined as the inverse transform of the transfer function, i.e.

$$g(t) = \mathcal{L}^{-1}[G(s)] \tag{12}$$

It is easy to show that, if a system with transfer function G(t) is excited by an impulse $\delta(t)$, its response will be given by g(t). Hence, the weighting function is also known as the system *impulse response*.

The Bode Diagram

Let us return now to the transfer function. It is easy to show that, if the input is a sine wave of frequency f_o, i.e.

$$v_i(t) = \sin \omega_o t = \sin 2\pi f_o t$$

then the steady-state response of the network will be given by

$$v_{o-ss}(t) = G(j\omega_o) \sin(\omega_o t + \phi) \tag{13}$$

i.e., the output is also a sine wave of the same frequency, but with a changed amplitude and phase. The amplitude change is determined by the magnitude of the transfer function evaluated at the frequency ω_o, and the phase shift is determined by

$$\phi = \tan^{-1} \frac{\text{Im } G(j\omega_o)}{\text{Re } G(j\omega_o)} \tag{14}$$

where Im and Re denote the imaginary and real parts of the transfer function, respectively. Eq. (13) is applicable to any linear system of arbitrary complexity whose transfer function is given by G(s). For the particular system shown in Figure App. 1-1, if we apply an input sine wave of unit magnitude, the magnitude of the transfer function is obtained by substituting $j\omega$ for s in Eq. (9):

$$|G(j\omega)| = \left| \frac{1}{1 + j\tau\omega} \right| = \sqrt{\frac{1}{1 + \tau^2\omega^2}} \tag{15}$$

and the phase angle is given by applying:

$$\phi = -\tan^{-1} \omega\tau \tag{16}$$

In order to obtain a graphical representation of these two expressions, it is customary to plot the magnitude term in decibels and the phase angle in degrees, both against the logarithm of frequency. (The decibel is defined as twenty times the logarithm to the base ten of the magnitude.) For our example system, amplitude ratio and phase as a function of frequency are given in Figure App. 1-2.

The phase and amplitude versus frequency diagram given in Figure App. 1-2 is known as a Bode diagram. An examination of this figure shows the behavior of the system as a function of the frequency of the impressed voltage. The amplitude curve shows that the example network is a "low-pass system," which allows low frequencies to pass through with very little amplitude change, but which progressively attenutates the amplitude of inputs of higher and higher frequency. The solid line represents the actual output-to-input amplitude ratio. It can be seen that, as the solid line approaches the dotted line asymptote, the network produces an attenuation of 6 dB for every factor of two increase in frequency of excitation. It is easy to verify that 6 dB corresponds to a factor of two decrease in amplitude. Thus, we can expect that a doubling of input frequency will produce a reduction in the amplitude of the output sine wave by a factor of two in the higher

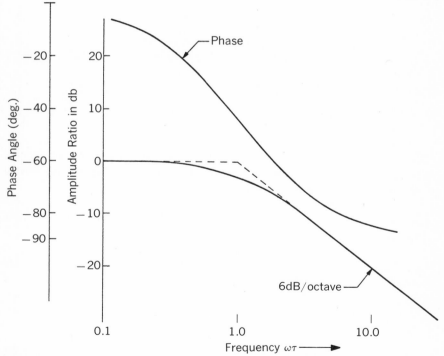

Figure App. 1-2. Bode diagram of system in Figure App. 1-1.

frequency regions covered by this diagram. The phase shift curve, on the other hand, shows that at very low frequencies the input and output sinusoids are nearly in phase, while as the frequency increases the output sine wave gradually lags further and further behind the input sine wave. Finally, as the solid phase curve becomes asymptotic to the minus 90° phase angle line, it becomes evident that, eventually at higher frequencies, the input and output sinusoids will become 90° out of phase.

In higher order systems, the Bode diagram not only provides us with a general amount of insight into system behavior as a function of frequency but it may also contain significant information regarding system stability, which will be discussed below.

Pole-Zero Description

Finally, an alternative means of presenting the information contained in the system transfer function is through an examination of the roots of its numerator and denominator. The values of complex frequency at which the numerator of the transfer function is equal to zero are known as the *transmission zeros,* or simply as the *zeros* of the system. When plotted on the complex plane, they are generally denoted by small circles. The values of complex frequency at which the denominator of the transfer function becomes zero clearly represent points at which the system response is unbounded. Hence, these points are denoted as the *poles* of the system, and they are designated by the letter x when plotted in the complex s plane (Fig. App. 1-3).

To illustrate these concepts more clearly, consider first our example system, described by the transfer function of Eq. (9). In order to facilitate the determination of the roots of the numerator and denominator of the transfer function when seeking a system description in terms of poles and zeros, it is customary to rewrite Eq. (9) as follows:

$$G(s) = \frac{1/\tau}{s + 1/\tau} \tag{17}$$

where the location of the single pole of this transfer function is now explicit. Recalling that the complex variable s is defined as $s = \sigma + j\omega$, it is easy to see that the denominator becomes zero when $\sigma = -1/\tau$ and $j\omega = 0$. The location of this pole is plotted on the s plane in Figure App. 1-3a. The system whose transfer function is Eq. (17) has a corresponding weighting function (impulse response) given by

$$g(t) = e^{-t/\tau} \tag{18}$$

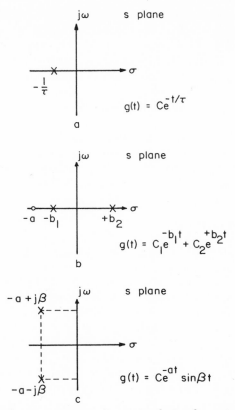

Figure App. 1-3. Pole-zero patterns for several simple systems.

This is evidently a stable response to initial conditions, which decays exponentially with a time constant τ. It is easy to see that pole locations along the negative real axis, as with this example, will always lead to exponentially decaying transient responses. On the other hand, a pole located along the positive real (σ) axis will lead to a rising exponential response which will eventually grow out of bounds. This can be illustrated by considering a system with a transfer function

$$G(s) = \frac{K(s + a)}{(s + b_1)(s - b_2)} \tag{19}$$

whose poles and zeros are plotted in Figure App. 1-3. There are two poles and a single zero, at the locations indicated. The transient response of this system is indicated by the weighting function

$$g(t) = c_1 e^{-b_1 t} + c_2 e^{+b_2 t} \tag{20}$$

which illustrates graphically the dependence of the time response on the pole locations in the s plane. The location of the zero will determine the magnitudes of the coefficients c_1 and c_2.

As a third example of the relation between pole location and time response, consider a system whose transfer function is given by

$$G(s) = \frac{K}{(s + a)^2 + \beta^2} \tag{21}$$

In this case the denominator has complex roots. The corresponding time response is now a sinusoid with an exponentially damping envelope

$$g(t) = Ce^{-at} \sin \beta t \tag{22}$$

Transfer Functions of Closed-Loop Control Systems

Consider now the feedback control system shown in Figure App. 1-4. The system reference input is indicated by r(t), the controlled variable (system output) by c(t), the system error by e(t), and the controller output by u(t). The plant or controlled elements are designated by the transfer function $G_p(s)$, the controller by the transfer function $G_c(s)$, and the feedback elements, which usually include the properties of the sensors and transducers associated with the measurement of the controlled variables, by H(s). The closed-loop transfer function is defined as the ratio C(s)/R(s). From the discussion in the previous section, it is evident that this transfer function, which we shall designate T(s), contains all the information needed to describe the system. The closed-loop transfer function can be derived as follows. The controller output is obtained from the system error and the controller transfer function

$$U(s) = E(s)G_c(s) \tag{23}$$

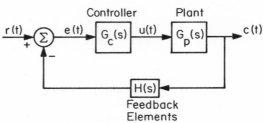

Figure App. 1-4. Closed-loop system.

The system output is obtained from the controller output and the plant transfer function

$$C(s) = U(s)G_p(s) \tag{24}$$

Substituting Eq. (23) into Eq. (24) we obtain

$$C(s) = E(s)G_c(s)G_p(s) \tag{25}$$

Thus, it is evident that cascaded dynamic elements can be represented by simply multiplying their transfer functions, which in part explains the simplification that results from the study of dynamic systems using Laplace transforms. The system error is given by

$$E(s) = R(s) - C(s)H(s) \tag{26}$$

If we substitute Eq. (26) into Eq. (25) we can eliminate the error transform and obtain

$$C(s) = [R(s) - C(s)H(s)]G_c(s)G_p(s) \tag{27}$$

Now, collecting terms, we can obtain the desired closed-loop transfer function:

$$T(s) = \frac{C(s)}{R(s)} = \frac{G_c(s)G_p(s)}{1 + G_c(s)G_p(s)H(s)} \tag{28}$$

This is the basic expression for the study of linear control system behavior. We are now in a position to examine quantitatively the differences between open- and closed-loop control with respect to sensitivity to plant parameter variations, disturbance rejection, and stability.

Sensitivity to Parameter Variations

In order to compare open- and closed-loop systems, consider the specific configurations in Figure App. 1-5. The two diagrams in this figure represent a plant or controlled element with negligible frequency dependence, which can be approximated by $G_p(s) = 1$. Assume for the moment that the disturbance $n(t) = 0$. The objective of our problem is to design a controller such that the output $C(t)$ will be ten times as large as the input, that is, we desire

$$C(s) = 10\,R(s) \tag{29}$$

The design of the open-loop controller for this problem is extremely easy. Since, in the absence of noise, the output can be written as

$$C(s) = G_{cl}(s)G_p(s)R(s) = G_{cl}(s)R(s) \tag{30}$$

and since we desire to satisfy the input-output relation in Eq. (29) it is evident that we obtain

$$G_{cl}(s) = 10 \tag{31}$$

Consider now the type of error in the controlled variable described by C(s) which would be obtained if our knowledge of the plant were inaccurate. Let us say that there is a 5 percent error in our measurement of the plant characteristics so that, in fact, $G_p(s) = 1.05$. Then the output will be given by

$$C(s) = (10)(1.05)R(s) = 10.5R(s) = 1.05C(s) \tag{32}$$

In other words, the 5 percent error in our description of the plant has resulted in a 5 percent error in the magnitude of the controlled variable.

Consider now the design of a controller to accomplish the same objective in the feedback configuration of Figure App. 1-5. If again we begin by assuming the disturbance to be absent, using Eq. (28) and substituting $G_p(s) = 1$, we obtain the requirement

$$\frac{G_{c2}(s)}{1 + G_{c2}(s)H(s)} = 10 \tag{33}$$

It can be seen that the feedback configuration has given us an additional degree of freedom, since we must choose both the feedback element

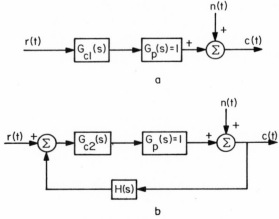

Figure App. 1-5. Comparison of: a. open-loop, and b. closed-loop control systems.

and the controller. As a possible solution, let us choose $G_{c2}(s) = 1000$ and $H(s) = 0.099$. It can be verified by substitution that these values do indeed satisfy Eq. 33. Consider now what happens if we again assume a 5 percent change or error in our estimate of the plant gain. Then the output is obtained by substituting in Eq. (28):

$$C(s) = \frac{(1000)(1.05)R(s)}{1 + (1000)(1.05)(0.099)} = 1.0005 \; C(s) \tag{34}$$

In other words, a 5 percent error in the plant gain has resulted in only a 0.05 percent error in the controlled variable. This simple example dramatically illustrates the fact that closed-loop systems can be designed in such a way that they are extremely insensitive to variations in plant parameters. Of course, in this example, only algebraic quantities were used and no frequency dependence was present. Nevertheless, the same type of relationship applies in the presence of dynamic elements. It now becomes intuitively clear why biological control systems are generally feedback systems. The closed-loop nature of the control systems makes it possible to maintain important physiological variables within very close tolerances, even in the presence of wide variations in the parameters describing the organisms.

Disturbance Rejection

A second important advantage of closed-loop over open-loop control systems is their ability to tolerate large magnitudes of disturbances without producing significant changes in the controlled variables. This can be illustrated by again examining Figure App. 1-5; assuming that a disturbance equal to one-tenth of the magnitude of the input is present:

$$N(s) = 0.1R(s) \tag{35}$$

In the open-loop case, shown in Figure App. 1-5a the output is obtained from the expression

$$C(s) = G_{cl}(s)G_p(s)R(s) + 0.1R(s) \tag{36}$$

which illustrates the fact that the disturbance simply adds linearly to the output produced by the controller-plant combination. In fact, with additive disturbances (noise) of the type shown here, there is no way of designing the controller to affect in any way the amount of noise, and corresponding error, produced at the system output.

In the closed-loop case, it is easy to show that the output transform in the presence of noise is

$$C(s) = \frac{G_{c2}(s)G_p(s)}{1 + G_{c2}(s)G_p(s)H(s)} R(s) + \frac{1}{1 + G_{c2}(s)G_p(s)H(s)} N(s) \quad (37)$$

Using the numbers which were employed in the design of the closed-loop system in the previous section, we obtain

$$C(s) = 10R(s) + \frac{N(s)}{100} \quad (38)$$

This equation illustrates the fact that the feedback has reduced the effect of the noise on the system response by a factor of 100. Once again, it is evident from a teleological point of view why biological systems have in general evolved as closed-loop systems; if this were not the case, living organisms would be unable to maintain any constancy of their internal environment in the presence of large variations in their external environment. It is easy to postulate that, if closed-loop feedback control systems had not been developed in the evolutionary process, living organisms would be confined to operation only in relatively constant environments, such as those present, e.g., in water of a certain temperature, pressure, and chemical composition.

To summarize, feedback control systems are less sensitive to parameter variations and environmental disturbances than are open-loop systems. However, a comparison of the controllers $G_{c1}(s)$ and $G_{c2}(s)$ used in our design illustrates the fact that the closed-loop system accomplishes this by requiring a much higher gain controller. In addition, the closed-loop system requires feedback sensors and some form of comparison process for operation. Finally, as an additional price paid for the advantages of closed-loop control, the stability of such systems cannot be taken for granted, even when the plant and controller individually display well-behaved and stable characteristics. The question of stability will be investigated in the next section.

Stability

A system is defined as stable if, when perturbed away from an equilibrium position, it has a tendency to return to equilibrium following the perturbation. Thus, a pendulum, when perturbed away from the vertical position, will tend to return toward equilibrium. Depending on the amount of friction at the support point, the resulting motion will be an oscillation about the equilibrium point with gradually decreasing amplitude. As long as the resulting motion gradually damps out, the system is considered *stable*. If the resulting motion grows, the system is considered *unstable*. Finally, if, as a result of a perturbation, the system continues to oscillate with a constant amplitude, it is sometimes called *neutrally stable*. The problem of stability is particularly critical in feedback control systems, since the nature of the controller and the feedback element will govern the stability properties of the system.

Among the common ways of studying the stability of linear control systems are the Bode diagram method and the root locus method. We shall examine both of these briefly.

The Bode Diagram Method. The Bode diagram was introduced as a plot of the amplitude ratio and phase of a system versus frequency when the system was excited with a sinusoidal input. In order to study stability, we plot the Bode diagram for the product of the controller, plant, and feedback transfer functions, $G_c(s)G_p(s)H(s)$. Consider, for example, the system shown in Figure App. 1-6 and its corresponding Bode diagram. The controller is indicated as having an adjustable gain K. If we examine the Bode diagram, we can see that the dotted line shows a frequency ω_c at which the system phase characteristic becomes equal to $-180°$. This implies that, at this frequency, the feedback signal (which in this case is equal to the controlled variable, since the feedback is equal to unity) now adds to the reference input, rather than subtracting from it. Hence, at this frequency, our system behaves as a positive feedback system rather than as a negative feedback system. It can be seen that, if the system is now perturbed, and if the perturbation contains any energy at the frequency ω_c, the resulting feedback signal at that frequency will add to the input. Whether such positive feedback leads to instability will now depend on whether the combined gain of the controller and plant at that frequency exceeds unity or not. It is intuitively evident that, if the superposition of input and feedback signals is now attenuated by the controller-plant combination, the system will still return to equilibrium. The amplitude ratio plot has a vertical axis which depends on the unspecified controller gain K. It is important that, at the frequency ω_c, the open-loop system gain be less than unity (less than zero dB) in order for this system to remain stable. By plotting the diagram to scale, the student can easily verify that, for this condition to apply, we must have K less than or equal to 6.

Thus, the Bode diagram contains more than simply the dynamic

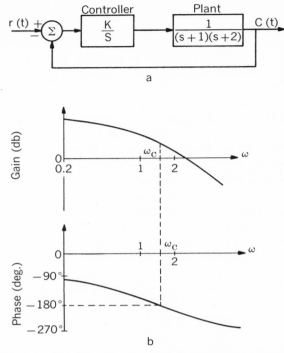

Figure App. 1-6. Third-order system: a. block diagram, b. approximate Bode diagram.

behavior of a particular system. We can also learn from it whether the addition of feedback to such a system will lead to stable operation.

The Root Locus Method. We have introduced the notion of describing a dynamic system by means of its pole-zero configuration in the s plane. Thus, the plant of Figure App. 1-6a is described by means of two poles, located on the negative real axis of the s plane at $s = -1$ and $s = -2$, respectively. However, when the loop is closed, it is now of interest to study the location of the poles and zeros of the closed-loop transfer function T(s). Just as an examination of the location of the poles of the particular system in Figure App. 1-6a has led us to conclusions regarding the time domain response, so an examination of the locations of the closed-loop poles will lead us to conclusions regarding the behavior of the entire feedback control system. In terms of Eq. (28), this means that we must find the roots of the equation

$$1 + KG_c(s)G_p(s)H(s) = 0 \qquad (39)$$

where the gain of the controller K has been brought out as a separate quantity. As we have seen in the previous paragraphs, stability often depends critically on the controller gain, thus making it desirable to examine the location of the roots of Eq. (39) as a function of the value of the controller gain. In general, if the system is higher than third order, an analytical evaluation of the roots of this polynomial is extremely difficult. However, standard computer programs exist for plotting the location of the roots of Eq. (39) as a function of the location of the individual poles and zeros of the three component transfer functions for various values of controller gain K. A plot of the location of the roots as a function of K is known as a *root locus diagram*. Consider again the simple system described by Figure App. 1-6a. The poles and zeros of the controller and plant are indicated in Figure App. 1-7. Since this system contains no finite zeros, none is seen on the diagram, and the three x's indicate the location of the controller pole at $s = 0$ and the plant poles at $s = -1$ and $s = -2$. To locate the closed-loop poles, we must find the roots of the equation

$$1 + \frac{K}{s(s + 1)(s + 2)} = 0 \qquad (40)$$

or

$$s(s + 1)(s + 2) + K = 0 \qquad (41)$$

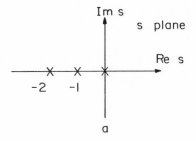

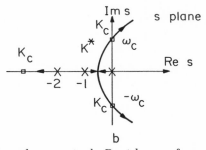

Figure App. 1-7. a. Open-loop roots; b. Root locus of example system.

This is a cubic equation in s, and by reference to standard books on applied mathematics the location of the roots can be found analytically. If these are plotted as a function of K, one obtains the diagram shown in Figure App. 1-7b, which shows the location of these roots as a function of K in the s plane. Note that, at any value of gain, there are three roots. The arrows indicate the movement of the roots as the gain increases. It is easy to verify that, for K = 0, the closed-loop poles are located precisely at the open-loop poles 0, −1, and −2. For low values of controller gain, the diagram of Figure App. 1-6 shows that the closed-loop roots are all real and negative. This means that, if the closed loop system is perturbed with a value of controller gain located in this region, its response will consist of a sum of three damped exponentials, and the response will be stable. At a certain gain indicated by K* in the figure, two of the roots of Eq. (40) become complex, and this behavior continues for all values of gain greater than K*. This means that, for those values of gain, the system response is now made up of a combination of two portions: (a) a damped exponential that arises from the root located on the negative real axis, and (b) an oscillatory response multiplied by an exponential that arises from the two complex roots. It is evident that, for values of controller gain between K* and K_c, at which point the root locus diagram crosses the imaginary axis, this oscillatory response will be damped and will gradually decay toward zero. Hence, a perturbation of the system with this value of controller gain will still result in stable operation. If the controller gain is equal to K_c, and if the system is perturbed, the response will contain a rapidly decaying exponential component, given by the root located on the negative real axis, and a constant amplitude sinusoidal oscillation, at the frequency ω_2. Hence, with this value of controller gain, our closed-loop system is neutrally stable. For all values of controller gain greater than K_c, the closed-loop poles are located in the right half of the s plane and the oscillatory response has an increasing envelope, thus resulting in an unstable system. If the diagram of Figure App. 1-7 is plotted accurately, it is easy to determine whether $K_c = 6$, as we would expect from the Bode diagram analysis in the preceding paragraphs.

Instability in Nonlinear Systems

The above discussion of the simple example has illustrated the nature of the stability problem for linear control systems. It can be seen, however, that such a discussion can never apply to biological systems, where no variable can increase arbitrarily in magnitude without bounds. In biological systems, as in all other real systems, all variables are constrained in magnitude by physical limitations. If an amplitude constraint is placed on a linear system, this constraint, which can be viewed as a saturation characteristic, makes such a system nonlinear. Since the amplitude will no longer grow without bounds, nonlinear systems are capable of oscillating with a constant amplitude determined by the constraint. The resulting oscillation will now not be sinusoidal in shape because of the distortions that arise precisely as a result of saturation. Such oscillations in nonlinear systems are known as *limit cycles* and are known to occur under certain conditions in biological systems as, for example, in connection with the phenomena of Cheyne-Stokes breathing which was mentioned in Chapter 2.

Appendix 2
Pattern Recognition

"How we understand what we are seeing" could be a subtitle to a chapter on pattern recognition. Thanks to recent advances in the field of neurophysiology, the mechanisms of vision are beginning to be understood. Experimental studies have yielded very valuable information on the eye and the retina.

But the retina is only half the story. It is a sensitive organ, one input of the nervous system. Without it we could not see at all. However, with the retina alone we could not understand anything of what we are seeing. The eye acts as an optical system; it intercepts light and focuses it through a natural lens, thus producing in its focal plane an inverse image of the object outside. This image acts as a stimulus to the retina.

The retina itself is complex; its first layer is a carpet of millions of light-sensitive cells; when light impinges upon them, they send electric impulses on to the brain. However, there are successive layers of interconnected neurons in the retina which perform certain information-processing operations on the pattern of electric impulses before sending it to the brain.

The result of this "preprocessing" of the image is then sent to the brain through the optical nerve, which is made up of about one million fibers. In comparing this number to the several hundred millions of sensitive cells of the retina, it is quite apparent that the retina has already done preprocessing on the image (the retina sometimes has been compared to a "small brain behind the eye").

Whatever the transformations performed on the pattern when it passes through the layers of the retina, it can reasonably be assumed that (a) if the *same* optical image hits the eye, the *same* pattern of electric impulses will be transmitted to the brain, and (b) if a *different* optical image hits the eye, then a *different* pattern of electric impulses will be produced (provided the difference between the two images is noticeable).

Thus, a mapping of the set of eye images corresponds to a hypothetical set of patterns transmitted to the brain. A very important operation (namely, pattern recognition) must now be performed by the brain before it can understand what it sees.

One and the same object can be seen in an infinity of different ways; for the same object, there is an infinity of different patterns of electric impulses coming to the brain. A baby never sees his mother in the same way twice. Thus, optical images of the mother are always different to him; his brain receives different patterns of electric impulses. But the baby could not long survive if, every time he received a different pattern, he thought he was seeing a different thing. He must learn to recognize, among all patterns coming to his brain, those particular ones which come from his mother.

How does he learn such a pattern? Other information is given to him at the same time, and this information along with many other sensations allows recognition of the mother by other means before he learns to recognize her visually. Hence, the brain is able to make an infinite set of patterns into a set of known "objects."

The assumption is made that this function is performed by "physical mechanisms" in the brain. Very little is known about the structure (be it nervous, chemical, or both) which allows the function of pattern recognition to be performed in the brains of living animals. It is easy to understand why: the brain is so complex, and its relevant phenomena are composed of so many microscopic events, that progress through direct, experimental observation is very slow. This is probably one of the reasons why another approach to the problem of organic pattern recognition is developing. This approach aims at understanding the "logic" of the process.

This is the approach of psychologists, who have asked the question "How does the brain understand what it sees?" without being able to test directly the hypotheses they put forward. Applications of a logical approach can be readily found in speech recognition, processing of electrocardiographic data, automatic photointerpretation, etc.

Objects and Classes of Patterns

A set of patterns of an infinite number can be classified so that every pattern belongs to one (and usually only one) class. For instance, the word "water" is a pattern, and it belongs to a class of all possible interpretations of the word "water."

A system has the function of pattern recognition with regard to a set of patterns if, and only if, given any pattern X as an input it can give as an output the class to which this pattern belongs. At least two possible outputs are necessary if the system is to be capable of recognizing patterns. Patterns must either belong to a class or not belong to it. The two outputs are then "A" and "non-A"; the system recognizes the patterns correctly if it gives output "A" when presented with a pattern of class A, and if it gives output "non-A" when presented with any other input. Actually, a certain percentage of errors is tolerated, depending on the problem and the accuracy required.

Classes are preexistent; the belonging or not belonging of a pattern to a certain class is known to the people who construct or use the pattern recognition system. For instance, if the patterns are the optical signals (images) sent by an object O_i, we shall define the class C_i as the set of all patterns which are images of O_i. Thus, the abstract notion of a class is complemented by a very concrete relation between any pattern of the class and the object which produced it. In more frequent cases, the limits of the classes are unknown, even if classes are not overlapping.

Pattern recognition problems can be divided into three main areas: (a) where the extent of the class is known, (b) where the number of classes is known, but not the relationships between classes, and (c) where the number of classes is unknown.

The set of all possible patterns is very large, but there is a limited number of classes which are actually subsets. Each pattern belongs to only one subset; the problem with any given pattern is to be able to tell to which subset (class) it belongs.

All problems of pattern recognition can be divided into two "families," according to the way a class of patterns is defined. Substantially different methods apply to these two families of problems.

The first family of problems is one in which a class is defined by means of a given prototype. Every pattern of the class can be deduced from the prototype by the application of one or more permissible transformations.

The second family of problems is defined by means of the properties of its members—or by characteristic features, where a feature of a pattern is like a property of a pattern, the only difference being that a property is either absent or present on a pattern, whereas a feature can have an intermediary degree of presence.

The Pattern Recognition System

All pattern recognition systems begin with a sensor or sensing unit the input of which is the physical phenomena that are associated with (produced by) the objects. In a very general sense, the sensors are the input devices of a system, and this is also true for living organisms. They act as the interface between "environment" and "system." The input to sensors are physical variables. The sensors perform a set of measurements, which means that they yield information relative to some characteristic of the physical variables. The set of the values yielded by the measurements constitutes the pattern. This set may possess a structure.

The patterns are very often noisy. For instance, the characters printed in a newspaper are very frequently blurred or smeared. Sounds can be heavily blanketed with noise.

One of the functions of preprocessing is to reduce or eliminate noise. This is generally done by recognizing noise, thanks to its high frequency (in sound patterns) or high "spatial frequency" in visual patterns, and providing a filter of some sort to remove it. Contrast is also a very important parameter which must be enhanced as much as possible before the measuring operation is performed.

An Example of Pattern Recognition

Let us assume that we have a reading machine with characters to be read from the same type of font. Take the letter "A." It will always produce, on the sensor of our system, a pattern similar to "A." However, the letter may be altered in various ways: (a) it may be *translated* or moved to one side of the space it should occupy, (b) it may be *rotated* in space, (c) it may be *homothetized,* made smaller or larger, or (d) it may have *noise* associated with the pattern.

Our first task is to normalize the pattern to a standard format. This can be done in a variety of ways. It is possible to look for the center of gravity of the pattern, and to make it coincide with the center of the sensor; this is translation.

The process of the normalization of translations can be called "centration." With some techniques (electronical filtering, optical filtering), this can be avoided.

Suppose all prototypes are inscribed in a box of fixed size.

Then a given pattern, after having been centered, can be expanded uniformly until it touches one of the borders of the box. This is *homotheses.*

We now assume that we can normalize a pattern in the modes just described. The normalized pattern of an unknown character "X" should match one of the prototypes; the problem now is to find which one. We must compare our normalized pattern to every prototype. The result can be summarized by a number which is a measure of the matching between our pattern "X" and a given prototype.

For instance, if our pattern "X" is a "C," one can find the following values of matching criteria m:

$$m(A,X) = 0,20$$
$$m(B,X) = 0,25$$
$$m(C,X) = 0,90$$
$$m(D,X) = 0,70$$
$$. . . . .$$
$$m(G,X) = 0,85$$
$$. . . . .$$
$$m(O,X) = 0,80 \quad \text{etc} . . .$$

Looking at the values of the matching criteria, we shall choose the highest one, $m(C,X) = 0,90$, and decide that "X" = C. Notice that the C, O, Q, etc. . . . would also yield high values of the criteria, because they are similar to C. A small noise can induce an error in the system, causing it to mistake a C for a G, an O for a Q, or vice versa.

Optical Methods for Pattern Recognition

Two-dimensional correlation, as described above, can be performed by optical methods in coherent light; the time required for the operation is the time necessary for the rays of light to cross an optical system.

Suppose we want to detect in a plane $F(x, y)$ a small object o located in (x_0, y_0). This small object can be a letter on a page. A "filter" is matched to the object o, that is, a slide whose transparency $s^*(p, q)$ is the conjugate of $s(p, q)$. When the filter is interposed in the frequency plane, the plane will appear black except at point of coordinates $(-x_0, -y_0)$, where a sharp maximum of light intensity can be seen. Luminous points corresponding to the locations of the o on the page appear in the image plane.

A "unit" of information can be as small as a few microns; a square of 10 cm $\times$ 10 cm can thus contain 30,000 bits on one channel and 30,000 channels, or 9×10^8 bits on one channel.

The ratio of signal to noise is also very high; if the page is considered as noise and the looked-for character as the signal, a signal-noise ratio of 48 dB can be obtained.

Optical information processing may well go beyond visual patterns; it is quite possible to transform an acoustical signal into a visual one (the sound track of films is recorded in this fashion). Visual signals can also be transformed from electronic signals for the optical processing of radar data.

Theoretical Problems in Pattern Recognition

A theoretical conception of pattern recognition is the representation of a pattern recognition problem in an n-dimensional space.

Imagine d number of classes that we want distinguished with n characteristic features, some being characteristic of class A, others of class B, others of class C, etc. The first step is to test the pattern to see if (and how much) it has the characteristic features 1, 2, ... n. Obtaining a set of n numbers for every pattern, let them be $x_1, x_2, \ldots x_n$, standing for "the results of the tests 1, 2, ... n on a given pattern."

With these n values x_i, resulting from n measurements, we can associate an n-dimensional space where each axis represents a variable x_k; a particular pattern will then be represented by a point in this space. We shall call this space the pattern space. The pattern can be denoted by

$$\overline{X} = (x_1, x_2, \ldots x_n)$$

We can also think of a pattern $\overline{X}$ as a vector extending from the origin to the point of coordinates $(x_1, x_2, \ldots x_n)$, with coordinates $\begin{pmatrix} x_1 \\ x_2 \\ | \\ x_n \end{pmatrix}$

A given class, A or B or C, is therefore a set of points in the pattern space. If the choice of characteristic features, i.e., axes, is good, the points of a given class will not be disposed throughout the space; on the contrary, they should form a "cluster."

One property allows the separation of two classes. For separating d classes, at least m properties, such that $2^m = d$, are required. Suppose we have chosen to made n dichotomic tests on the patterns. This ordered set of tests yields for any pattern an ordered set or "list" of binary values:

$$xi = 0 \text{ or } I$$

Each class corresponds to a "regular" series of values; for instance:

$$\text{"A"} = (000000)$$
$$\text{"B"} = (000III)$$
$$\text{"C"} = (IIIIII)$$

This means only that a regular pattern of class "C" should give $xi = I$ as an answer to all the six tests.

If, for example, having measured the presence or absence of six properties on a pattern, we find the series (000III), then the pattern belongs to class B. If (000I0I) is obtained, the answer may also be "class B."

Generally speaking, the Hamming distance can be found between a given series of values and the regular series corresponding to classes A, B, and C (the Hamming distance between two binary numbers of k digits is the number of digits of some rank that are different).

On our example, if $P = 000I0I$

$$d(A,P) = 2$$
$$d(B,P) = 1$$
$$d(C,P) = 4$$
So we chose B.

In the case of a retina with binary (ON-OFF) photocells, any image that is projected on this retina gives rise to a pattern of ON and OFF cells, which in turn can be represented as a series of 0 and I.

The property list method is more general than template matching, precisely because it can operate on elaborate properties. For instance, in the case of letters, one property can be the presence or absence of more than one point of intersection by a horizontal line passing through the middle of the letter; C, M, N, and L, would give 0's as results of this test; A, B, H, E, etc., would give I's.

The difficulty is to find a set of properties that would be present or absent for all members of a class, and that would give the same value

for all members of the set. With a well-defined set of intraclass transformations, those properties common to all patterns of a class are
invariant under the set of intraclass transformations. Topological properties are invariant in the set of linear transformations: translations,
rotations, and homotheties.

Another difficulty is to correctly define the "distance" between a
pattern and a regular series. The criterion of the Hamming distance
is not good, because it gives exactly the same importance to all properties. In series 000I0I above, the pattern was determined to be of class
B-000III. If this is right, why did a zero occur in the fifth place instead
of the regular I? Obviously, a statistical error had occurred, and it is
necessary to know the correct number. Suppose the two classes C and
D are described by C-00 and D-II, and results of the tests are a 0I
pattern. If it is known, by statistical experience, that errors are much
more frequent on the second property than on the first, it is then more
probable that this pattern belongs to C than to D.

The mathematical conclusion is that, since the first property is more
reliable, it should be given a heavier weight in the decision. This weight
seems to be directly related to the frequency of good performance by
this property. All this can be expressed in mathematical terms in what
is called the Bayes approach to decision.

However, one usually must go further in decision making, and
introduce the new concept of a risk or loss function. This is actually
outside the scope of pattern recognition as a theory, but will be encountered time and again when dealing with concrete problems such
as automatic diagnosis of diseases, weather prediction, signal detection,
etc.

Learning Pattern Recognition

The best decision surface is usually not unique; in probabilistic
cases, clusters interfere, i.e., the same point may belong to two classes,
although it has a much stronger probability of belonging to one. In such
cases, there should be an optimal decision surface. Learning starts with
an arbitrary surface and adjusts coefficients to move closer and closer
to the "optimal surface," this closeness being measured by reference
to the decisions produced by the surface. This is a particular case of
a very general method called "identification," which is widely used in
modern automatic control, and in bioengineering model making. Identification applies when a physical system (be it a muscle, a nerve, an
oven, or a chemical plant) has input-output pairs from a black box that
cannot be opened. In such a case, general steps can be taken:

1. Select a set of inputs.
2. Construct a model of the system, in the form of a set of relations between the input variables (and their derivatives) and the output variables (and their derivatives).
3. Find a criterion of how good the model is; this criterion is a *comparison* between outputs of the system and outputs of the model for the same inputs. The goal is to find a particular set of values $a_1^*, a_2^* \ldots a_n = 1^*$ of the structural parameters a_i, such that the criterion I is minimum for *this* set of values. If I $(a_1^*, a_2^*, \ldots a_{n \div 1}^*) = 0$, then we have a perfect model of the system for that particular class of considered inputs.
4. Step 4 is the computing step: finding a minimal value of a function $I(a_1, a_2, \ldots a_{n+1}$, of which only points are known.) There are general methods for doing this.

Appendix 3
Introduction to Analog/Hybrid Programming

The ultimate goal of the engineer is the economical synthesis of safe useful devices or systems, whether these be bridges, disk brakes, airplanes, computers, or surgical monitoring systems. Classically, the engineer attempts to incorporate discoveries from the basic sciences into his synthetic process; and, conversely, he quite often suggests areas for investigation to those involved in the basic sciences.

Synthesis must, however, go hand-in-hand with analysis, since limitations in the analytical tools at the disposal of the engineer will inevitably result in limitations in the safety and usefulness of the devices and systems which arise from his attempts at synthesis.

The purpose of this chapter is twofold: to suggest in explicit terms a way for the student to view complex systems in order to have some basis for the analysis/synthesis procedures, and to provide an introduction into simulation as an analytical tool for both the life scientist and the engineer.

The Simulation Process

Simulation, or modeling, is the process by which one observes some system in operation and attempts to evolve a mathematical description of the behavior of the system. This evolution is an iterative procedure whereby one makes a first guess as to the nature of the equations governing the behavior of the system. These equations are solved and their solutions are compared to the behavior of the real system. Changes are then made in the equations to account for any discrepancies between the previous model and the real system. The behavior of the revised model is then examined and revised further if necessary. At some point in this process, the modeler will feel that his needs are now fulfilled by the current model, that further revision is unnecessary, and that the model may be put to use.

The worth of this current model can be judged only according to how well it meets the requirements of the modeler and not how well it may solve the problem of someone else. Any model does, however, have certain intrinsic characteristics which must be specified by the modeler before its applicability for other uses can be evaluated. Foreknowledge of these requirements for specification will aid the student in understanding the nature of simulation.

The Model

First, the exact physical system being modeled must be carefully specified. Second, the operational modes over which the performance of the system was considered must be delineated. Third, system characteristics which were not taken into account in the derivation and verification of the model must be specified.

For example, a model of the human circulatory system conceived for the purpose of studying heat dissipation or drug distribution may not be at all applicable to the study of gravitational effects on the circulation by merely adding force terms to the blood flow. The original modeler probably did not verify his model under the appropriate conditions, so that there is no assurance that the added operational mode of the system (e.g., going from a prone to a standing position under gravity) can be successfully imposed on the model by the simple addition of force terms on the blood.

In addition to the above considerations, verification of the model proceeds on several levels, and each level must be documented by the modeler. Individual components within a model, e.g., heart, aorta, lungs, etc., must be "isolated" and verified by isolated organ experiments in physiology. These components are then appropriately interconnected, and the overall system model is verified by physiological experiments on the intact organism.

Omission of verification at the component level leads to excessive "tuning" of the model. Uniqueness cannot be assured, since inaccuracies in one component might result in compensating inaccuracies tuned into other components to obtain what would appear to be accurate overall system performance.

The First Guess

Having read this rough overview of the simulation process and some of the considerations involved in specification and verification, the student may well question what mystic processes are entailed in coming up with the first guess. Quite often this "guess" takes the form of an earlier model to which refinements and additions are to be made. Similarly, one may wish to combine two existing models, such as the respiratory control system and the baroreceptor heart rate controller, to study more complex phenomena. Conversely, however, one may be forced to start from scratch with basic considerations from biophysics. In any case, the steps to implementing the first guess are similar and will now be described.

Considerations of Measurement Domains

Hopefully, the question "What purpose is my model to fulfill?" has been answered. Implicit in this answer are hints concerning the time domain in which the model must satisfactorily perform and the degree of precision needed in the model.

To illustrate the importance of the above considerations, let us digress momentarily and consider a typically complex biosystem, the human being.

It has been suggested that the complicated physiological organism be viewed as a set of interacting control subsystems organized in a hierarchical control scheme whereby successive levels of control operate in separate frequency domains. This hypothesis, in addition to being consistent with available data, is quite convenient in that overall system stability can be explained without having to resort to new and unknown principles of control theory.

For example, consider the heart rate as a system variable which is driven primarily by the baroreceptors to achieve blood pressure regulation. This control loop operates rapidly in the 0.5- to 10-cps range. Influencing the action of this loop, however, are the respiratory control mechanisms that operate in the 1- to 0.01-cps range. Carotid and aortic chemoreceptors become active during hypotension and influence the vasomotor center to introduce 20 to 40 second-per-cycle oscillations, particularly during hypoxia. Blood volume control by the atrial receptor-diuretic hormone loop operates in cycles per hour, and these resulting blood volume changes are reflected in heart rate. Other, even slower, hormone controls operate over periods of days and can also be seen in heart rate patterns.

Thus, heart rate, though a single variable, reflects the operation of many interacting subsystems whose characteristic response times range from tenths of seconds to days.

All of these subsystems are interacting with one another to some degree; thus, all necessitate acknowledgement to some extent, regardless of the specific purpose of the cardiovascular system model. The crucial consideration can be seen to lie in the time period over which the model must be valid to meet its purpose. It also follows that the accuracy required of long-term integrations will be influenced by the above.

Mechanization of the Solution

Once a set of equations has been obtained, the task of familiarization with the behavior of these equations remains. The word "familiarization" is used in order to point out that one does not obtain a single solution, or even a whole family of solutions. Rather, for complex systems, months may be spent investigating general properties of the derived equations, during which period the iterative modeling process mentioned earlier takes place.

The hardware used as a solution tool during these months of familiarization will, of course, vary among installations. One finds entire large-scale digital computers devoted solely to a single simulation effort. Historically, however, large-scale simulation efforts have utilized analog computers for their computing speed and economy. The aerospace age brought the hybrid computer into popular usage for large simulations of systems having widely varying time constants and contrasting ranges of precision requirements.

Much literature, including several texts, has been devoted to the pros and cons of analog vs digital, or hybrid vs pure digital, computation. This chapter will not attempt a critique of the current status of these discussions, but will merely point out that circumstances alter cases.

For the purposes of this brief chapter, it was felt that a more meaningful tutorial could be given on the analog solution methods than on the digital methods. The example problems and the exercise problems yield useful insight into both analog computing and certain types of physiological systems, and they can be quickly mechanized on the smallest of analog computers. Some problems may even be combined for more advanced systems study, should the computing power be available.

Comments on System Representation

While a particular physical system may be unique, its mathematical representation may take on one of a number of different forms.

The body of knowledge accumulated in classical control theory using the transfer function representation of systems made that approach initially an attractive vehicle for representing biological systems. There are, however, two features of this approach which limit its applications to biological systems. First, only linear systems are amenable to analysis by transfer functions, and biological systems are nonlinear. One might argue that adequate analysis can be accomplished using a linear approximation to the nonlinear system, but one important characteristic of biological control systems is their ability to exhibit sustained oscillations (respiration, ovulation, etc.) or to "hunt" (chemo-receptor-heart rate loop). Since no physically realizable linear system can have this property, linearizing the system model would be self-defeating.

Second, transfer functions basically describe a unique relationship for a system with one input and one output, and are not explicitly concerned with what happens in between. Rarely, however, are life scientists confronted with single input-output systems; and frequently, the intermediate stages are of great interest.

In summary, although transfer function techniques have proven very useful in dealing with some biological systems, much excellent text material already appears on the subject; moreover, this approach is not well suited for many classes of biological systems.

The presentation here will deal with the state function approach in which the system is represented by a set of first-order differential equations. The use of state variables allows one to define nonlinear, time-varying systems which may then be subjected to stability and optimality analysis via the state functions of Liapunov and Pontryagin.

For most practical systems of any consequence, the equations obtained cannot be solved analytically; thus, the computer is a mandatory tool for obtaining numerical solutions. We shall now turn to the analog computer as one means for solving the equations of the system.

The following step-by-step example illustrates the *necessary* procedure for analog programming. It is not intended to be an exhaustive treatment of each phase of analog programming, but rather an introduction to some of the terminology and operational aspects of the subject.

Example

System description: Consider a block of mass "a" suspended by a viscous damper of damping "b" in parallel with a spring whose force constant is "c." The system operates in a frictionless, gravity-free environment, and motion is only allowed axial to the spring-damper and is assumed to occur only in the linear portion of the spring and damper. An external force F(t) may be imposed on the block. Let x(t) designate the position of the block in time where x(t) = 0 at equilibrium with F(t) = 0. A simulation of the motion of the block in time is required.

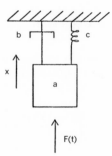

Step 1:   Write the system equations:

$$a \frac{d^2x}{dt^2} + b \frac{dx}{dt} + cx = F(t) \qquad (1.1)$$

$$\dot{X}(0) = \dot{X}_0 \qquad (1.2)$$

$$X(0) = X_0 \qquad (1.3)$$

Step 2:   Rewrite the system equation with the highest order derivative on the left and all other terms on the right:

$$\frac{d^2x}{dt^2} = -\frac{b}{a} \frac{dx}{dt} - \frac{c}{a} \times x + \frac{1}{a} F(t) \qquad (2.1)$$

$$\dot{X}(0) = \dot{X}_0 \qquad (2.2)$$

$$X(0) = X_0 \qquad (2.3)$$

Step 3:   Use the following "boxes" to draw a machine diagram, first forming the highest order derivative:

3.1   Boxes—General Definitions

3.1.1   Summer

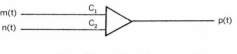

$$C_1 m(t) + C_2 n(t) = -p(t)$$

3.1.2   Integrator

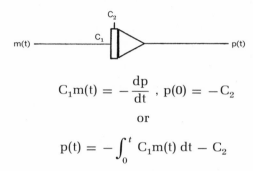

$$C_1 m(t) = -\frac{dp}{dt} \ , \ p(0) = -C_2$$

or

$$p(t) = -\int_0^t C_1 m(t) \ dt - C_2$$

The constant C_2 here is called the initial condition.

3.1.3   Summer—Integrator

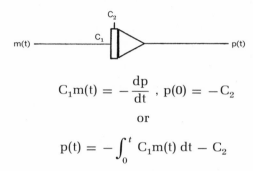

$$C_1 m(t) + C_2 n(t) = -\frac{dp}{dt}, \ p(0) = -C_3$$

or

$$p(t) = -\int_0^t \left[C_1 m(t) + C_2 n(t) \right] dt - C_3$$

3.1.4   Coefficient Multiplier (Pot)

$$am(t) = p(t), \ a = \text{constant}$$
$$0 \le a \le 1$$

3.2   Machine Diagram

3.2.1   Form the highest order derivative term as indicated by equation (2) and integrate, assuming that the necessary variables are available:

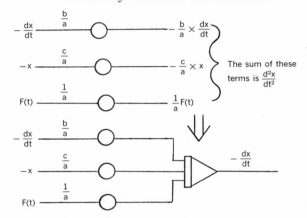

The sum of these terms is $\frac{d^2x}{dt^2}$

3.2.2   Adding an additional integrator:

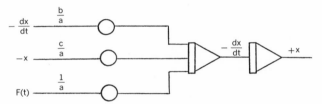

3.2.3   Feed back the terms which were needed to form the highest order derivative and insert initial conditions.

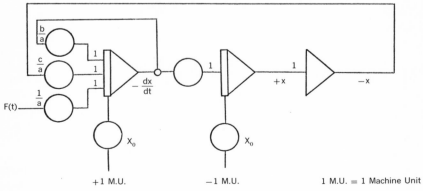

+1 M.U.   −1 M.U.   1 M.U. = 1 Machine Unit

We now have an unscaled machine diagram.

Step 4: Scaling: The boxes used above will overflow when their output exceeds a value of 1 Machine Unit. (1 Machine Unit = 10 volts, 20 volts, or 100 volts, depending upon the particular computer being used.) Hence, it is necessary to scale the system equations.

    4.1 Write down the *expression* for the output of each box— not including pots. A simple inversion ($+x$ to $-x$ in this case) is redundant.

Unscaled Output
$\dfrac{dx}{dt}$
x

    4.2 Assign constants to represent the maximum value of each *system* variable, and then evaluate the output expressions for their maximum values, e.g., S_{dx} = maximum value or upper limit of dx/dt.

Unscaled Output	Maximum Value
$\dfrac{dx}{dt}$	S_{dx}
x	S_x

    4.3 Write a scaled *machine* variable in brackets with an associated scale factor.

Unscaled Output	Maximum Value	Machine Variable	Scale Factor
$\dfrac{dx}{dt}$	S_{dx}	$S_{dx}\left[\dfrac{1}{S_{dx}}\dfrac{dx}{dt}\right]$	$\dfrac{1}{S_{dx}}$
x	S_x	$S_x\left[\dfrac{1}{S_x}x\right]$	$\dfrac{1}{S_x}$

NEVER REMOVE THE BRACKETS

    4.4 Insert each of the expressions in the machine variable column *in their entirety* into the machine equation, using the following rules:

4.4.1 Scale the highest order term the same as the next highest order term. Example II shows how the highest order term may be scaled separately.

$$\frac{d^2x}{dt^2} = S_{dx}\left[\frac{1}{S_{dx}}\frac{d^2x}{dt^2}\right]$$

4.4.2 The coefficient multipliers (pots) may only have values from 0 to 1. Thus, it may later be necessary to redistribute the total coefficient (that part outside of the brackets) so that the pot is properly scaled. Temporarily, however, consider the total quantity outside of the bracket to be the pot coefficient. Place ( )'s around these coefficients.

$$S_{dx}\left[\frac{1}{S_{dx}}\frac{d^2x}{dt^2}\right] = -\frac{b}{a} \times S_{dx}\left[\frac{1}{S_{dx}}\frac{dx}{dt}\right]$$
$$-\frac{c}{a} \times S_x\left[\frac{1}{S_x}x\right] + \frac{1}{a}F(t)$$

$$S_{dx}\left[\frac{1}{S_{dx}}\dot{X}(0)\right] = \dot{X}_0$$

$$S_x\left[\frac{1}{S_x}X(0)\right] = X_0$$

$$\left[\frac{1}{S_{dx}}\frac{d^2x}{dt^2}\right] = -\left(\frac{b}{a}\right)\left[\frac{1}{S_{dx}}\frac{dx}{dt}\right]$$

$$-\left(\frac{c}{a}\frac{S_x}{S_{dx}}\right)\left[\frac{1}{S_x}x\right] + \left(\frac{1}{S_{dx} \times a}\right)F(t)$$

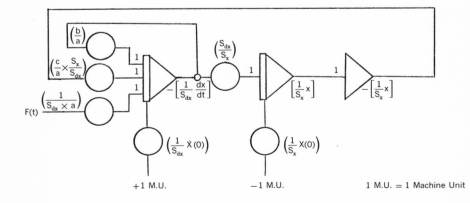

+1 M.U.   −1 M.U.   1 M.U. = 1 Machine Unit

Note: The (S_{dx}/S_x) valued pot between the integrators was introduced by the scaling process. It merely serves to equate

$$\left[\frac{1}{S_x}x\right] = +\int\frac{S_{dx}}{S_x}\left[\frac{1}{S_{dx}}\frac{dx}{dt}\right]dt$$

To scale the pots, first evaluate their expressions.

Suppose $a = 0.5$ slug $\dot{X}_0 = 0$ $S_{dx} = 5$
  $b = 2$ lb/ft $X_0 = 10$ $S_x = 20$

then

$$\left(\frac{b}{a}\right) = 4 \quad \left(\frac{1}{S_{dx}}\dot{X}_0\right) = 0 \quad \left(\frac{S_{dx}}{S_x}\right) = 0.2500$$

$$\left(\frac{c}{a}\times\frac{S_x}{S_{dx}}\right) = 1.2 \quad \left(\frac{1}{S_x}X_0\right) = 0.5$$

$$\left(\frac{1}{S_{dx}\times a}\right) = 0.4$$

Thus, the two pots with values greater than 1 must take the form:

$$\left(\frac{b}{10a}\right) = 0.4 \quad \left(\frac{c}{10a}\times\frac{S_x}{S_{dx}}\right) = 0.12$$

These input gains to the integrator must each be increased by a factor of 10.

The machine equation is now rewritten with the machine variables in brackets, the pot expressions in parentheses, and the input gains open:

$$\left[\frac{1}{S_{dx}}\frac{d^2x}{dt}\right] = -10\left(\frac{b}{10a}\right)\left[\frac{1}{S_{dx}}\frac{dx}{dt}\right]$$

$$-10\left(\frac{c}{10a}\times\frac{S_x}{S_{dx}}\right)\left[\frac{1}{S_x}x\right]+\left(\frac{1}{S_{dx}\times a}\right)\left[F(t)\right] \quad (4.1)$$

$$\left(\frac{1}{S_{dx}}\dot{X}(0)\right) = \frac{\dot{X}_0}{S_{dx}} \quad\quad (4.2)$$

$$\left(\frac{1}{S_x}\dot{X}(0)\right) = \frac{X_0}{S_x} \quad\quad (4.3)$$

The machine equations are now fully scaled.

Note: Machine Variables are in [Brackets]
       Pot Coefficients are in (Parentheses)
       Amplifier Input Gains are Open

As a check on the scaling procedure, remove all brackets and parentheses. Scaling terms should all cancel out, and the original system equations 1.1, 1.2, and 1.3 should be recovered.

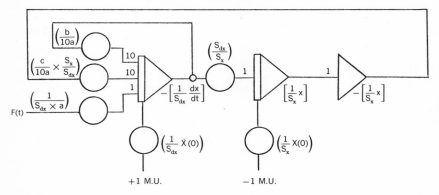

Step 5:   Once a scaled machine diagram is made, a box on the machine must be assigned to each box on the machine diagram for patching and running the problem.
All boxes are labeled by an alpha character followed by 3 B.C.D. digits.
Summers and Integrators use an AXXX; Pots use a PXXX. The machine diagram is labeled by placing the component address *inside* the box:

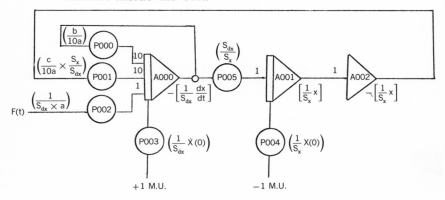

Step 6:   For documentation purposes, Pot Sheets and Amplifier Sheets are filled out (see Step 10).

P-	Expression	Run #1 Coeff	Run #1 Value
000	$b/10 \times a$	0.4000	−0.0800
001	$(c/10 \times a)(S_x/S_{dx})$	0.1500	−0.0045
002	$1/a \times S_{dx}$	0.1000	
003	$\dot{X}_0/S_{dx}$	0.2000	0.0000
004	X_0/S_x	0.5000	−0.5000
005	S_{dx}/S_x	0.2000	−0.0400

A-	FCN	Expression	Run #1
000	Int.	$-\left[\dfrac{1}{S_{dx}}\dfrac{dx}{dt}\right]$	−0.2000
001	Int.	$+\left[\dfrac{x}{S_x}\right]$	+0.0300
002	Inv.	$-\left[\dfrac{x}{S_x}\right]$	−0.0300

D-	Expression	Run #1
000	$(-b \times \dot{X}_0/a - c \times X_0/a + F/a)/S_{dx}$	0.1250(D/10)
001	$\dot{X}_0 S_{dx}/S_x$	0.0400(D)

Step 7:   The machine diagram from Step 5 is patched on the removable patchboard.

Step 8:   The patchboard is placed on the computer and the Analog Mode is put into Pot Set.

Step 9:   The Pot Sheet is read and the precalculated coefficients of the pots are set up.

Step 10:   The Analog Mode is changed to Static Check. In the Static Check Mode, the following quantities are available to be read out.

10.1   The Integrator Outputs have their initial conditions.

10.2   The Summer Outputs have the inverse of the sum of their inputs.

10.3   The Pot Output *values* are available (as opposed to reading their coefficient settings).

> *10.4*   The sum of the Integrator Inputs may be read (D or D/10
> for derivative).

Thus, the program is "alive" with all boxes working except
for the Integrators which are in the Initial Condition. The
Static Check quantities produced from the machine equations
and the patching diagram in Step 6 (Pot Sheet, Amplifier Sheet,
and Derivative Sheet) are now compared with the quantities
(10.1 to 10.4 above) produced by the patched computer. Thus,
the following types of errors can be detected.

> *10.4.1*   Diagramming errors or patching errors.
>
> *10.4.2*   Arithmetic errors in pot calculations.
>
> *10.4.3*   Machine component tolerances.

Step 11: The analog computer is placed in the Initial Condition (sometimes called Reset) Mode. With all readout devices now operating, the computer is placed in the Operate (compute) Mode.
The integrators are now free to operate on their derivatives
to produce the solutions to the patched equations.

Repetitive Operation

Modern high-speed analog computers are equipped with electroni-
cally controlled I.C. and Operate switches which are driven by digital
counters to produce I.C. and Operate times as short as 10 μsec. Automatic cycling between I.C. and Operate at high speed is called Rep.
Op. (Repetitive Operation). Typically, with I.C. = 100 μsec and Operate = 900 μsec, 1000 complete solutions to a complex system of nonlinear differential equations can be produced each second and displayed on an oscilloscope. During the 100-μsec I.C. period, parameter
values may be changed, and one can observe an entire family of solutions displayed "simultaneously." Particular applications where such
speed is helpful are in statistical studies of system parameter variations,
Monte Carlo-type simulations, and mixed-end condition boundary-
value problems where one must search a parameter domain to optimize
certain performance criteria.

Analog Computation—The Practical Case

At this point, the student may be somewhat overwhelmed with the
amount of bookkeeping and documentation involved in even the simplest of the preceding examples. In fact, most texts on analog computing

omit these details and concern themselves with implementing more sophisticated mathematical expressions. One should note, however, that even a relatively simple nonlinear system presents horrendous difficulties when any method other than computer solution is employed for its investigation. Also, never having encountered large simulations of any form, the student might assume that less bookkeeping is required for all-digital solutions. It should therefore be pointed out that the bookkeeping for a problem setup is inherently related to the complexity of the problem; only the difficulty of performing this bookkeeping varies with the type of machine.

It was therefore felt both appropriate and necessary to explicitly introduce the student to the rigorous structuring which becomes mandatory when solving larger problems of practical significance and complexity. Chemical and aerospace industrial simulations using a thousand amplifiers are not uncommon.

As for the more sophisticated mathematical aspects of analog programming, the student may turn to a number of texts on the subject.

References

1. Iberall, A. S., and Cardon, S. Z.: *Regulation and control in biological systems.* Proceedings of IFAC Tokyo Symposium on Systems Engineering for Control System Design, 1965, pp. 463–473.

2. Schultz, D. G., and Melsa, J. L.: *State Functions and Linear Control Systems.* McGraw-Hill, New York, 1967, p. 435.

3. Korn, G. A., and Korn, T. M.: *Electronic Analog and Hybrid Computers.* McGraw-Hill, New York, 1964, p. 584.

4. Milsum, J. H.: *Biological Control Systems Analysis.* McGraw-Hill, New York, 1966, p. 466.

5. Fein, L.: *The structure and character of useful information-processing simulations.* Simulation 5:417, 1965.

Index

421